W9-AMV-334

ELECTRICAL WIRING FUNDAMENTALS

CONTEMPORARY CONSTRUCTION SERIES

DEVELOPED AND PRODUCED BY
VOLT INFORMATION SCIENCES, INC.

WRITTEN BY JOSEPH H. FOLEY

GREGG DIVISION
McGRAW-HILL BOOK COMPANY

New York / Atlanta / Dallas / St. Louis / San Francisco
Auckland / Bogotá / Guatemala / Hamburg / Johannesburg / Lisbon
London / Madrid / Mexico / Montreal / New Delhi / Panama
Paris / San Juan / São Paulo / Singapore / Sydney / Tokyo / Toronto

SPONSORING EDITOR: CARY BAKER
EDITING SUPERVISORS: ALICE V. MANNING, PAT NOLAN
DESIGN SUPERVISOR: NANCY AXELROD
PRODUCTION SUPERVISOR: KATHLEEN MORRISSEY

COVER DESIGNER: AMPERSAND STUDIO

Library of Congress Cataloging in Publication Data

Volt Information Sciences (Firm)
 Electrical wiring fundamentals.

 (Contemporary construction series)
 Includes index.
 1. Electric wiring, Interior. I. Foley, Joseph H.
II. Title. III. Series.
TK3285.V64 1981   621.319'24   80-20097
ISBN 0-07-067561-9

ELECTRICAL WIRING FUNDAMENTALS
Copyright©1981 by McGraw-Hill, Inc. All rights reserved. Printed in the United
States of America. No part of this publication may be reproduced, stored in a
retrieval system, or transmitted, in any form or by any means, electronic,
mechanical, photocopying, recording, or otherwise, without the prior written
permission of the publisher.

ISBN 0-07-067561-9

· CONTENTS ·

CHAPTER 16
WIRING LOW-VOLTAGE CIRCUITS   261

Low-Voltage Transformers   Low-Voltage Wiring   Antenna Installation and Wiring   Antenna Mounting   Remote Control   Review Questions

GLOSSARY   279

INDEX   283

· PREFACE ·

Electrical Wiring Fundamentals introduces the beginning student to the materials and the methods used in residential wiring. No prior knowledge of the subject is assumed.

The material is organized in a sequence that allows the student to move confidently through the subject matter. Each chapter is built upon, and follows logically from, those that precede it.

The introduction to each chapter describes how the material in the chapter can be applied to practical situations in electrical wiring. When appropriate, this introduction also relates the subject of the chapter to other subjects in the book.

Review questions follow each chapter. The questions are designed to reinforce the learning process and test the student's grasp of code rules and trade practices. The questions summarize the main points the student must learn from each chapter. They also allow the student and the instructor to measure progress and to identify areas needing further work. Some questions are intended to stimulate discussion; others require the student to consult the National Electrical Code. The author believes students should become familiar with the organization of material in the NEC itself and learn to locate information in it. For this reason, exact code references in the text have been kept to a minimum to induce the student to make direct use of the NEC.

Units of electrical measurement used throughout this book are those recommended by the *Metric Guide for Educational Materials*, published by the American National Metric Council. In areas other than electrical units, however, this book employs traditional measurements to conform with the primary usage in the National Electrical Code and in the catalogs and pricing sheets of major manufacturers of electrical material.

The structure and content of this book have been carefully scrutinized by knowledgeable reviewers. Prior to the writing of the text, a detailed outline of the proposed book was reviewed and approved by a group of educators familiar with the needs of vocational training. During preparation, the manuscript and illustrations were reviewed by experienced educators. We are greatly indebted to all the reviewers for their guidance and their comments.

ACKNOWLEDGMENTS

Many manufacturers and organizations provided printed matter and product illustrations that were the basis for illustrations in this book. Special thanks are due Midland-Ross Corporation, General Cable Corporation, Slater Electric, Inc., Ideal Industries, General Electric Company, Amprobe Instruments, Square D Company, Underwriters' Laboratories, Inc., and the National Fire Protection Association. Thanks are also due the New York State Division of Housing and Community Renewal for permission to use material from the New York State building code manual.

Finally, I wish to thank my wife, Christine, for her valuable editorial suggestions and for the care with which she edited my manuscript.

Joseph H. Foley
for Volt Information Sciences

1
BASIC ELECTRICITY

• INTRODUCTION •

Energy gets work done.

Electricity is a form of energy.

Electricity can provide light when and where you need it, can produce heat for warmth and cooking, and can make motors run to do work. Electricity does these jobs when it is *under control*.

You have seen bolts of lightning split the summer sky and perhaps damage buildings or cause fires. During heavy rain or wind you may have seen sparks and flashes of light from electric power lines. From time to time you may have felt a slight but unpleasant shock when you touched an electrical appliance. These are just a few of the things that happen when electricity is *out of control*.

Well designed and correctly installed electrical systems keep electricity under control. A good electrical system conserves energy, too. When wire, cable, switches, fixtures, and power outlets are properly used, when wasteful wiring practices are avoided, electrical energy is delivered efficiently when it is needed. The best way to learn to do a good job and to work safely with electricity is to learn, first, what electricity is and how it behaves.

You can see what electricity does, but you cannot see electricity itself. Still, to know how electricity behaves, you must learn to think about electricity as if you could see it. Luckily, the best explanation of electrical energy, the *electron theory*, is easy to understand. It helps you form pictures in your mind of how electricity flows, what makes it flow, and what stops it from flowing. This chapter tells you about the electron theory. Read it carefully; it will help you understand electricity and the rules of electrical wiring that electricians must know.

• STATIC ELECTRICITY •

The principles of a form of energy called "static electricity" can be used to demonstrate how electrical charges act. Static electricity is the energy which, for example, causes the shock we feel when we touch an auto door handle after sliding across a car seat. Static electricity is also the energy that makes some kinds of clothing stick to our bodies.

For this demonstration we will use simple materials to generate small static charges. We will then transfer these charges to bits of lightweight material. These bits of material will then be made to move, without any visible force being applied to them. By noting how the bits of material move, we can learn something about how all electrical charges act.

Static electricity can be demonstrated with many different kinds of material. Some suggestions are listed below.

1. Two small pith (paper pulp) balls suspended by thread from movable supports.
2. Static-generating materials. One set consists of a glass rod and a piece of silk cloth; the other set, a hard rubber rod and a piece of fur.

To get the best results the demonstration should be done on a dry (not humid or rainy) day. Pick an area that is protected from drafts or breezes that might affect the results.

Now follow the steps below.

Step 1. Set the supports holding the pith balls on a table or bench. Separate the supports by about a foot.

Step 2. Rub the glass rod briskly with the silk cloth for 15 or 20 seconds.

Step 3. Hold the rod near one of the pith balls. The ball will swing toward the rod and stick to it.

Step 4. Move the rod away slowly until the ball pulls loose and hangs free. Step 2 generated a static charge on the glass rod. Step 3 transferred this charge to one of the pith balls.

Step 5. Rub the rubber rod with the fur (Fig. 1-1).

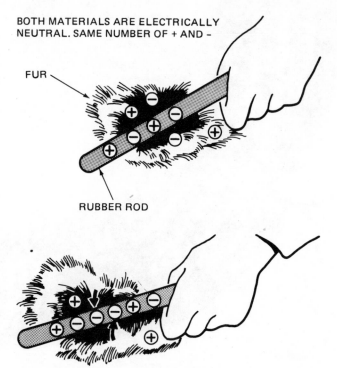

BOTH MATERIALS ARE ELECTRICALLY NEUTRAL. SAME NUMBER OF + AND –

FUR

RUBBER ROD

AFTER RUBBING, RUBBER ROD HAS MORE – THAN +

WHEN GLASS ROD AND SILK CLOTH ARE USED, GLASS ROD HAS MORE + THAN –

Figure 1-1. Producing static charge by friction.

Step 6. Hold this rod near the other pith ball. The ball will swing toward the rod and stick to it.

Step 7. Move the rod away so the ball hangs free. Step 5 generated a static charge on the rubber rod. Step 6 transferred this charge to the pith ball.

Step 8. Slide the supports holding the pith balls closer together. As the supports get closer the balls will swing toward each other. When close enough, they will touch (Fig. 1-2).

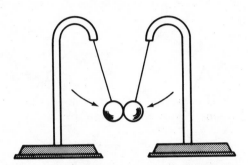

Figure 1-2. Pith balls attracted to each other.

Static charges of two kinds can be represented by positive (+) and negative (−) signs. The materials used to generate the charges caused each rod to have a different charge. The glass rod became positively charged and the rubber rod became negatively charged. These charges were transferred to the pith balls (Fig. 1-3). When oppositely charged, the balls were attracted to each other. This demonstrates a basic electrical law: unlike charges attract.

Now use the same materials to perform these steps.

Step 1. Touch each of the pith balls with your finger for a moment or two. This removes the charges applied earlier and the balls are now uncharged.

Step 2. Set the pith ball supports as close together as possible.

Step 3. Rub the glass rod with the silk cloth to recharge it.

Step 4. Touch one pith ball with the glass rod, then move the rod away. Recharge the rod and touch the second pith ball.

Step 5. The pith balls will swing away from each other as much as possible. Move the supports about. The balls will swing in all directions to avoid each other (Fig. 1-4).

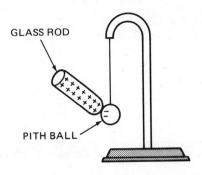

GLASS ROD

PITH BALL

NEGATIVE CHARGES MOVE FROM BALL TO ROD

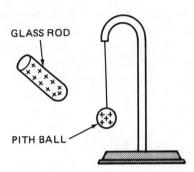

GLASS ROD

PITH BALL

BALL THEN HAS POSITIVE CHARGE

a

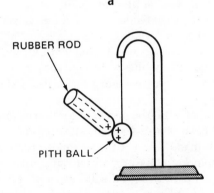

RUBBER ROD

PITH BALL

NEGATIVE CHARGES MOVE FROM ROD TO BALL

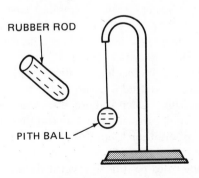

RUBBER ROD

PITH BALL

BALL THEN HAS NEGATIVE CHARGE

b

Figure 1-3. Transferring the static charge to the pith balls; (*a*) using glass rod; (*b*) using rubber rod.

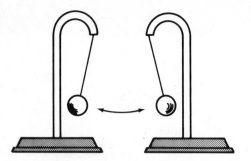

Figure 1-4. Pith balls repel each other.

This time both pith balls were given the same charge and they repelled each other. This demonstrates the other part of this basic law: like charges repel.

In the next section when we look at what atoms are made of, you will learn what positive (+) and negative (−) charges mean and what the force is that caused the pith balls to move.

• SMALL, SMALLER, SMALLEST — MOLECULES, ATOMS, ELECTRONS •

The world we live in is made of many millions of different things that take up space and have weight. The general name for all these things is *matter*. Matter includes the buildings we live and work in, the clothes we wear, the automobiles we ride in, the air we breathe, the water we drink, even our own bodies. Everything, whether solid, liquid, or gas, is matter.

As different as all the materials around us may seem, scientists have established that all things in our world are really made up of a fairly small group of basic building blocks of nature. By physical and chemical actions in laboratories, all substances, whether solid, liquid, or gas, can be broken down into smaller and smaller bits. This process of separating substances into bits can be continued until the smallest particle is obtained that still has all of the chemical characteristics of the larger pieces of the substance. These small particles are called molecules (Fig. 1-5).

Once the tiny molecule was discovered, scientists asked "What do we find if we break the molecules into still smaller bits?" Chemical processes were devised to break down molecules and it was found that molecules were made of another kind of smaller particle, called the atom.

Some molecules contained several different atoms. For example, things like wood, rubber, water, steel, and plastics were in this group (Fig. 1-6). Molecules of some other substances, however, contained only one atom. Copper, gold, iron, and sulfur were in this group.

It was found that there were only about 100 different kinds of single-atom molecules. The molecules of all other substances in the world contain various combinations of these atoms.

Those substances whose molecules contain only one atom are called elements. All the things in our world that have weight and take up space are composed of combinations or mixtures of these different elements.

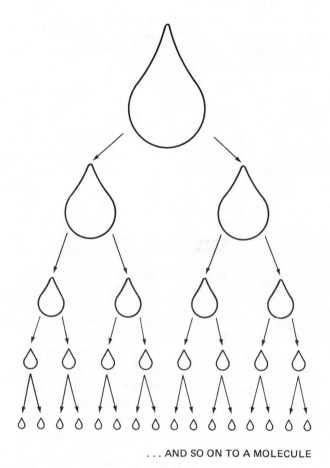

. . . AND SO ON TO A MOLECULE

Figure 1-5. Dividing a drop of water.

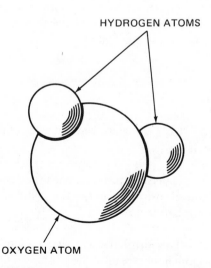

HYDROGEN ATOMS

OXYGEN ATOM

Figure 1-6. Hydrogen and oxygen atoms combined in a water molecule.

Atoms are so small that they are difficult even to imagine. The tiniest speck of matter that can be seen contains billions and billions of atoms. The most powerful microscopes cannot make atoms visible; yet we know a great deal about them and their structure.

Although the things we know about the structure of the atom are based on a theory, it is a theory that has been proven in many spectacular ways. The most spectacular are atomic explosions. The fact that these explosions occur is one proof of the theory of atomic structure. The use of atomic energy in electric generating plants is another.

What Atoms Are Made Of

Atoms consist of three kinds of particles held together by a natural force. The force, a form of energy, that binds atomic particles is the true source of electric energy.

The three particles in the atom are neutrons, protons, and electrons. The neutrons and protons are grouped together at the center of the atom to form a core or nucleus. The electrons move in circular paths (called orbits) around the nucleus (Fig. 1-7). Each of the three particles in the atom has weight, but the electrons are the lightest. It would take 1840 electrons to equal the weight of one proton or neutron.

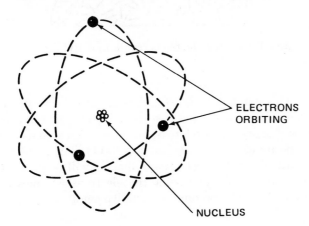

ELECTRONS
ORBITING

NUCLEUS

Figure 1-7. Atomic model.

Neutrons contribute to the weight of the atom, but carry no electrical charge. Neutrons play a big part in the work of releasing and controlling the energy locked up in atoms. In fact, the word we hear most for atomic energy, *nuclear energy,* refers to the neutron and the nucleus of the atom. However, the source of electrical energy is the proton and electron, so we will concern ourselves with these particles.

With this picture of the atom in mind, let's review the experiment described at the beginning of this chapter. The experiment showed us how one form of electricity, called static electricity or static charge, could be generated by friction.

The materials that become charged by friction are materials whose atoms can give up electrons when the surface is rubbed. The heat and motion of rubbing transfers electrons from one material to the other. When this happens there are some atoms on one surface that are short of electrons and there are atoms on the other surface that have a surplus of electrons.

When the glass rod was rubbed with the silk cloth, electrons moved from the glass rod to the silk cloth. This caused a shortage of electrons on the glass rod. Rubbing the rubber rod with the fur caused electrons to move from the fur to the rubber. This caused a surplus of electrons on the rubber. The energy that binds atoms together then exerts a force to restore the natural balance. This is the force that caused the pith balls to be attracted to each other in the first experiment (Fig. 1-8a).

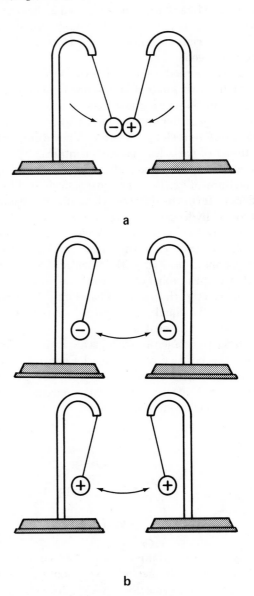

a

b

Figure 1-8. Static charges. (a) Opposite charges attract pith balls; (b) like charges repel pith balls.

The force is generated because two of the particles, protons and electrons, have opposite electric charges. The charge on the proton is positive and is shown as +. The charge on the electron is negative and is shown as –. The symbols + and – represent *unlike charges* and the attraction they show is stated as the electrical law that "unlike charges attract."

The second part of the static electricity experiment showed how, when the same charge was applied, it caused the pith balls to repel each other. When we charged the materials we generated a positive (+) charge on each and then (+) and (+) were brought near each other. When this was done we saw that a force existed that tried to keep these materials apart (Fig. 1-8b). This is a demonstration of the second part of that important electrical law "like charges repel."

The natural force that causes unlike charges to be attracted and like charges to be repelled is the source of all electrical energy. Not just the energy generated by rubbing materials together, but *all* electrical energy, however it is generated.

We have learned that there are only about one hundred different kinds of atoms, one kind of atom for each element. Now we know that all atoms are composed of three different particles. How, then, do the atoms of different elements differ? They differ only in the number of particles that each contains.

All neutrons are alike, all protons are alike, all electrons are alike. But they are joined in different numbers to form the different elements. For example, an atom of hydrogen, the lightest known substance, contains only one proton and one electron; an atom of copper has 29 protons in the nucleus and 29 electrons orbiting around it (Fig. 1-9). When nothing disturbs the balance, the number of electrons and protons in *any* atom is exactly the same. Therefore, the positive charges exactly equal the negative charges and the atom is electrically neutral.

The force that holds atoms together is the attraction between the positive protons in the nucleus and the surrounding negative electrons. Remember, the number of positive charges and negative charges are the same, and opposite charges attract.

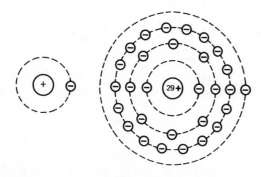

Figure 1-9. Atoms of hydrogen and copper.

Free Electrons

Because the positive charge is concentrated in the core or nucleus, while the total negative charge is made up of many orbiting electrons, the electrons tend to stay at fixed distances from the nucleus as they spin around it. However, to maintain the proper balance some electrons orbit close to the nucleus, others farther away. The electrons that are close in can be thought of as *locked in* to the atom. It would take a tremendous force to dislodge these electrons. The electrons in orbit farther from the nucleus are not as securely locked in. These electrons are called *free electrons*. They can drift from atom to atom in a random way (Fig. 1-10). If all or most of the free electrons in some material could be forced to drift toward the same point, that point would soon have a surplus of electrons. Because electrons are negatively charged, a surplus of electrons is another way of saying that point would have negative charge. Of course, if there is a surplus of electrons at one point, there must be a shortage of electrons at another point. This second point would then have a positive charge because the positive charge in the nucleus would not be offset by an equivalent negative charge.

Figure 1-10. Free electrons (random drift).

This point is basic to understanding electricity. If some force causes the free electrons in a material to move in one direction, two points can be created having opposite electrical charges (Fig. 1-11). The point toward which the electrons are moving will become the negative point. The point the electrons are moving away from will become the positive point.

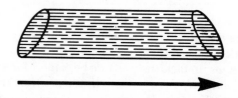

Figure 1-11. Free electrons (non-random drift).

This movement of electrons and the charges that result are what electrical energy is all about (Fig. 1-12). The main things to remember from this section are:

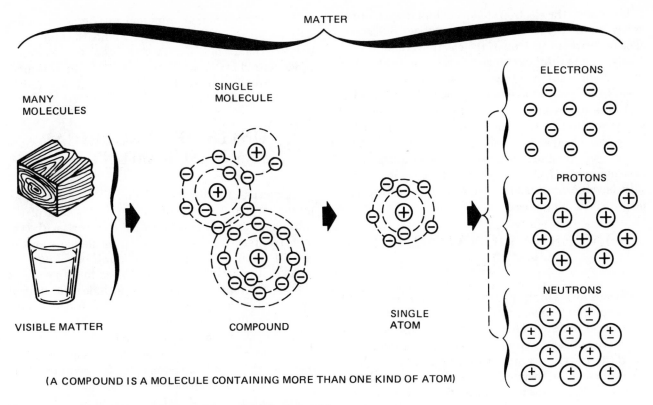

Figure 1-12. Breakdown of matter into electrical particles.

1. Every substance, whether it is liquid, solid or gas, is composed of atoms.
2. Atoms contain particles called protons that have a positive charge and particles called electrons that have a negative charge.
3. The protons are clustered at the center of the atom; the electrons revolve or orbit around the center.
4. In the atoms of some materials, the electrons farthest from the center are only loosely bound to the atom and can drift from the home atom to one nearby.
5. If all or most of the free electrons in some material can be forced to drift in one direction, the point toward which they are drifting will acquire a negative charge and the point the electrons are leaving will acquire a positive charge.

To say that two points have positive and negative charges is another way of saying that a force exists between these points. The force is a form of the energy stored in atoms. The force wants to restore the natural balance of electrons and protons in the atom. This force is called *voltage*. The greater the unbalance of protons and electrons (the greater the number of atoms that have lost electrons) the stronger the force will be; tha is, the greater the voltage will be between the two charged points (Fig. 1-13).

If an easy path is provided for the electrons to move toward the protons, the electrons will follow the path. The movement of electrons along this path is called *current flow*. The number of electrons that move depends on the strength of the force (voltage) that is acting on them and the ease with which they can move along the path. If the electrons can move freely along the path, the electron flow will be heavy even though the force acting on them is small.

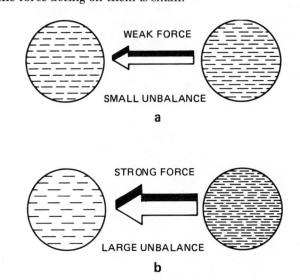

Figure 1-13. Electron unbalance. (*a*) Light; (*b*) heavy.

If the path the electrons follow allows them to move freely, we can say that the path offers "low" *resistance*. The opposite is also true. If the electrons cannot move freely, the path offers "high" resistance. If the path offers high resistance the electron flow will be light, even if the applied force (voltage) is high (Fig. 1-14).

All the statements made about voltage, current, and resistance suggest that electrical force, electron flow, and ease of electron movement are closely related. A bit farther on, in Ohm's law, we will see how this relationship works. But, before we do that we must consider how these important electrical characteristics can be described in numbers.

• INTERNATIONAL METRIC SYSTEM •

Various forms of the metric system have been in use for about 200 years. A modern simplified system was established in 1960. This system is now the standard international language of measurement, abbreviated SI. Basic electrical units of measurement have long been expressed in metric terms, so little change is required in this area to conform to the new metric standard.

Converting sizes and units of measure of electrical products and materials to new sizes and units of measure involves considerable expense to manufacturers. For this reason, the changeover to metric units will come about slowly in this field. It is probable that for some time to come, electrical manufacturers will retain present product sizes, but may include metric equivalents on packages and in catalogs.

Measurements in traditional units are converted to metric units by means of decimal quantities called conversion factors. If a conversion factor is less than one, you must divide to get the metric equivalent; if the factor is more than one, you must multiply.

The conversion factor for feet to meters is 0.3048. To convert 6 feet, for example, to meters, divide 6 by 0.3048. Answer: 1.83 meters. The conversion factor for inches to millimeters is 25.4. To convert 5 inches, for example, to millimeters, multiply 5 by 25.4. Answer: 127 millimeters.

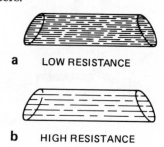

a LOW RESISTANCE

b HIGH RESISTANCE

Figure 1-14. Resistance to electron flow. (*a*) Light; (*b*) heavy.

In this book metric units and metric notation are used for all electrical quantities. Standard trade units, abbreviations, and names are used for electrical products and materials. When a metric unit or symbol will eventually replace the trade term, this new unit is explained in the text.

• UNITS OF ELECTRICAL MEASUREMENT •

To describe the quantity of a force or substance, we must first define a unit of the force or a unit of the substance. To be able to measure and work with current, resistance, and voltage, we must define a unit of current, a unit of resistance, and a unit of voltage.

Current is electron flow. So we need to define a unit of *flow*. A definition of flow must include some unit of time, as in gallons per hour or liters per second. A unit of current, then, can be defined as the movement of some number of electrons in a given amount of time.

Some things are best defined by noting the effect that they produce. For example, the temperature of air causes more or less expansion in a column of mercury in a thermometer. By dividing the column into equal units we can define changes in temperature in degrees. Resistance to electron flow in any material raises the temperature of the material. A unit of resistance can be defined in terms of the amount of heat generated in the material by the flow of a unit of current.

If we can define a unit of current and a unit of resistance, the third characteristic, voltage, can be defined in terms of those two.

Defining each of the units of electrical energy were major milestones in physics and so the units were named for their discoverers.

Measurement of Current

The unit of measurement for electron flow, or electrical current, is the *ampere*. The unit is named for a French scientist, André Marie Ampère, who lived from 1775 to 1836 and made many important discoveries about electron flow. He defined one unit of current as a flow of 6,250,000,000,000,000,000 electrons past a point in 1 second (Fig. 1-15). That big number is needed to measure electron flow because the electrical charge on each electron is small and a great many must move to make the combined electrical charge large enough to measure. You may see that number of electrons written as 6.25×10^{18}. The term 10^{18} is simply a shorthand way of writing a one followed by eighteen zeros. Six and one-quarter times one followed by eighteen zeros will give the number shown above.

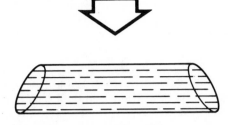

1 AMPERE = 6,250,000,000,000,000,000 ELECTRONS PAST THIS POINT IN ONE SECOND

Figure 1-15. Electron flow and time.

This large quantity of electrons is known as a *coulomb* (pronounced "koo-lomé"). The practical unit of current flow that is used in electrical work, however, is the ampere. The standard abbreviation for ampere is A. You will hear electricians speak of amps as units of current flow. This is trade slang and should not be confused with the standard abbreviation. In mathematics when current flow is referred to as an electrical characteristic, the symbol I is used. I in this usage means *intensity* of flow. Do not confuse this with A, which represents *units* of flow.

Measurement of Resistance

The unit of measurement for resistance is the *ohm*. Again the unit is named for an early scientist who made important electrical discoveries. The resistance unit is named for Georg Simon Ohm, a German physicist who lived from 1787 to 1854. The unit of resistance is really a measure of heat. When the flow of electrons is resisted, the temperature of the material through which the electrons are flowing rises, that is, it becomes hotter (Fig. 1-16). All materials through which electrons can flow offer *some* resistance. The resistance may be quite small but it is never zero. Ohm defined a small unit of heat* and then said that any material whose temperature could be raised that amount by the flow of one ampere had one unit of resistance. In this book when units of resistance are mentioned in text or shown in figures, the word ohm is always used. The metric symbol for ohms is Ω, a Greek letter called omega. You will often see this symbol used on products and drawings to represent electric resistance. When resistance is referred to as an electrical characteristic—as in mathematics—the symbol R is used.

Measurement of Voltage

The unit of measurement for the force that exists between positive and negative points is the *volt*. This

*Ohm's small unit was approximately one-fourth of a calorie. One calorie is the heat required to raise the temperature of 1 gram of water by 1°C.

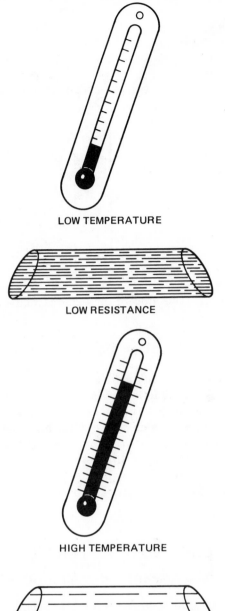

LOW TEMPERATURE

LOW RESISTANCE

HIGH TEMPERATURE

HIGH RESISTANCE

Figure 1-16. Resistance and heat.

unit, too, was named for an early experimenter, Count Alessandro Volta, who lived and worked in Italy (1745-1827). The volt is the amount of pressure required to cause 1 ampere of current to flow through a resistance of 1 ohm (Fig. 1-17). The volt is a unit of electrical pressure caused by a difference in electric potential. The standard abbreviation for volts is V. The symbol V is the preferred symbol for voltage as an electrical characteristic in mathematics. You may see the symbol E used for voltage in some mathematical texts. Voltage is sometimes called "electromotive force" or emf, so E was used as the symbol. The symbol E is now obsolete.

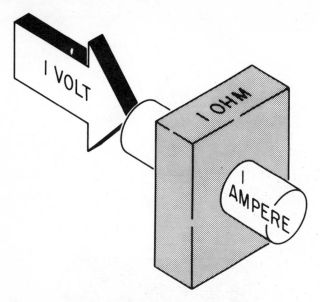

Figure 1-17. One ampere through one ohm equals one volt.

Larger or Smaller Measurements

The quantity of voltage, electron flow, and resistance that is used in practical situations is often either too large or too small to be conveniently stated in units: volts, amperes, and ohms. To solve this problem, prefixes are added to the basic units. Prefixes are something put before (or in front of) words to change the meaning. There are many possible prefixes that can be used, but in electrical wiring the most common are those in Fig. 1-18.

$$1,000 \times VOLT = 1 \; KILOVOLT$$
$$1/1000 \; AMPERE = 1 \; MILLIAMPERE$$
$$1,000,000 \times OHM = 1 \; MEGOHM$$
$$1,000 \times OHM = 1 \; KILOHM$$

Figure 1-18. Prefixes as multipliers.

Volts

Large voltages are used to transport power over utility company lines. These voltages are so large that it is easier to talk about units of 1000 volts than just 1 volt. The prefix meaning 1000 is *kilo*. So the statement "That is a 12-kilovolt line" means "The line voltage is 12,000 volts." Kilovolt is abbreviated kV.

Amperes

Under certain conditions, quite small quantities of electron flow can occur in power wiring. For this situation it is convenient to divide the ampere into a

thousand parts. The prefix for this is *milli*. The trade term "8 milliamps" or sometimes just "8 mils" means "eight one-thousandths of an ampere." Milliampere is abbreviated mA.

Ohms

Many electrical devices offer great resistance to electron flow, so ohms need a prefix to mean a large quantity. In conversation you will hear the term *megohm* or just *meg*. The prefix mega means a million. The statement "That insulation is good for 50 megs" means "That insulation offers a resistance to electron flow of 50 million ohms." The standard abbreviation for electrical resistance—Ω—can be combined with metric prefixes to represent thousands of ohms—kΩ—or millions of ohms—MΩ. These symbols are used primarily on electrical drawings.

• OHM'S LAW •

The relationship of voltage, resistance, and current flow was discovered many years ago by the physicist for whom the resistance unit was named. He stated the ways that current, voltage, and resistance affect each other in a basic law of electricity that bears his name, Ohm's law.

Ohm's law states that a simple mathematical relationship exists among the three characteristics of electricity. When a force (voltage) exists between two points and a path (resistance) for electron flow (current) is created between the two points, the voltage will cause current to flow and the relationship of the three characteristics, stated in words, will be:

1. The voltage in volts will be equal to the electron flow in amperes multiplied by the resistance in ohms.
2. Electron flow in amperes will be equal to the voltage in volts divided by the resistance in ohms.
3. Resistance in ohms will be equal to the voltage in volts divided by the electron flow in amperes.

Using the symbols for voltage *V*, electron flow *I*, and resistance *R*, the same three formulas can be written:

1. $V = I \times R$.

2. $I = V/R$.

3. $R = V/I$.

Because this law is so important in electrical work, it is often useful to have some aid to help you remember these formulas. One way that many people find easy is the Ohm's law triangle (Fig. 1-19). The positions of the symbols tell you how to find the missing quantity.

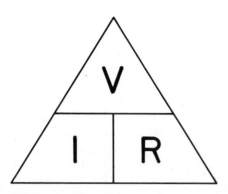

Figure 1-19. Ohm's law triangle.

1. To find V you must know I and R. I and R are on the same line, so they must be multiplied, as in formula 1 above.
2. To find I, V and R must be known. V is over R, so V must be divided by R.
3. To find R, V and I must be known. V is over I, so V must be divided by I.

Summary

1. Voltage is a force that is created when the balance of electrons and protons in atoms is changed by causing electrons to leave some atoms. It is measured in volts.
2. Current is the flow of electrons that takes place when both a voltage and a flow path are available. Electron flow is measured in amperes.
3. Resistance is a characteristic of the electron flow path that resists or holds back the movement of electrons. Resistance is measured in ohms.
4. Voltage, electron flow, and resistance are closely related. Any one of the three can be found if the other two are known. The mathematical relationship of voltage, electron flow, and resistance is called Ohm's law.

• CONDUCTORS AND INSULATORS •

So far we have examined how electrons flow along what we called a path. The actual paths along which electrons flow are wires. Now that we know something about voltage, electron flow, and resistance, it will be easier to see why some materials make good paths for electron flow and others do not.

We know that the atoms of each element are different in the number of electrons and protons that they contain. Each atom can be identified by an atomic number, a number that tells the quantity of electrons and protons the atom contains. We also know that the electrons orbit in groups or rings at various distances from the nucleus.

Each of these groups or rings of electrons can contain a maximum number of electrons. When the ring does contain the maximum number the ring is said to be "stable" and it will neither accept nor give up electrons. This condition can be thought of as *balance*. When a ring has its maximum number of electrons, the spinning mass is well balanced and therefore stable.

Atoms are so formed that the rings are successively filled to the maximum from the inner ring outward. The outermost ring then holds whatever electrons are left. An atom of copper has 29 protons in the nucleus and, therefore, in a neutral state, 29 electrons in orbit. If we look at a diagram of that atom (Fig. 1-20), we see that the inner ring has 2 electrons, its maximum number. The next ring has 8, its maximum number. The third ring also holds its maximum of 18. That accounts for $2 + 8 + 18 = 28$ electrons. Only one electron is left for the outer ring. This outer ring would like to have eight electrons, its maximum number. With only one electron where eight are needed for balance, this ring is *unstable*. That means that the electron in this ring can readily drift into another outer ring. Electrons can move into the now vacant ring, and the process continues.

It is this unstable outer ring that makes copper such a good path for electron flow. Other materials that have this characteristic are silver, gold, aluminum, and iron. In other words, all metals have atoms whose outer rings contain much fewer than the maximum number of electrons. Materials in this group are called *conductors*. Because electrons can flow easily in these materials they have low resistance. The words "easy

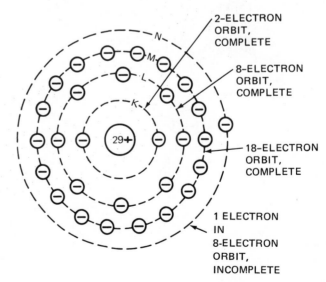

Figure 1-20. Copper atom showing electron rings.

electron flow" and "low resistance" mean exactly the same thing.

Let us look now at the opposite condition in the outer ring of an atom. In this case the outer ring has the maximum number of electrons, or close to the maximum. This causes the outer ring to be balanced or stable. There are no free electrons to drift from atom to atom. Materials with stable atoms are called *insulators*. Insulators include most plastics, rubber, cloth, wood, and paper.

A distinction must be made clear at this point. Electron *flow* in response to an applied voltage is quite different from electron movement that can take place in any material.

Rubber, glass, fur, and silk were used in the experiment at the beginning of this chapter to demonstrate how electric charges work. All of these materials are insulators, yet they became electrically charged. Is this a contradiction? The answer is "No." Electron transfer between the insulating materials occurred as a result of heat and friction, not an applied voltage. The rubbing action literally forced electrons from the surface of one material to the surface of another. A *force* then existed between the charged materials, but no electron flow took place. That is why electric charges on insulating materials are called "static"; they tend to stay put. The point to remember, then, is that electrical conduction takes place only when free electrons are present in a material and a voltage is applied.

For electrical wiring, the characteristics of conductors and insulators make them an ideal combination: a good conductor covered by a good insulator. The conductor allows electrons to flow readily and the insulator covering it prevents the electron flow from finding other paths by touching objects.

In practice, the metal most widely used as a conductor is copper. Copper has the best combination of qualities needed for current flow. First, low resistance; next, reasonable cost (compared to gold and silver); third, good strength for size; and fourth, it is light enough to work with but still strong enough to stand the sometimes hard use it gets on the job.

Wire does its job because the discoveries about atomic structure and electron theory have been put into practical use to manufacture a product needed to make electricity available when and where it is needed. The fact that electrons can move easily in some materials, called conductors, and cannot move readily in other materials, called insulators, provides the ideal practical electron flow path: insulated wires.

Wires and bundles of wires called cables are the material that carries electrical energy from the huge generators of electric utility companies right up to the wall receptacle in your home (Fig. 1-21).

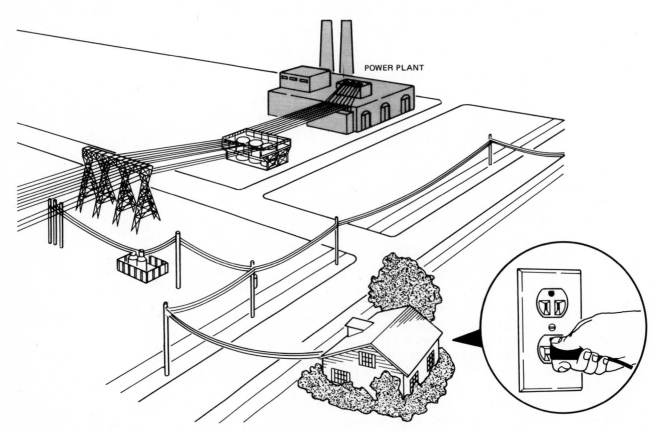

Figure 1-21. Wires carry electrical energy from the utility company to the wall outlet.

1. Atoms contain three particles; protons, neutrons, and electrons. Two of these particles are the source of electrical energy. Which two?

2. Electrical charges are represented by positive (+) and negative (-) signs. Which sign applies to each particle?

3. The static experiment demonstrated two laws of electricity. Which law does each of these pictures represent?

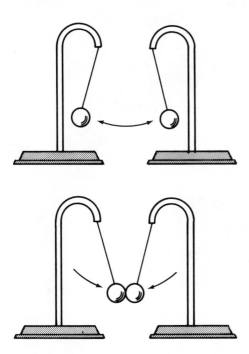

4. Some atomic particles can leave one atom and drift to another. What are they called?

5. Electrical pressure is called voltage and is measured in units called volts. What units are used to measure current? Resistance?

6. Sometimes prefixes must be added to electrical units to represent larger or smaller quantities. Which prefixes represent which number?

   a. milli   × 1,000,000 _____

   b. meg   × 1000   _____

   c. kilo   ÷ 1000   _____

7. Ohm's law describes the relationship of voltage, current, and resistance. If we know any two of these quantities we can find the missing one. Refer to the Ohm's law triangle (Fig. 1-19), and write the formula for finding each quantity.

8. What do we call materials in which current can flow easily?

9. What do we call materials in which current cannot flow?

2
ELECTRICAL CIRCUITS

• INTRODUCTION •

The subject of this chapter, electrical circuits, is an area of electrical knowledge with which an electrician *must* be familiar.

Electrical circuits are installed in buildings so electricity can do work. Circuits combine a source of power, wires, switches, outlets, fixtures, and other electrical devices to carry electricity where it is needed and to provide convenient places to connect lamps and appliances.

This chapter tells you various ways circuits can be wired and what happens to voltage, resistance, and current in each circuit arrangement. Chapter 1 covered what electricity really is. With that background you should have no difficulty with the material in this chapter.

• BASIC ELECTRICAL CIRCUIT •

To control electricity and make it available where we need it, we must combine voltage, current, and resistance with conductors.

Let us review the main points of Chapter 1. If the free electrons in some material can be made to move in the same direction, we can create a point having an electron shortage and a point having an electron surplus. A force will exist between these points which exerts pressure to restore the electron balance. If a path is provided between the two points, electrons will flow from the surplus point to the shortage point. If we direct this electron flow through the proper devices we can generate light or heat or make motors run.

The practical combination of electrical devices and hardware which puts electron theory to work is called a circuit. All electrical circuits have four basic parts (Fig. 2-1).

1. A *source* of power. The source provides the two points of electron surplus and shortage. The points are usually identified as positive (electron shortage) and negative (electron surplus). The more electrons at a point, the more negative the point is. The greater the electron unbalance, the stronger the pressure between the points.
2. *Conductors.* The conductors connected to the + and – points and then to other points in the circuit provide a path for electron flow.
3. A *load.* For the present we will consider a load to be any electrical device that makes electron flow do work: a light bulb, a motor, a heater. In the next section we will see *how* electron flow functions.
4. *Control.* To be useful, a circuit must have some way to control electron flow. Switches turn

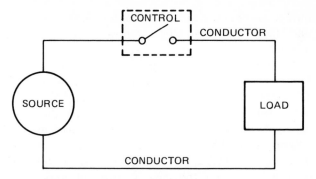

Figure 2-1. Parts of a basic circuit.

current on or off and make electron flow heavier or lighter by changing the flow path. As we will see in Chapter 3, changes that occur within the load can also affect electron flow.

• ELECTRICAL POWER — WATT'S LAW •

If any two of the values of voltage, current, or resistance of a basic circuit are known, Ohm's law can be used to calculate the unknown quantity. We can then analyze how the circuit functions (Fig. 2-2).

If we assume a source voltage of 10 volts and a resistance of 5 ohms, we can find the current flow by dividing the voltage (10 volts) by the load resistance (5 ohms):

$$I = \frac{V}{R}$$

The current flow is 2 amperes.

From this calculation we can conclude that a 5-ohm load across a source of 10 volts produces a current flow of 2 amperes. It takes voltage *and* current to do work.

Another basic electrical law describes the combination of voltage and current as *power* required by this 5-ohm load. Known as Watt's law, it states that the power required by a load is equal to the product of the current through the load and the applied voltage. The

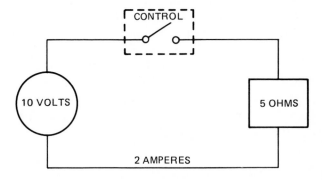

Figure 2-2. Circuit for a 20-watt load.

unit of power is the *watt,* abbreviated W. Both the law and the unit are named for James Watt, a Scottish inventor.

The unit of power is important to electrical work because it provides a uniform basis for measuring the time rate at which electrical power is being consumed or the rate at which work is being done. Neither voltage alone nor current flow alone can be used for this measurement. The watt combines voltage and current in a single unit. According to Watt's law a 100-watt lamp will consume the same quantity of power connected to either circuit in Fig. 2-3. If the lamp is illuminated for 1 hour, the power consumed in each circuit is: 100 watts for 1 hour, or 100 watt-hours.

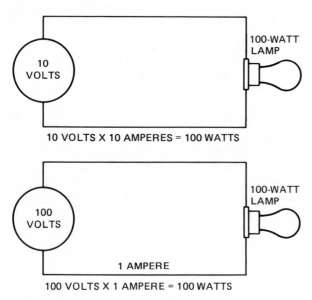

Figure 2-3. Two 100-watt circuits.

NOTE: A 10-volt source will not *properly* illuminate a standard 100-watt lamp. This is an example of the unit of power only.

The symbol for watts is W and the formula is $W = V \times I$. This formula, like Ohm's law, can be manipulated so that watts can be determined if we know any two of the basic values.

From Ohm's law we know $V = I \times R$. This can be substituted in Watt's basic formula to get $W = (I \times R) \times I$. I appears twice as a multiplier, so the formula can be written as $W = I^2R$. The symbol I^2, read as "I squared," means I is multiplied by itself.

By a similar process we can substitute V/R for I. $W = V/R \times V$ or $W = V^2/R$. In the circuit of Fig. 2-2, these calculations would be:

$$W = V \times I$$
$$= 10 \times 2$$
$$= 20 \text{ watts}$$

$$W = I^2 \times R$$
$$= (2 \times 2) \times 5$$
$$= 4 \times 5$$
$$= 20 \text{ watts}$$

$$W = \frac{V^2}{R}$$
$$= \frac{10 \times 10}{5}$$
$$= \frac{100}{5}$$
$$= 20 \text{ watts}$$

• DIRECT CURRENT •

Electrical energy can be generated in the following ways, which are discussed here and in Chapter 3.

1. *Friction.* The static electricity project in Chapter 1 showed how friction could move electrons from one piece of material to another. This resulted in an electrical charge on each piece of material. This principle is used in industry when paint is sprayed on metal surfaces such as automobile bodies. The particles of paint acquire a small negative charge as a result of friction encountered in the spray nozzle. The base metal of the auto body is positively charged. This causes the particles of paint to be attracted to the metal surface and not to remain suspended in the surrounding air. In addition, as the negatively charged paint coats the surface, it repels the negative particles being sprayed. This causes the particles to fall on bare metal, providing a more uniform coat.
2. *Thermoelectricity.* When some materials are heated they emit electrons. This type of electron emission is used, for example, in television picture tubes. A loop of wire called a filament in the base of the tube is heated by passing an electrical current through it. It then emits a beam of electrons that illuminates the screen of the tube.
3. *Piezoelectricity.* This form of electricity is generated when pressure is applied to certain crystalline substances. Its most common form is the crystal microphone. Pressure from sound waves causes a voltage to be developed between the opposite faces of a sliver of crystal.
4. *Photoelectricity.* Some materials emit electrons when light falls on them. Photovoltaic cells or solar cells use silicon to generate an output voltage from light.

Friction, thermoelectricity, piezoelectricity, and photoelectricity all have uses in industry, but at the present time none of these generating techniques can produce power in large enough quantities for use by electric utility companies. Solar cells show the most promise as energy sources for the future, but additional development work must be done.

Two forms of electricity generation have practical value at present. They are magnetoelectricity and chemical action. Magnetoelectric generation is described in Chapter 3. Chemical action is described below.

Electrical energy is produced in two additional forms known as direct current (dc) and alternating current (ac). As an electrician you will work with ac power and devices almost all the time. The study of dc power and circuits is useful, however, as a means of understanding ac power and circuits. Direct current circuits are simpler because dc power sources produce a constant voltage. Alternating current sources produce a constantly changing voltage. At any instant in time, with the same voltage applied, electron flow is exactly the same in a circuit regardless of whether the power source is producing dc or ac. We will consider dc power and circuits first; then we will use this information to understand how ac power and circuits work.

Direct Current Sources

CELLS AND BATTERIES • The simplest and most common dc sources are dry cells and batteries. Dry cells use chemical action to cause electrons *within the cell* to move from one piece of metal to another. Chemical action maintains this electron unbalance. Electrons flow in a circuit *outside the cell* to restore the balance.

The terms *cell* and *battery* have different meanings. A cell is a combination of two electrodes and a chemical solution. The materials used determine the voltage produced. The size of a cell determines how long current can be drawn from it and how large the current drain can be without a significant decrease in voltage.

Cells can be joined together so that the voltages add. This grouping of cells is properly called a battery. There are two kinds of cells, primary and secondary.

PRIMARY CELLS • When two different substances, such as zinc and copper or zinc and carbon, are placed a little distance apart in certain acid solutions called electrolytes, a voltage exists between them. As a result of this voltage, electron flow takes place from the zinc to the copper (or carbon) electrode when they are connected externally by a conductor. The combination of the two plates, electrolyte, and container is called a *primary cell* (Fig. 2-4). An example is the ordinary "dry" cell (strictly speaking, it isn't dry), which uses a

paste containing ammonium chloride as the electrolyte. Its force is 1.53 volts. With other electrolytes and electrodes the force may be from 0.7 to 2.5 volts.

As the chemical action continues, the materials are used up in the production of electricity. When the materials are used up, the device is useless and cannot be used over again. Such a device is known as a primary cell because it is a primary, or original, source of electricity, and it cannot be restored to usefulness after the materials have been used up. The voltage produced depends on the materials used and not on the size of the cell.

A dry cell, which is a primary cell, produces about 1.5 volts whether it is a tiny flashlight cell or a large No. 6 dry cell (Fig. 2-4), and provided no more than a very small current is drawn from the cell. The current which can be drawn from a cell depends on the size of the cell. If you attempt to draw more current from a cell than it is capable of supplying, the voltage will decrease to a low value.

Carbon-zinc cells are best suited to short periods of use at low current drain. Other types of cells can be used when other service is required. Mercury cells produce a voltage of about 1.4 volts. These cells are combined to form various battery voltages. Mercury cells can provide a fairly constant voltage even when used for long periods of time and with changing loads.

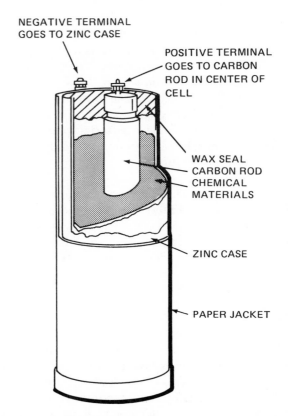

NEGATIVE TERMINAL GOES TO ZINC CASE

POSITIVE TERMINAL GOES TO CARBON ROD IN CENTER OF CELL

WAX SEAL
CARBON ROD
CHEMICAL MATERIALS

ZINC CASE

PAPER JACKET

Figure 2-4. Primary dry cell.

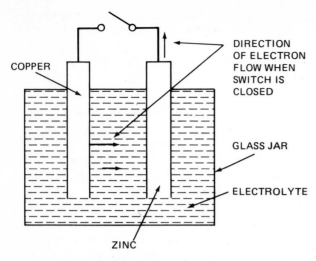

Figure 2-5. Typical wet cell.

SECONDARY CELLS • These are chemical cells that can be restored to their original condition by charging them from an outside source of electricity (Fig. 2-5). They are used in automobiles and are known as storage batteries. The cells are usually used in batteries made up of either three or six cells. The output of each cell is about 2 volts. The voltage would be either 6 or 12 volts, depending on the number of cells connected so that the voltages add. Secondary cells can be recharged by passing an electric current through them in a direction opposite to their discharge direction. Other secondary dry cells are alkaline cells and nickel-cadmium cells.

Alkaline cells produce a voltage of 1.5 volts. Alkaline batteries are available in several sizes up to 15 volts. These batteries are especially well suited to use in circuits when current drain is high. Alkaline cells are also available in non-rechargeable form.

Nickel-cadmium cells produce a voltage of 1.25 volts and these, too, are available in batteries up to 12 volts. Nickel-cadmium cells can withstand hard use over a wide range of temperatures.

GENERATORS • Direct current can also be produced by dc generators. Generators convert mechanical energy to electrical energy. Internally, all generators produce alternating current. Direct current output is obtained by adding an automatic switching device, called a commutator, to the generator outputs. The operation of generators is covered in Chapter 3.

Circuits

We will use some simple electrical diagrams to describe how basic circuits operate. A complete description of the various types of electrical diagrams is given in Chapter 12. It will be easier to understand how to read and use more complex electrical diagrams after you have covered the material in Chapters 1 through

11. For the present, two schematic symbols are all you need to know.

1. *Voltage Sources.* The symbols for cells and batteries are shown in Fig. 2-6. Other symbols shown in Fig. 2-6 are more general and may represent any voltage source according to the notation.
2. *Circuit Loads.* Devices called resistors can be used to represent other kinds of dc loads, such as lamps, heating elements, or motors. The symbols and the devices they represent are shown in Fig. 2-6.

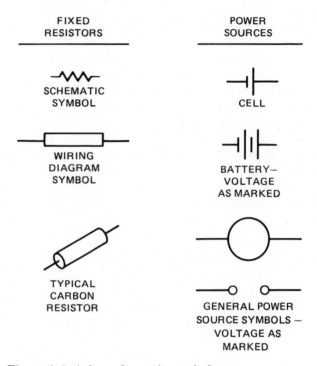

Figure 2-6. A few schematic symbols.

KIRCHHOFF'S LAWS • So far we have covered two sets of laws that tell us how electricity behaves. First, we looked at Ohm's law, which tells us how voltage, current, and resistance act in a circuit. Next, we looked at Watt's law. This extended the voltage and current relationship to show how to calculate electrical power.

To understand dc circuits we must consider another set of laws. These laws were first described by Robert Gustav Kirchhoff, a German physicist, and they now bear his name.

Kirchhoff's first law deals with current flow, *Kirchhoff's current law.* In any circuit the total amount of current (electron flow) leaving any point in the circuit must be exactly equal to the total current approaching that point.

Figure 2-7 shows a circuit that illustrates this law. Resistors R_1 and R_2 are connected so that current from the battery flows through both of them. According to

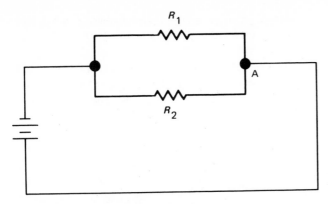

Figure 2-7. Kirchhoff's current law.

Kirchhoff's law, the current at point A must be equal to the sum of the current through R_1 and the current through R_2. We will see in the discussion of series and parallel circuits how this simple observation can be used to solve circuit problems.

Kirchhoff's second law deals with voltage, *Kirchhoff's voltage law*. In any circuit, the sum of the voltage drops around any complete path is exactly equal to the voltage applied to the path. This means that all voltage applied to a circuit will be used to cause current to flow. None will be "left over."

Figure 2-8 shows two simple circuits. Circuit A has a single load of 50 ohms and a source voltage of 100 volts. Circuit B has 3 loads totaling 80 ohms and also has a source voltage of 100 volts. From Kirchhoff's voltage law we know that voltage across the single 50-ohm load must be equal to the applied voltage. Similarly, the sum of the voltages across the three loads in circuit B must also be equal to the applied voltage. Ohm's law can be used to calculate the current flow in each circuit. From this, the voltage across each resistor can be calculated.

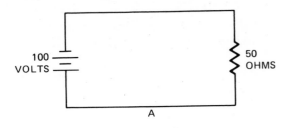

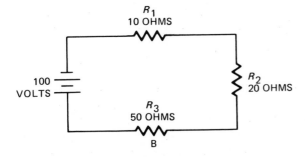

Figure 2-8. Kirchhoff's voltage law.

These calculations show that 2 amperes will flow in circuit A and 2 amperes through 50 ohms requires 100 volts. The current flow in circuit B is 1.25 amperes and the voltage across each resistor is, respectively, 12.5 volts, 25 volts, and 62.5 volts, or a total of 100 volts.

Regardless of the number or value of the loads in a circuit, the sum of the voltages across each load will equal the applied voltage. The portion of the applied voltage that is "used up" by each load is called the *voltage drop* across that load.

There are three ways of connecting loads to a source.

In series
In parallel
In series-parallel combination

We will consider what happens to the load resistance, the current, and the voltage in each of these circuits.

SERIES CIRCUITS • In a series circuit all loads are connected one to the other in a continuous loop. Current leaving the negative side of the source must flow through each load in turn to reach the positive side of the source. In accordance with Kirchhoff's current law, the current will be the same at every point in the circuit. In other words, the same amount of current will flow through each *load regardless of the load resistance*. If current flow is constant, then more source voltage will be required to push this amount through high resistance loads and less source voltage will be required for low-resistance loads. In accordance with Kirchhoff's voltage law, the sum of the voltage drops across the loads will equal the source voltage.

Figure 2-9, a simple series circuit, shows how this works. The total resistance (R_T) in a series circuit is the sum of the individual resistances:

$$R_1 + R_2 + R_3 + R_4 = R_T$$
$$8 + 15 + 20 + 17 = 60 \text{ ohms}$$

Ohm's law can be used to calculate the current flow in the circuit.

$$I = \frac{V}{R_T} = \frac{120}{60} = 2 \text{ amperes}$$

We know that 2 amperes will flow through each resistor, so we can now calculate the voltage drop across each load.

$$V = I \times R_1 = 2 \times 8 = 16 \text{ volts}$$
$$V = I \times R_2 = 2 \times 15 = 30 \text{ volts}$$
$$V = I \times R_3 = 2 \times 20 = 40 \text{ volts}$$
$$V = I \times R_4 = 2 \times 17 = 34 \text{ volts}$$

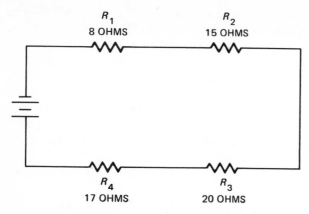

Figure 2-9. Simple series circuit.

The total voltage drop, then, is 16 + 30 + 40 + 34, or 120 volts, the source voltage.

There are three important points to remember about series circuits.

1. Current flow is the same through each load in a series circuit.
2. Voltage drop across each load is proportional to the load **resistance**.
3. The sum of the voltage drops is equal to the applied voltage (Kirchhoff's voltage law).

PARALLEL CIRCUITS • By far the most common circuit arrangement is the wiring of all loads in parallel. The word parallel describes the electrical relationship of the loads. When parallel circuits are shown schematically the loads (if drawing layout permits) are shown across parallel legs of the circuit. For this reason, parallel circuit schematic diagrams are sometimes referred to as "ladder" diagrams. It is often convenient to show the two lines to the voltage source along the top and bottom of the drawing. The loads are then shown on parallel lines between the power lines. The overall arrangement resembles a ladder with the power lines being the siderails and the loads being the rungs.

Virtually all residential and commercial power circuits are wired in parallel. Parallel circuits have several practical advantages over series circuits.

The parallel arrangement applies the same voltage to every across-the-line leg of the circuit so all circuit devices can be designed for a known voltage. Switches can be added to each across-the-line leg and any leg can be turned on and off without affecting the voltage applied to the other legs. Similarly, each leg can be a connection point (for example, a wall outlet) and loads can be connected and disconnected as desired.

The voltage, current, resistance, and power in a parallel circuit can be analyzed using Ohm's, Kirchhoff's, and Watt's laws.

A simple parallel circuit is shown in Fig. 2-10. The voltage across R_1 and R_2 is the same because both ends of R_1 and both ends of R_2 are connected directly to the source voltage.

The current flow through R_1 and R_2 however, depends on the resistance. The current through R_1 is:

$$I = \frac{V}{R_1} = \frac{120}{10} = 12 \text{ amperes}$$

The current through R_2 is:

$$I = \frac{V}{R_2} = \frac{120}{15} = 8 \text{ amperes}$$

Note that the values of 12 amperes and 8 amperes represent the current flow through each individual resistor. But at point A there must be enough current to supply both legs of the circuit. Therefore, 20 amperes must flow to point A—by Kirchhoff's current law — and 20 amperes must leave point A. The 20 amperes divides according to our calculations; 12 amperes flows through R_1 and the remaining 8 amperes flows through R_2. At point B the current through the two legs joins and 20 amperes flows back to the source.

A parallel circuit can contain any number of branches or legs, but the pattern of current flow is the same. Current flow divides at each parallel leg, in accordance with the resistance of the leg, and current from all legs merges in the return line.

Total current flow in a parallel circuit can be calculated two ways. Ohm's law can be used to calculate the current flow through each load and these values can be added together. This method was used in the preceding example. In the second method a resistance value is calculated that is equivalent to the effective resistance of the parallel loads. This resistance value can then be used to calculate the total current that will be drawn from the source.

To understand the importance of equivalent resistance and how it is calculated, let us compare series and

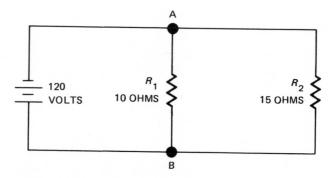

Figure 2-10. Simple parallel circuit.

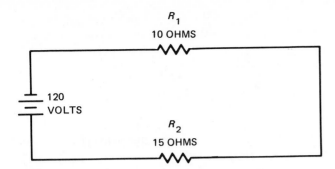

Figure 2-11. Comparative series circuit.

parallel load arrangements and see what general rules apply to current, voltage, and resistance in each circuit. Figure 2-11 is a series circuit which has the same resistance values and the same applied voltage as the parallel circuit of Fig. 2-10. To begin, we will compare the current flow in each circuit.

We have previously calculated that 20 amperes flows in the parallel circuit. The series circuit current formula is $I = V/R_T$, where R_T is the sum of the resistance in the circuit $R_1 + R_2$).

$$I = \frac{120}{25} = 4.8 \text{ amperes}$$

Clearly, this indicates that for the same load resistance and source voltage, current flow is greater in a parallel circuit. The reason for this difference in current flow is simple, but it is not readily apparent.

In a series circuit every load resistance simply adds to the other loads in the circuit. The more loads, the greater the resistance and, therefore, the less current flow. Each added load reduces the current flow.

When loads are wired in parallel, each load provides another current path. That is, each load *increases* the current flow from the source.

To look at this from another point of view, let us calculate the value of *series* resistance that would draw 20 amperes from the source.

$$R = \frac{V}{I} = \frac{120}{20} = 6 \text{ ohms}$$

This tells us that for 20 amperes to flow when R_1 and R_2 are in parallel, R_1 and R_2 must "look like" 6 ohms to the power source. The value of 6 ohms is the effective resistance of a 10-ohm load and a 15-ohm load in parallel. The effective resistance of loads in parallel can be calculated directly, rather than by the roundabout method of comparing series and parallel circuits.

Because each added resistor reduces the effective total resistance, the "reciprocal of reciprocals" method applies:

$$R_E = \cfrac{1}{\dfrac{1}{R_1} + \dfrac{1}{R_2} + \dfrac{1}{R_3} \cdots \dfrac{1}{R_N}}$$

This method can be used for any number of parallel loads. For our two-resistor circuit, the calculation is:

$$R_E = \cfrac{1}{\dfrac{1}{10} + \dfrac{1}{15}}$$

$$= \cfrac{1}{\dfrac{5}{30}}$$

$$= \frac{30}{5}$$

$$= 6 \text{ ohms}$$

A somewhat quicker calculation can be made when only two loads are involved. This is the "product divided by the sum" method.

$$R_E = \frac{R_1 \times R_2}{R_1 + R_2}$$

$$= \frac{10 \times 15}{10 + 15}$$

$$= \frac{150}{25}$$

$$= 6 \text{ ohms}$$

This method can be used for any number of parallel loads by successively calculating groups of two.

To summarize the subject of equivalent resistance in parallel circuits, both methods of calculation are given below for the multiple load circuit of Fig. 2-12.

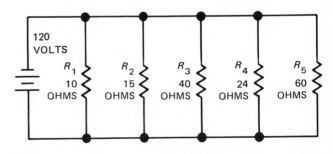

Figure 2-12. Multiple parallel loads.

1. Reciprocal of reciprocals method:

$$R_E = \cfrac{1}{\dfrac{1}{R_1} + \dfrac{1}{R_2} + \dfrac{1}{R_3} + \dfrac{1}{R_4} + \dfrac{1}{R_5}}$$

$$= \cfrac{1}{\dfrac{1}{10} + \dfrac{1}{15} + \dfrac{1}{40} + \dfrac{1}{24} + \dfrac{1}{60}}$$

$$= \cfrac{1}{\dfrac{30}{120}}$$

$$= \dfrac{120}{30}$$

$$= 4 \text{ ohms}$$

2. Successive combinations method (Fig. 2-13):

Step 1. Find equivalent resistance for R_1 and R_2 (Fig. 2-13*a*).

$$R_E = \dfrac{R_1 \times R_2}{R_1 + R_2}$$

$$= \dfrac{10 \times 15}{10 + 15}$$

$$= \dfrac{150}{25}$$

$$= 6 \text{ ohms}$$

Step 2. Find equivalent resistance for R_3 and R_4 (Fig. 2-13*b*).

$$R_E = \dfrac{R_3 \times R_4}{R_3 + R_4}$$

$$= \dfrac{40 \times 24}{40 + 24}$$

$$= \dfrac{960}{64}$$

$$= 15 \text{ ohms}$$

Step 3. Find equivalent resistance for (R_3, R_4) and R_5 (Fig. 2-13*c*).

$$R_E = \dfrac{(R_3, R_4) \times R_5}{(R_3, R_4) + R_5}$$

$$= \dfrac{15 \times 60}{15 + 60}$$

$$= \dfrac{900}{75}$$

$$= 12 \text{ ohms}$$

Step 4. Find equivalent resistance for (R_1, R_2) and (R_3, R_4, R_5) (Fig. 2-13*d*).

$$R_E = \dfrac{(R_1, R_2) \times (R_3, R_4, R_5)}{(R_1, R_2) + (R_3, R_4, R_5)}$$

$$= \dfrac{6 \times 12}{6 + 12}$$

$$= \dfrac{72}{18}$$

$$= 4 \text{ ohms}$$

When using the successive combinations method, loads may be combined in any order. In cases where two or more loads of equal resistance are connected in parallel, the equivalent resistance is simply the resistance of one load divided by the number of loads. In the circuit of Fig. 2-14, three 10-ohm loads are connected in parallel, so the equivalent resistance is:

$$R_E = \dfrac{\text{Resistance of 1 load}}{\text{Number of loads}} = \dfrac{10}{3} = 3.3 \text{ ohms}$$

This method can be combined with either or both of the other methods to calculate equivalent resistance of multiple parallel loads.

Power calculations for parallel circuits are made by applying the power formulas to the total circuit or to any part of it.

For the five-load circuit (Fig. 2-12) the current through each load is found by Ohm's law: $I = V/R$. With V equal to 120 volts and the resistor values as shown in Fig. 2-12, the current through each load is:

R_1 = 12 amperes

R_2 = 8 amperes

R_3 = 3 amperes

R_4 = 5 amperes

R_5 = 2 amperes

We know the voltage, current, and resistance of each load, so the power consumed can be calculated by Watt's law: $W = V \times I$ or $W = I^2 R$.

For each load the power consumed is:

R_1 = 1440 watts

R_2 = 960 watts

R_3 = 360 watts

R_4 = 600 watts

R_5 = 240 watts

The total power is 3600 watts.

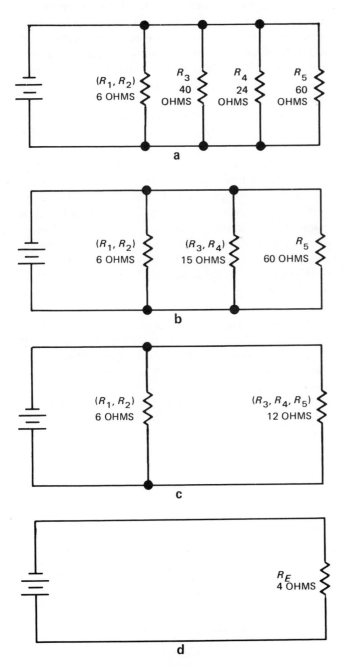

Figure 2-13. Finding equivalent resistance by successive combinations.

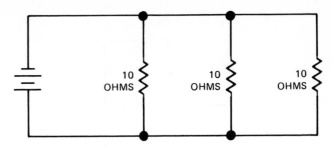

Figure 2-14. Equivalent resistance of equal loads.

Our calculation of equivalent resistance for this circuit gave us a value of 4 ohms. Current flow in terms of equivalent resistance is:

$$I = \frac{V}{R} = \frac{120}{4} = 30 \text{ amperes}$$

Total power is 120 × 30 or 3600 watts. The various methods of calculation provide a way of double checking answers.

SERIES-PARALLEL CIRCUITS • Both series and parallel load arrangements can be used in the same circuit. This combination is known as a series-parallel circuit.

To calculate current flow, equivalent resistance, or power in a series-parallel circuit, the procedures used for series alone and parallel alone are applied to the appropriate parts of the circuit.

In the circuit of Fig. 2-15a, if the equivalent resistance of R_1 and R_2 is calculated, that resistance plus R_3 will form a simple series circuit.

In a more complex series-parallel circuit, the circuit should be examined to determine the best way to group the loads for calculation. In the circuit shown in Fig. 2-15b, the following procedure will yield an equivalent resistance value in four steps.

Examination of the circuit suggests that if the equivalent resistance of R_1, R_2, R_3 and R_4, is found, and then if the equivalent resistance of R_5, R_6 and R_7 is found, those two equivalent values will form a simple series circuit with R_8. The procedure is as follows:

Step 1. Find the equivalent resistance of R_1 and R_2.

$$R_E = \frac{R_1 \times R_2}{R_1 + R_2}$$

$$= \frac{5 \times 20}{5 + 20}$$

$$= \frac{100}{25}$$

$$= 4 \text{ ohms}$$

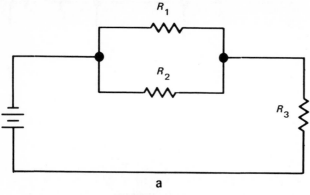

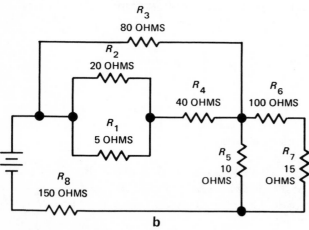

Figure 2-15. Series-parallel circuits.

Step 2. Find the equivalent resistance of $(R_1, R_2) + R_4$ in parallel with R_3.

$$R_E = \frac{[(R_1, R_2) + R_4] \times R_3}{[(R_1, R_2) + R_4] + R_3}$$

$$= \frac{(4 + 40) \times 80}{(4 + 40) + 80}$$

$$= \frac{3520}{124}$$

$$= 28.4 \text{ ohms}$$

Step 3. Find the equivalent resistance of $(R_6 + R_7)$ in parallel with R_5.

$$R_E = \frac{(R_6 + R_7) \times R_5}{(R_6 + R_7) + R_5}$$

$$= \frac{115 \times 10}{115 + 10}$$

$$= \frac{1150}{125}$$

$$= 9.2 \text{ ohms}$$

Step 4. Add equivalent resistance of steps 2 and 3 to R_8.

Step 2 = 28.4

Step 3 = 9.2

R_8 = 150.0

R_E = 187.6 ohms

SHORT CIRCUITS AND OPEN CIRCUITS • The two most common problems in electrical work are known as short circuits and open circuits. The words short and open refer to a change in circuit resistance as a result of an accident, equipment failure, or faulty wiring. The change in resistance caused by a short or an open is usually extremely large and usually occurs suddenly.

When a short circuit occurs there is a low (almost zero) resistance path for current flow in parallel with the load.

In normal operation of the circuit shown in Fig. 2-16, current flow through the 100-watt lamp when the switch is closed is about 0.83 ampere. If a short circuit occurs, such as path A, the low resistance in parallel with the lamp resistance results in an effective parallel resistance of less than 1 ohm. This causes current flow to jump to 120 amperes or more.

When this happens in an actual circuit, current flow is automatically cut off by an overcurrent protective device, as described in Chapter 11.

If an open circuit occurs—for example, a break in the conductor at point B—an extremely high resistance will be placed in series with the 100-watt lamp. The effect is the same as opening the switch: current flow drops to zero and the lamp goes out.

When shorts or opens occur, the cause must be found and corrected. The procedures for finding and correcting these problems are described in Chapter 14.

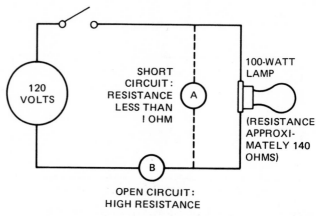

Figure 2-16. Short circuit and open circuit.

1. Name the four elements of a circuit.

2. Which of the four elements of a circuit does the work?

3. Power in watts is the product of _____ times _____ .

4. Why is the value of current squared (I^2) in the power formula $P = I^2R$?

5. Why is the value of voltage squared (V^2) in the power formula $P = V^2/P$?

6. The *chemicals* used to produce electricity in a cell determine the output _____ .

7. Cells to form batteries are combined to increase _____ or _____ .

8. How do primary and secondary cells differ?

9. The circuit shown in Question 8 is a series circuit. Answer the following.
   a. How much current flows through each resistor?
   b. According to Kirchhoff's voltage law, what will the sum of the voltages around the circuit equal?
   c. How much of the source voltage will be used by R_3?

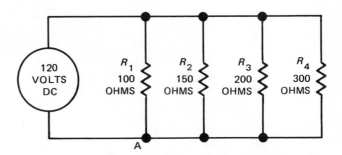

10. The circuit shown above is a parallel circuit. Answer the following.
   a. How much current flows through each resistor?
   b. What is the current flow at point A?
   c. What is the equivalent resistance of R_1, R_2, R_3, and R_4?
   d. How can the answer to question b be used to check the answer to question c?

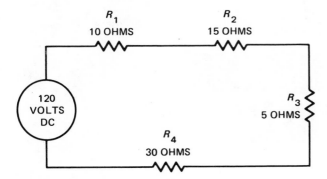

3
ALTERNATING CURRENT

• INTRODUCTION •

The form of electricity that utility companies generate and distribute is called alternating current, or ac. Virtually all the circuits installed by electricians use alternating current. This chapter tells you what special characteristics alternating current has. You will learn the things that alternating current can do and how it does them. Of course, there is more to electrical circuits and alternating current than this chapter and the previous one can cover, but the essentials provided here will give you a solid base on which to build. The effort you put into gaining a full understanding of electrical circuits and alternating current will give you a significant advantage in your future work.

• ALTERNATING CURRENT MACHINES •

Magnetism

Some materials, such as iron, nickel, and cobalt, have a unique property that permits them to become magnetized. Iron and iron alloys, in particular, show strong magnetic properties. Magnetism is a force that causes materials having magnetic qualities to be attracted or repelled in accordance with a definite set of rules.

The earth itself is a giant magnet (Fig. 3-1) and is the source of the names given to the two poles of force in magnets. One of the earth's magnetic poles is near the true geographic North pole and the other magnetic pole is near the earth's geographic South pole. If a small length of magnetic material is allowed to move freely while suspended by a thread or floating on liquid, the material will align itself so that each end points to one of the earth's magnetic poles. The end that points toward the north magnetic pole is called the north-seeking pole and the end that points toward the south magnetic pole is the south-seeking pole. On some magnets and on drawings these ends are marked N and S, respectively.

DEMONSTRATION OF LINES OF FORCE • A few simple demonstrations can show the shape of the zone of magnetic force and the way magnetic poles interact. The materials needed are:

Two permanent bar magnets
A magnetic compass
About 1/2 cup of fine iron filings
A piece of chalk

If the ends of the bar magnets are not marked N and S, the compass can be used to identify the poles and the magnets can be marked with the chalk. Work with the bar magnets one at a time.

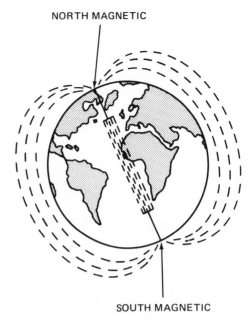

Figure 3-1. Earth as a magnet.

Step 1. Position one bar magnet and the compass as shown in Fig. 3-2.

Step 2. Move the compass close to the bar magnet until the compass indicator turns to a position approximately parallel to the bar.

Step 3. The north end of the indicator is pointing toward the south end of the magnet. Mark that end S.

Step 4. Mark the other end of the magnet N.

Step 5. Repeat this procedure with the other bar magnet and mark the ends of it also.

The marked magnets can be used to "feel" the magnetic lines of force. When the end of one magnet marked N is brought near the end marked S of the other magnet, the attraction of these poles can be felt

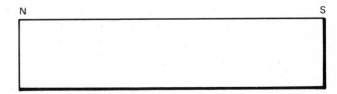

Figure 3-2. Identifying magnetic poles using a compass.

as the magnets snap together. The repelling force can also be felt when an attempt is made to bring together ends with the same polarity.

The magnetic force that was felt in this demonstration can be seen, indirectly, by using the iron filings.

Step 1. Place one magnet on a flat surface. Put a sheet of paper over the magnet.

Step 2. Sprinkle the iron filings slowly on the paper. The filings will move into the pattern in Fig. 3-3. This pattern is called the magnetic field of the magnet.

Step 3. Lift the paper from the magnet and return the filings to the container.

Step 4. Place both magnets on a horizontal line with opposite poles facing, but far enough apart that the magnets will not pull together.

Step 5. Again place the paper over the magnets and sprinkle the filings on the paper. The filings will move into the pattern in Fig. 3-4.

Step 6. Repeat this procedure except place the magnets with like poles facing. The pattern will be as shown in Fig. 3-5.

Four important points to remember from the magnet demonstration are:

1. Opposite magnetic poles attract.
2. Like magnetic poles repel.
3. Lines of force follow a curved path from one pole to the other.

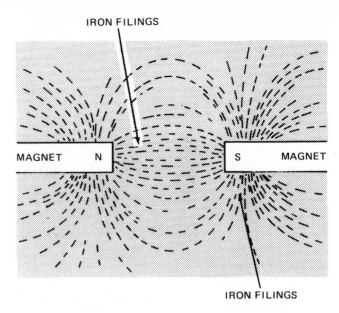

Figure 3-4. Lines of force between opposite poles.

4. The pattern of the filings showed that magnetic fields are strongest close to the poles.

The force that was felt in the magnet demonstration and the force that caused the filings to form patterns is explained by a theory called the molecular theory of magnetism. According to the molecular theory, each molecule of magnetic material is itself a tiny magnet. You will recall from Chapter 1 that a molecule is the smallest bit of any material that still has all the characteristics of the material.

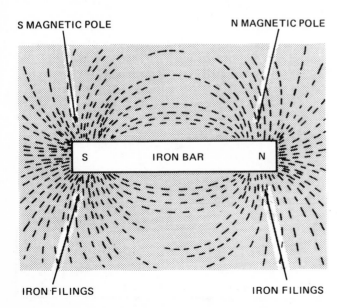

Figure 3-3. Magnetic field of a bar magnet.

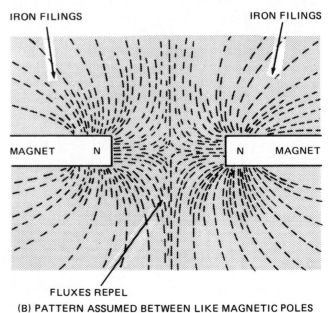

Figure 3-5. Lines of force between like poles.

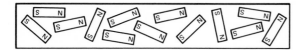

Figure 3-6. Molecular pattern in non-magnetized material.

When a piece of material—iron, for example—is not magnetized, these tiny molecular magnets are arranged in a haphazard way such that the magnetic force of each molecule tends to be cancelled by the force of another molecule (Fig. 3-6). The overall effect, then, is that the piece of material shows no magnetic properties.

If some force can be applied to this material to cause the molecules to be aligned so that all the N's point in one direction and all the S's point in another (Fig. 3-7), the material will then show the magnetic properties of attraction and repulsion of other magnets.

In some materials, such as the magnets that were used in the demonstration, once the molecules are aligned they will stay aligned for long periods of time. In many cases these magnets last for years without a noticeable decrease in magnetic strength. This, of course, is the reason they are called permanent magnets.

In other materials the molecules remain aligned only as long as a force is present to hold them in alignment. As soon as this force is removed, the magnetism drops to a low level and soon disappears. The force that aligns the molecules is electricity, and these magnets are called *electromagnets.*

Induction

ELECTROMAGNETS • You have probably noted by now that the action of electrical charges and the action of magnetic poles follow similar rules. Unlike charges and poles attract; like charges and poles repel.

This similarity was noted, of course, by early electrical experimenters and so they looked for a link between electricity and magnetism. The first link was discovered when it was found that an electric current flowing through a wire causes a magnetic field around the wire. The shape of the magnetic field that exists around a current-carrying wire is different from the field around a magnet. The lines of force around the wire are circular and have direction (Fig. 3-8).

The magnetic field that exists around a single conductor is too weak to be of any practical value, so a way

Figure 3-7. Molecular pattern in magnetized material.

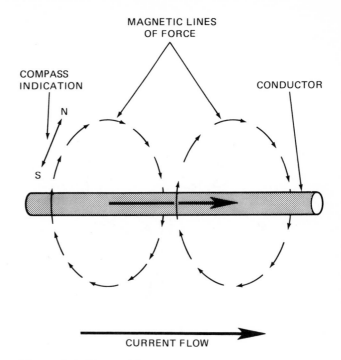

Figure 3-8. Lines of force around a straight wire.

had to be found to strengthen it. First, it was found that if the wire is wound into a coil, the magnetic lines of force will all pass through the center of the coil and reinforce each other (Fig. 3-9). Next, it was noted that magnetic lines of force pass more readily through magnetic material than through air, so an iron bar was added to the coil of wire. When current flows through the wire, the bar becomes magnetized. It has a north and a south pole, and can attract or repel other magnetic materials. This action is exactly the same as that of the bar magnets we used earlier, except for one important difference. The magnetic action of the bar in the coil can be started and stopped by turning the electric current through the coil on and off (Fig. 3-10).

Force becomes usable when it can be controlled. The combination of the coil of wire and the iron bar or core made magnetic force controllable. The electrical principle of induction is put into practical use in transformers, generators, motors, solenoids, relays, and vibrators.

When discussing magnetism and induction the term polarity is often used. A material or a device has polarity when it has points at which opposite forces exist. A magnet has polarity because it has a north pole and a south pole. A battery has polarity because it has a positive terminal and a negative terminal. Electromagnetic devices have both positive and negative electrical poles and north and south magnetic poles. A special property of electromagnetic devices is that these poles can be interchanged or reversed under certain conditions described later in this chapter. If we say that the polarity of one voltage is opposite to that of another voltage, we simply mean that the direction of current

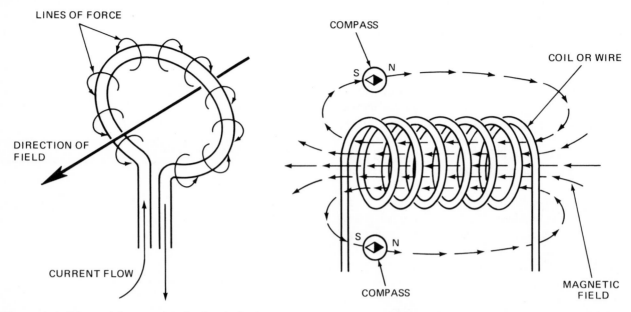

SINGLE LOOP

LINES OF FORCE

DIRECTION OF FIELD

CURRENT FLOW

COIL

COMPASS

COIL OR WIRE

S N

S N

COMPASS

MAGNETIC FIELD

Figure 3-9. Lines of force around a loop of wire.

flow produced by one voltage is opposite to the direction produced by the other. If we say the polarity of a voltage is reversed, we mean the positive and negative poles are reversed and consequently the direction of current flow is reversed.

To gain a better understanding of the effect induction has on a circuit, let us examine what happens in an electromagnet when current flow is turned on and off. Assume we have an electromagnet, a 120-volt dc source, and a switch connected as shown in Fig. 3-10. Coils used in electromagnets offer some resistance to the flow of direct current. We will assume that the coil in our test circuit has a resistance of 15 ohms. From Ohm's law we can calculate the current flow in the circuit:

$$I = \frac{V}{R} = \frac{120}{15} = 8 \text{ amperes}$$

When the switch is closed current starts to flow through the coil. As soon as current begins to flow, a magnetic field starts to expand around the coil. While the field is expanding, the lines of force around each loop of wire cut through other loops and induce a voltage. When the lines of force are expanding outward, polarity of the voltage induced in the loops of the coil is opposite to the battery voltage. This induced voltage causes current to flow in opposition to the battery current, which in turn causes the field to expand more slowly. The lines of force, then, cut more slowly, thus reducing the opposition current. In a short, but measurable, amount of time, the battery voltage

overcomes the opposition of the coil. Current flow in the circuit now is 8 amperes. At this time the field around the coil is no longer expanding and no voltage is induced to cause current to flow in opposition to the battery current. A steady condition of voltage, current, and resistance will exist in the circuit as long as the switch remains closed.

When the switch is opened, however, a similar action occurs in reverse. Current flow from the battery stops and the field around the coil begins to collapse. As the

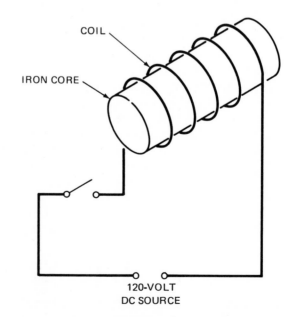

COIL

IRON CORE

120-VOLT DC SOURCE

Figure 3-10. Basic electromagnet circuit.

lines of force again begin to cut the loops of wire, a voltage is induced and current continues to flow. This induced voltage, however, is opposite to that of the expanding field; it has the same polarity as the battery voltage. In effect, the coil tries to keep current flowing as it was in the steady condition. Of course, with no energy input to the circuit, the field around the coil collapses and current flow ends.

Note that a voltage is induced in the coil only when the current flow is changing. Inductance can be defined as the ability of a conductor to induce a voltage in itself when current flow changes.

The unit of inductance is the *henry,* named for Joseph Henry, an American physicist. The symbol for inductance is *L.* A coil has an inductance of 1 henry when a voltage of 1 volt is induced by a uniform rate of current change of 1 ampere per second.

In studying the test circuit (Fig. 3-10), it is important to remember that when the switch was closed, the coil acted to oppose the flow of current. When the switch was opened, the coil acted to oppose the interruption in current flow. Inductance is that property of a circuit that opposes any change in current. This characteristic of inductive devices is important in understanding the subject of power factor, which is discussed in the section **Characteristics of Alternating Current**.

Generators

An English chemist and physicist named Michael Faraday (1791-1867) discovered another link between magnetism and electricity when he found that he could cause an electric current to flow in a wire by moving the wire through the field of a magnet.

A bar magnet can be bent into a U-shape to form a so-called horseshoe magnet. In this form, the north and south poles are brought closer together and the magnetic lines of force are concentrated in the space between. If a conductor is moved down through the space between the poles a voltage is induced in the wire and current will flow. Current will flow only as long as the conductor is moving. Moving the conductor up through the same space will again induce a voltage and current will flow. The current flow, however, will be in the opposite direction (Fig. 3-11). If the conductor is continuously moved up and down between the poles of the magnet, current will flow first in one direction, then in the other.

To induce a voltage in the conductor, the conductor must move through the magnetic lines of force. The conductor must, in effect, cut the lines. If the conductor is moved sideways so that it does not cut any lines of force, no current will flow. The real importance of Faraday's discovery lies in the fact that it provided a practical way to generate electricity.

Moving a conductor through a magnetic field requires mechanical energy. Such energy can be provided by coal, oil, or water power. In this way the energy in conventional fuels can be converted to electrical energy. In recent years, nuclear energy has been added to the list of fuels that can generate electricity.

Although the basic principles remain unchanged, many advances and refinements were necessary to develop Faraday's observations into practical electric generators.

Instead of a single loop of wire moving through the magnetic field, a device called an armature was developed (Fig. 3-12). An armature is a shaft on which many loops of wire are wound. The armature is shaped to fit

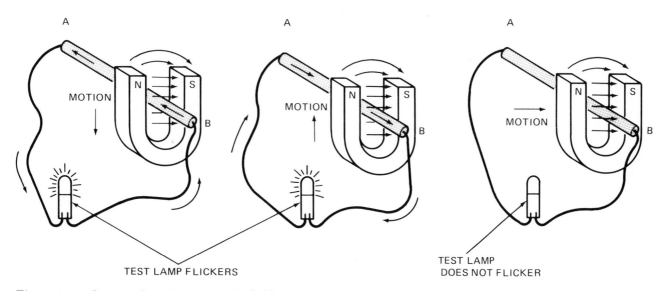

Figure 3-11. Current flow in a magnetic field.

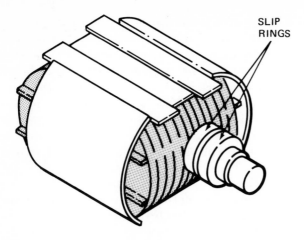

Figure 3-12. Typical ac generator armature.

closely between magnetic poles. The ends of the poles are curved to allow the armature to rotate even closer. This allows the loops on the armature to cut through the lines of force at their strongest point.

Because the electrical energy is induced in the armature, sliding contacts—called sliprings—are used to allow this energy to flow from the rotating armature. The armature loops are connected to metal rings on the end of the armature. Spring-type contacts mounted on the stationary part of the generator press against these rotating rings, conducting the electrical energy out of the generator. Because the stationary contacts brush against the sliprings, they are called *brushes*.

In present-day electric utility generating plants huge generators are used. In some the generator armatures

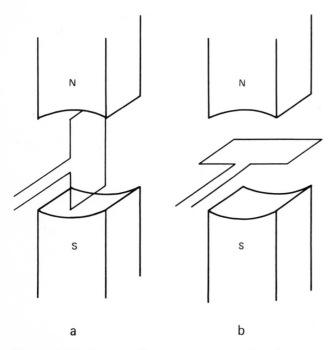

Figure 3-13. Current/flow in a single rotating loop.

are as large as railroad cars. In most generating plants, coal, oil, or nuclear fuels are used to make steam. The steam in turn drives a turbine that rotates these giant armatures. Of course, smaller generators are made for lower power outputs. These range from diesel-powered generators which supply emergency power for hospitals and office buildings down to generator-powered flashlights that can be held in one hand. Whatever the size or purpose of a generator, the action of the rotating armature produces electric power in a distinctive form.

THE FORM OF ALTERNATING VOLTAGE AND CURRENT • To understand why alternating current has its distinctive form, we will learn how voltage is induced in a single loop of wire as it rotates between the magnetic poles of the generator.

First, we will consider the general form of the generated power and then we will trace the generating process in more detail to develop the output waveform.

As a loop of wire moves into and through the position shown in Fig. 3-13a, the lines of force are being cut almost at right angles. In this area the voltage induced in the winding and the resultant current flow are greatest. When the loop has moved about 90° (Fig. 3-13b), both sides of the loop move almost parallel to the lines of force. At this location little or no voltage is induced in the loop.

During the next 90° of rotation, each leg of the loop moves into the position that causes the largest induced voltage. There is a difference, however. Both legs of the loop are cutting the lines of force in the opposite direction. This causes the polarity of the induced voltage to be reversed and current flows in the opposite direction.

From this it can be seen that the output of a generator will not be a constant voltage, as is the output of a battery. Generator output changes in magnitude and polarity constantly as the armature rotates.

To see what form this output takes, we will plot the output at several points as a loop of the armature is rotated. To simplify, we will represent the armature position by a single line. On the line to the right we will indicate the voltage and polarity. Of course, it takes time for the armature to rotate, so we will plot the angle of the armature on the line to the right to show the time relationship of the changes. As the line representing the loop is rotated counterclockwise, the height of the arrowhead above the starting line represents the strength of the induced voltage. If we plot this point to the right above the appropriate angle, these plotted points will show the way the voltage varies through one revolution. Assume the magnetic lines of force run from the top to the bottom of the page (north pole of the magnet above the armature circle of rotation; south pole below it). The voltage induced in the loop at each position is as follows (Fig. 3-14).

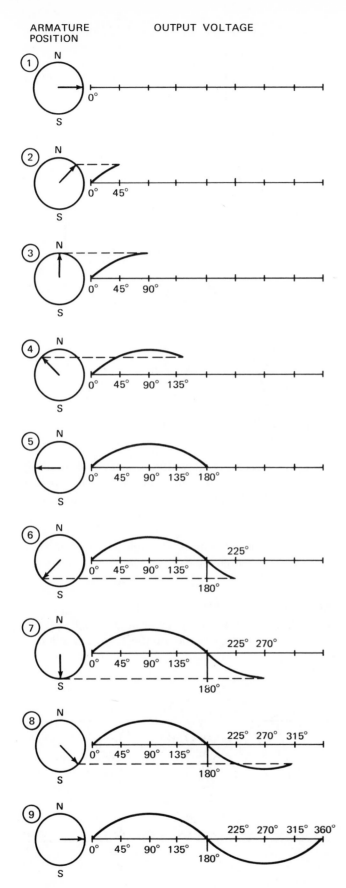

ARMATURE POSITION   OUTPUT VOLTAGE

Figure 3-14. Generation of one cycle of alternating current.

1. Lines of force pass through the loop. No voltage is induced.
2. At 45° the loop cuts the lines of force, but has not yet reached the maximum cutting angle. Some voltage is induced.
3. At 90° the loop cuts the lines of force at the sharpest angle. Maximum voltage is induced.
4. At 135° the condition is similar to 45°. The voltage drops from maximum to a lower level.
5. At 180° the lines of force again pass through the loop. No voltage is induced.
6. At 225° the loop again cuts the lines of force somewhat, but now the loop moves through the lines of force in the opposite direction. Voltage of opposite polarity is induced.
7. At 270° the loop cuts the lines with maximum efficiency, as at 90°, but still opposite in polarity to 90°.
8. At 315° the induced voltage has dropped from the peak, but some voltage is still induced.
9. At 360° or 0°, we are back where we started. No voltage is induced. One complete rotation of the armature represents one cycle of output voltage. The speed of rotation of the armature determines how rapidly the cycle is repeated.

Although we plotted only nine points, the points are connected by a curved line, rather than a straight one. If you place a ruler between the 0° and 90° points, you will see that the output at 45° does not fall on a straight line. A curved line must be drawn to connect the 0°, 45°, and 90° points. If more points were plotted, the overall shape of the curve would be as drawn.

A curve of this form is called a sine (pronounced "sign") curve. The name refers to one of the trigonometric functions of a right triangle. The sine of an angle is the ratio of the side opposite the angle to the hypotenuse of the right triangle. The angle in this case is the angle of rotation of the armature. The hypotenuse is the line that represents the armature. This line remains the same length at every angle. The ratio of the side opposite to the hypotenuse then varies as the side opposite varies. The side opposite is a vertical line drawn from the arrowhead to the base line (Fig. 3-15). At 0° the side opposite is 0 length, so the sine of 0° is 0. At 90° the hypotenuse and the side opposite are the same length, so the ratio is 1:1, or 1. The ratio at some points in between is shown in Fig. 3-15.

DIRECT CURRENT GENERATORS • Direct current generators operate exactly the same as ac generators operate. The dc output is obtained by switching the armature output every half cycle to reverse the polarity. The device that does the switching is called a commutator.

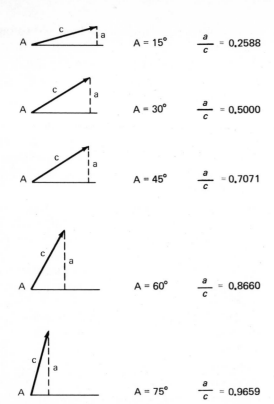

A = 15°	$\dfrac{a}{c}$ = 0.2588	
A = 30°	$\dfrac{a}{c}$ = 0.5000	
A = 45°	$\dfrac{a}{c}$ = 0.7071	
A = 60°	$\dfrac{a}{c}$ = 0.8660	
A = 75°	$\dfrac{a}{c}$ = 0.9659	

Figure 3-15. Sine wave values.

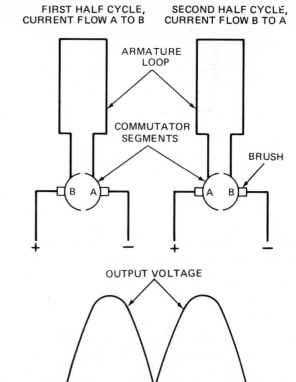

FIRST HALF CYCLE, CURRENT FLOW A TO B   SECOND HALF CYCLE, CURRENT FLOW B TO A

ARMATURE LOOP

COMMUTATOR SEGMENTS

BRUSH

OUTPUT VOLTAGE

Figure 3-16. Commutator switching.

The output of an ac generator is obtained from the armature through sliprings. Brushes mounted on the generator frame make continuous contact with the sliprings as the armature rotates. To obtain a dc output the sliprings are replaced by separate semicircular segments with insulating material between the segments. Each segment is connected to one end of the rotating loop. The segments are positioned on the armature so that the output is switched every 180° of rotation (Fig. 3-16). Consequently, current flow through the brushes is always in the same direction.

Actual armatures have many loops of wire, and actual commutators have many segments. However, the principle is the same. The two segments of the commutator that are making contact with the brushes always have the same polarity at the time of contact.

The output of a dc generator is not a steady dc, as is the output of a battery. Rather, it is pulsating dc. The polarity (direction of current flow) of the output remains the same, but the voltage level varies. This variation in output can be smoothed out by a process called filtering. This is described in the paragraphs on capacitance at the end of this section.

Transformers

Our study of generators showed that the energy produced by a rotating armature in a magnetic field constantly changes in amplitude and periodically changes in direction of flow. If this energy—alternating current—is applied to inductive devices, the applied energy can be changed or transformed in a number of ways. These transformations enable us to do many things that make electricity easier to use and easier to distribute over power lines. The inductive devices that make these things possible are, of course, transformers.

MUTUAL INDUCTANCE • The magnetic lines of force that expand and contract around a single coil not only induce a voltage in the coil itself, but can induce a

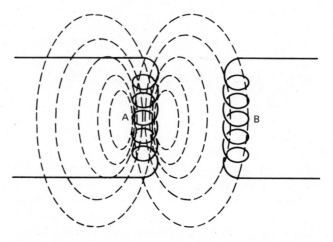

Figure 3-17. Mutual inductance between coils.

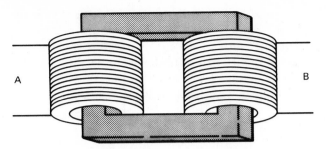

Figure 3-18. Iron core transformer.

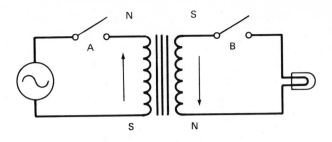

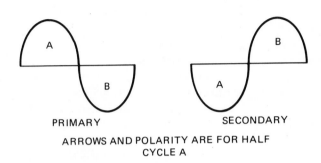

PRIMARY   SECONDARY

ARROWS AND POLARITY ARE FOR HALF
CYCLE A

Figure 3-19. Mutual inductance circuit.

voltage in another coil as well (Fig. 3-17). For maximum efficiency, both windings are wrapped on a common core (Fig. 3-18). Let us consider how a transformer works in a simple circuit (Fig. 3-19). We will connect one coil to a power source and the other coil to a light bulb. Figure 3-19 shows the conventional symbol for a two-coil transformer. The straight lines between the coil symbols tell you that the transformer has a solid iron core. The coil—or winding—that is connected to the voltage source is called the primary winding. The coil connected to the load is called the secondary winding.

When switch A is closed, alternating current is applied to the primary winding. Consider how the lines of force change in one complete cycle of source voltage. During half cycle A, the polarity of the voltage applied to the primary winding is as shown in Fig. 3-19 and current flows in the direction of the arrow. This causes the magnetic field around the primary winding to expand. The change in magnitude of applied voltage causes the field to expand slowly at first, then more rapidly as the peak is approached, then less and less rapidly, until at 0 voltage, the field has expanded to maximum.

The changing rate at which the primary field is expanding means the voltage induced in the secondary field increases slowly at first, then more rapidly as the peak is approached, then less and less rapidly. In other words, the voltage induced in the secondary field has the same form as the voltage in the primary field. However, the direction of current flow in the primary remains the same, and the lines of force in the primary have the same polarity during this half cycle. Note that the lines of force inducing the secondary voltages have the opposite polarity. This means the secondary voltage has opposite polarity and secondary current flow is opposite to primary current flow. The half cycle of voltage and current in the secondary is opposite in direction of flow and amount to that of the primary voltage and current. The second half cycle repeats the action of the first with direction of current flow reversed. Polarities are reversed and the second half cycle in the secondary winding is again a mirror image of the source voltage. This action will continue as long as the switch is closed.

If the secondary winding of the transformer has exactly the same number of turns as the primary, the voltage in the secondary is the same as the primary voltage. If we assume that the source voltage is 120 volts and the secondary load is a 100-watt lamp, we can calculate current flow in the circuit.

$$I = \frac{W}{V} = \frac{100}{120} = 0.83 \text{ ampere}$$

Current flow in the secondary circuit then is 0.83 ampere. This secondary load determines the current flow in the primary circuit. The primary current flow is also 0.83 ampere. If switch B is opened to turn off the lamp and interrupt current flow, current flow in the primary will also drop to zero. Similarly, if we add a second 100-watt lamp in parallel in the secondary, the secondary will require a current flow of 1.66 amperes and the current flow in the primary will also increase to this value. Current flow in primary and secondary transformer circuits is determined by the load requirements placed on the secondary circuit.

There are, of course, some practical aspects that must be considered in transformer circuits. Transformers are not perfect devices. There is some power loss in transformers, so that current flow in the primary and secondary circuits is never *exactly* the same. When the secondary circuit is opened, the current flow in the primary does not actually drop to zero. A small primary current flows even with no secondary load. In addition, transformers, like other devices, are manufactured to handle different maximum loads. For example, the size of the wire used in the primary and secondary windings must be suitable for the current and voltage with which the transformer is to be used. Transformers are manufactured in many sizes and shapes, to handle 350,000 volts or more on utility company power lines, or to handle as little as 8, 12, or 24 volts to ring doorbells.

STEP-UP AND STEP-DOWN TRANSFORMERS • The transformer we have considered so far had the same number of turns in both the primary and secondary windings. This type of transformer is called a *one-to-one* or *isolation* transformer. One-to-one means that voltage in the secondary is equal to voltage in the primary. Isolation describes the main use of a one-to-one transformer. There is no direct wire connection from the secondary to the power source. Transformers of this type are sometimes used to prevent shock and damage to instruments by isolating the secondary load from the power source.

It is far more common, however, for power transformers to have different numbers of turns in the primary and secondary windings. The reason is that the voltage in the secondary winding can be increased or decreased by changing the number of turns. If the secondary has twice as many turns as the primary, the voltage induced in the secondary will be twice that of the primary. If the secondary has half as many turns, the secondary voltage will be half that of the primary. To put this another way, the voltage induced in the secondary winding of a transformer is proportional to the ratio of the turns in the primary to the turns in the secondary. This is generally referred to as the *turns ratio*. If the secondary has more turns than the primary, the device is called a *step-up transformer*. If the secondary has fewer turns than the primary, the device is called a *step-down transformer*.

Note that the terms step-up and step-down apply only to voltage. Current flow in the secondary, in fact, moves in the opposite direction. Assuming the circuit load remains the same, if the voltage is doubled the current will be reduced to half. If the voltage is reduced to half, the current will be doubled. In other words, in a step-up or step-down transformer power available in the primary and secondary circuits is the same.

Our transformer test circuit can be changed to show how this works. First, we replace the one-to-one transformer with a one-to-two step-up transformer (Fig. 3-20). With 120 volts applied to the primary, the secondary voltage becomes 240 volts. With the same 100-watt load, we calculate the secondary current:

$$I = \frac{W}{V} = \frac{100}{240} = 0.416 \text{ ampere}$$

However, for the primary to supply 100 watts of power to the secondary, there must be 0.83 ampere at 120 volts in the primary.

If a step-down transformer is used (Fig. 3-21), the power requirements of the primary and secondary will remain the same. The secondary current will be 1.66 ampere at 60 volts. The primary current will remain 0.83 ampere.

At this point it may appear that no advantage is gained by the use of step-up and step-down transformers. However, if we recall some basic relationships, the advantage of power transformation will be easy to understand. Power is the product of voltage times current ($P = V \times I$). Voltage is electrical pressure. Current is electron flow. Transformers allow us to choose the predominant term in the power equation.

From Kirchhoff's voltage law we know that the voltage drop across any load is equal to the current flow through the load times the load resistance ($V = I \times R$). Electrical power must be carried long distances from utility generating stations to the towns and cities where power is used. For distances measured in miles the resistance of the conductor cables becomes a significant power load. If power had to be transmitted at high current levels, large amounts of power would be wasted in line voltage drop. Transformers allow the voltage level to be raised and current reduced so that energy is conserved by keeping line loss as low as possible. Step-up transformers at the generating station raise voltage levels to 350,000 volts or more. The corresponding decrease in current flow that accompanies this increase in voltage allows large amounts of power to be transmitted with little line loss.

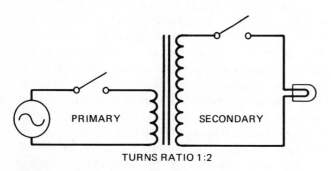

Figure 3-20. Step-up transformer circuit.

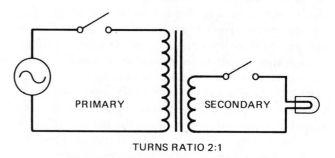

Figure 3-21. Step-down transformer circuit.

Of course, a voltage this high cannot be used in homes or stores, so step-down transformers are used at the end of the transmission line to reduce voltage to a few hundred volts. The current increase that accompanies the reduction in voltage provides the power needed by the end users.

MULTIPLE WINDING TRANSFORMERS • In addition to increasing or decreasing voltage between primary and secondary windings, transformers can provide two or more secondary voltages by multiple secondary windings or tapped windings.

The power distribution system that provides both 120- and 240-volt power for homes uses a multiple secondary winding. The pole transformer that delivers utility company power to a neighborhood location usually has about 4000 volts as the primary. The step-down secondary may be either two independent windings or one center-tapped winding, depending on service provided. If houses or apartments are provided with two-wire 120-volt service, each secondary winding can supply power to a different group of users (Fig. 3-22).

When houses or apartments are provided with three-wire 240-volt service, the two secondary windings are connected together and one wire common to both windings, plus a wire from the other end of each winding, make up the three-wire input power lines (Fig. 3-23).

AUTOTRANSFORMERS • In all the transformers we have considered so far the primary and secondary windings are connected only by induction. There is no direct-wire connection. However, this is not the only form of transformer that is used. A single winding with taps at various points can also step up or step down voltage. This type of transformer is called an autotransformer, because the voltage induced by the input power induces in turn a voltage in the other winding segment (Fig. 3-24).

Autotransformers are somewhat less expensive to manufacture, but do not provide electrical isolation between windings. This transformer is used principally on utility transmission lines.

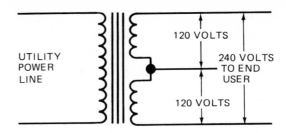

Figure 3-23. Three-wire 240-volt power transformer.

Motors

A detailed discussion of electric motors is beyond the scope of this text, but general background information on the subject is provided. The functional parts of motors are described first. This is followed by a summary of the main features of the most common types of motors. Special wiring requirements for electric motors are covered in Chapter 13.

The principles of induction make it possible to convert mechanical motion to electrical energy. This conversion is accomplished by generators. This action can be reversed; electrical energy can be converted to mechanical energy. This conversion is accomplished by motors.

The output of a generator is voltage and is expressed in volts. Motor output is mechanical movement and is expressed in terms of *torque* (pronounced "tork"). This word refers to the ability of the motor to overcome turning resistance. A motor with high starting torque is able to turn a heavy load quickly after power is applied.

Because of the similarity in action, motors and generators have the same major functional parts. In motors, the parts operate differently.

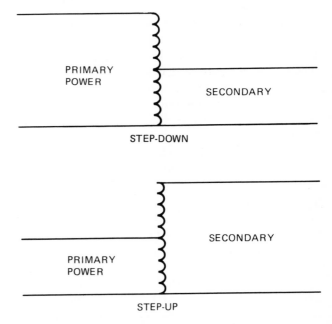

Figure 3-24. Autotransformer connections

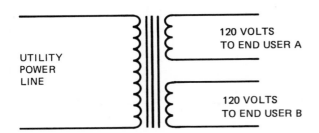

Figure 3-22. Two-wire 120-volt power transformer.

INDUCTIVE WINDINGS • The electromagnetic windings that produce electricity in a generator and motion in a motor are the field winding and the armature. The field winding produces the lines of force that are cut by the armature loops. In a motor the field winding must be connected to the power source to produce the necessary magnetic field. The armature must also be connected to the power source that drives the motor. As in a generator, the armature is made in the form of a drum containing many loops of wire.

In both ac and dc motors, rotation is produced by changing the polarity of induced fields in such a way that the force of attraction of opposite poles and repulsion of like poles "pushes" the armature around. In ac motors the change in polarity is produced by half-cycle alternation of the power source. In dc motors a switching device produces the change in polarity.

SLIPRINGS • These are the same devices used to conduct energy out of a generator armature. In ac motors sliprings are the rotating half of the electrical connection of source power to the armature.

BRUSHES • Brushes are the stationary half of the electrical connection of source power to the armature. The brushes rub against the spinning sliprings to complete the electrical circuit to the armature. Brushes are commonly made of graphite, a relatively soft material, and are shaped to fit the contour of the sliprings to assure continuous contact. Wear produced by the constant motion is compensated for by pigtail springs in the brush assembly. As the brushes wear down, the spring pressure keeps them in firm contact with the sliprings. After a time, of course, the brushes must be replaced.

COMMUTATORS • Commutators on dc motors are the equivalent of sliprings on ac motors. Instead of being solid rings of metal, commutators are made of metal segments with insulating material between the segments. The segments are connected to loops in the armature. As the commutator rotates, it makes contact with the brushes, just as sliprings do. As the brushes make contact with different pairs of segments, the dc power is switched to provide the polarity change necessary to make the armature rotate.

Types of DC Motors

SERIES MOTOR • The armature and field winding are wired in series (Fig. 3-25). This arrangement provides high starting torque for heavy loads. Automotive starter motors are series-wound. This type of motor must be operated with a load. If allowed to run free, the speed will continue to increase or "run away" until the motor is damaged.

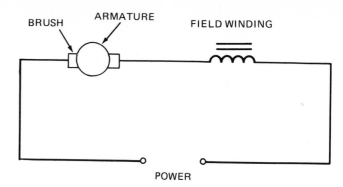

Figure 3-25. Series-wound dc motor diagram.

SHUNT MOTOR • The armature and field winding are wired in parallel (Fig. 3-26). This type of motor can maintain a constant speed even though the load varies.

COMPOUND MOTOR • The features of the series motor and shunt motor are combined to provide high starting torque and constant speed. The field winding is split so that part is in series and part is in parallel (Fig. 3-27).

AC OR DC UNIVERSAL MOTOR • These motors are generally used on small appliances and power tools. They are series-wired (Fig. 3-25) and they provide a large output for their size. Speed can be controlled by varying the voltage to either the field or the armature.

Types of AC Motors

SPLIT-PHASE MOTOR • This motor is used for easy-to-start loads such as fans and table saws. The motor contains a starting field winding and a running field winding (Fig. 3-28). The starting winding causes a voltage shift to get the rotor moving. A switch which is operated by centrifugal force that increases as the speed of the motor increases, switches out the starting

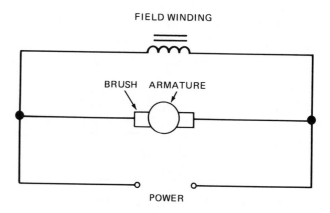

Figure 3-26. Shunt-wound dc motor diagram.

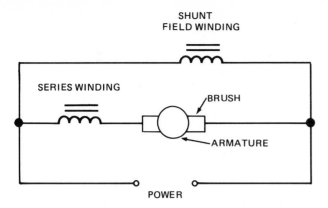

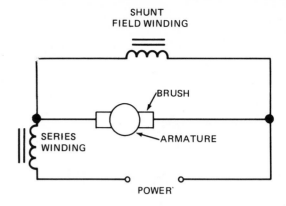

Figure 3-27. Compound dc motor diagrams.

winding when the motor reaches running speed. Flywheel effect keeps the rotor moving as long as power is applied.

CAPACITOR-START MOTOR • This motor is similar to the split-phase type, except that a device called a capacitor is in series with the starting winding (Fig. 3-29). Capacitors are described later in this section. The function here, however, is to increase the voltage shift. This increases the starting torque so that this motor can be used for heavier loads than the split-phase motor. As in the split-phase motor, a speed-activated switch cuts out the capacitor and the starting winding when the motor reaches running speed.

SHADED-POLE MOTOR • The field winding of this motor contains additional heavy copper coils. The copper coils produce a secondary magnetic field that acts with the field winding to make the motor rotate. This motor is used for loads that are easy to start, such as fans and blowers.

REPULSION-START MOTOR • This motor uses the additional force of repulsion to provide a strong initial "kick" to start heavy loads. During the starting

period, the field windings induce a voltage in the rotor. Secondary magnetic fields produced by rotor current are repelled by the field winding to provide high starting torque. A centrifugal switch cuts out the repulsion windings when the motor approaches full speed.

SYNCHRONOUS MOTOR • This is a simple induction motor designed to rotate in step with the alternating power source. The rotation of the armature is an even multiple of the source voltage frequency. This motor is used in clocks and other timing devices when constant speed is important.

Other Inductive Devices

SOLENOIDS • These are simple, dependable devices which produce back-and-forth mechanical movement. Solenoids consist of a coil of wire wound on a tube (Fig. 3-30), which contains an iron bar that is free to move in and out of the coil. A spring attached to one end of the bar holds the bar partly out of the coil. With the coil connected to a power source, the magnetic field of the coil pulls the iron bar into the tube. The magnetic field

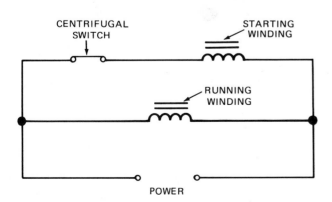

Figure 3-28. Split-phase ac motor diagram.

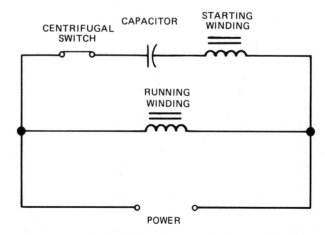

Figure 3-29. Capacitor-start ac motor diagram.

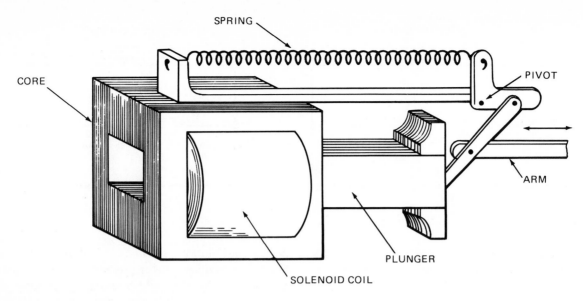

Figure 3-30. Solenoid operation.

produced by the coil holds the bar centered in the tube. When current to the coil is turned off, the spring pulls the bar out of the tube. Solenoids can operate directly from line voltages. They are used to control water valves, for example, on washing machines and dishwashers.

RELAYS • These are used primarily for remote switching. Relays consist of a coil with a fixed iron core. Stationary and movable switch contacts are mounted above the core in such a position that the magnetic field of the core attracts the movable switch contact. Current flow in the relay coil creates a magnetic field that causes the movable contact to make or break an electrical connection with the fixed contact.

Because the field of the relay coil must maintain the same polarity to hold the movable contact in position, most relay coils are designed to be operated from low voltage dc. Special types of relays, however, are made for use on ac. The coils of these relays have a laminated core and a shading coil. A shading coil is a single closed loop of copper that covers about half the core. The shading coil and the core laminations act to hold in the relay during current alternations. Drawing symbols for various types of relays are shown in Fig. 3-31.

VIBRATORS • Vibrators operate on the same principle as the solenoid, but they are designed to produce rapid, short-stroke movement (Fig. 3-32). When a switch is closed, current flows through an internal switch to a coil. The magnetic field attracts an iron arm. When the arm moves it opens the internal switch. The field collapses and the arm is pulled back by a spring. This closes the internal switch and the cycle repeats as long as the external switch is closed. Vibrators are used primarily as doorbells and buzzers.

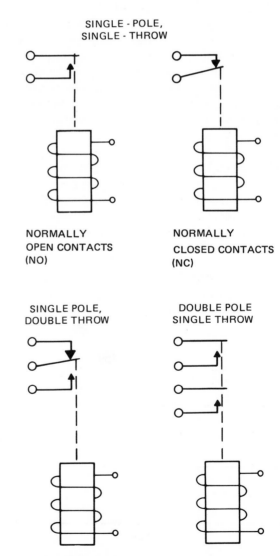

Figure 3-31. Relay symbols.

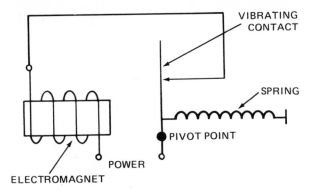

Figure 3-32. Vibrator operation.

Capacitance

From our study of induction we know that an electromagnetic field is created when current flows in a wire. Another kind of electrical field is created when a voltage difference exists between two points. The two points might be, for example, the positive and negative terminals of a battery or the power lines from a generator.

An important characteristic of this electrical field is that no current flows across it. Because no current flows, the field is called *electrostatic*. In this case "static" means stationary. There is no change in the electrical field as long as the voltage across the field remains the same.

An electrostatic field consists of lines of force, just as an electromagnetic field does. The form of an electrostatic field, however, is different. The lines of force in an electrostatic field represent the path that electrons would follow if they were free to move in the field. Because like charges repel each other and electrons are negatively charged, an electron in an electrostatic field will move away from the negative point toward the positive point. The lines of force, then, extend from the negative point to the positive point (Fig. 3-33).

When this force is present between points such as battery terminals and power lines, the force is quite

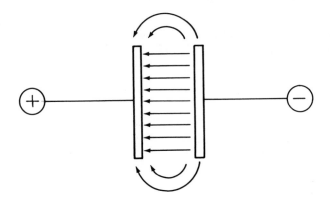

Figure 3-33. Electrostatic field.

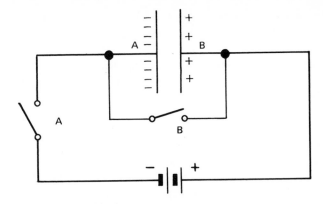

Figure 3-34. Capacitor test circuit.

small and of no practical value. The field must be made stronger to be useful. To increase the strength of an electrostatic field, the positive and negative charges must be spread over large areas and the space between these areas must be as small as possible.

The devices that provide this strong electrostatic field are called capacitors. The large area for each charge is provided by thin sheets of metal foil. To keep the distance between the sheets of foil as small as possible and still block current flow, insulating material, called a *dielectric*, is placed between them. The sheets of foil act as storage points for charged particles.

A simple test circuit (Fig. 3-34) shows how capacitors work. Assume that there is no charge on the capacitor. At the moment switch A is closed, electrons begin to flow from the negative battery terminal to plate A. As the electrons flow onto plate A, the increasing negative charge repels electrons from plate B. Thus, current flows in all parts of the circuit even though no current flows *through* the capacitor. During the period of time that electrons are flowing to plate A and away from plate B, there is high current flow and zero resistance in the circuit. With no apparent resistance in the circuit, there is no voltage drop across the capacitor. At this time current flow is maximum and voltage is zero. As electrons flow to plate A and away from plate B, a charge builds up on the plates; plate A becomes more and more negative, plate B becomes more and more positive. When the charge across the capacitor equals the battery voltage, current flow stops. The circuit remains in this condition indefinitely as long as the battery voltage remains constant. If switch A is opened, the capacitor will still retain the charge for a long period of time. Closing switch B shorts the capacitor. Electrons flow from plate A to plate B until the capacitor charge drops to zero. Capacitors, then, provide a means for short-term storage of an electrical charge.

Capacitors tend to oppose changes in voltage. We saw that when the capacitor was uncharged and voltage was applied to the circuit, the action of the capacitor slowed down the change in the circuit from zero voltage to the battery voltage. Another way of saying this is that the voltage change lagged the current change. When switch A was opened, current could no longer flow in the circuit, but the voltage on the capacitor remained equal to the battery voltage. Again, the voltage change lagged the current change.

The ability of a capacitor to hold a charge is its capacitance, which is measured in units call farads. The symbol for a farad is F. One farad is the amount of capacitance that can be charged to 1 volt by a current flow of 1 ampere for a period of 1 second. The farad is an extremely large unit. Practical capacitance is measured in millionths of a farad, or microfarads (μF). In electronic work a still smaller unit, the picofarad (pF) is used. This unit is a millionth of a microfarad.

When connected to an alternating current source, capacitors charge and discharge with each half-cycle change in polarity. Current flow in the circuit is then continuous. However, the charge on the capacitor opposes the changing source voltage, so that voltage peaks occur later than current peaks. In other words, the voltage continuously lags the current.

You will recall that inductive devices in a circuit had exactly the opposite effect. Inductive devices caused current peaks to occur later than voltage peaks. The action of these devices is an important part of the subject of power factor, discussed in **Characteristics of Alternating Current**. There are three types of capacitors in general use. Some typical capacitors and the circuit symbols that represent them are shown in Fig. 3-35. One widely used type of capacitor contains foil sheets with a thin sheet of wax paper or plastic between them. Because the dielectric (insulating material) is paper or plastic, these are often referred to as paper or plastic capacitors. A "sandwich" is formed by

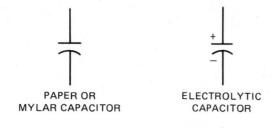

PAPER OR MYLAR CAPACITOR   ELECTROLYTIC CAPACITOR

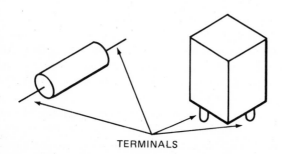

TERMINALS

Figure 3-35. Capacitor types and circuit symbols.

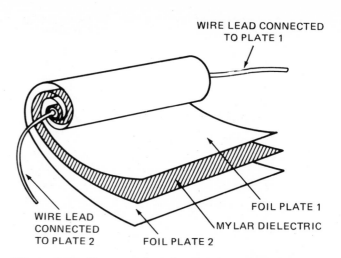

Figure 3-36. Construction of a plastic capacitor.

the foil sheets and the dielectric. A short length of wire is attached to each piece of foil, and the sandwich is rolled and enclosed in a paper or plastic tube (Fig. 3-36). Capacitors of this type are suitable only for light duty.

For heavier duty, capacitors known as oil-filled are made with oil as the insulating material. Foil sheets with spaces between them are placed in a metal can. The can is then filled with oil and sealed. The oil fills the space between the foil to act as the dielectric.

A third type of capacitor is called electrolytic. This type provides the largest capacitance for physical size and can be thought of as a special type of rechargeable battery. You will recall that the acid used in batteries is called an electrolyte. A similar chemical is used in the electrolytic capacitor. The electrolyte allows the capacitor to store a large charge in a physically small package. Like batteries, electrolytic capacitors have positive and negative terminals, and these polarities must be observed when electrolytic capacitors are wired into a circuit. Electrolytic capacitors are frequently used to filter or "smooth" the output of pulsating dc voltage, such as the output of a dc generator. The capacitors become charged on the voltage peaks and discharged in the "valleys."

• CHARACTERISTICS OF ALTERNATING CURRENT •

The Sine Wave

The interaction of the rotating magnetic fields in an ac generator results in an output that varies in amplitude and periodically changes polarity. If many instantaneous values of the generator output voltage and current are measured and plotted on a graph, and a smooth line is drawn through the plotted points, the result would look like Fig. 3-37. There are a number of

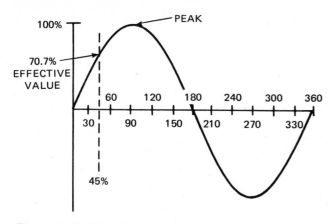

Figure 3-37. The sine wave.

characteristics of this sine wave variation that are important in electrical work.

Peak and RMS Values

First of all, how is this constantly varying voltage described? A simple average value could be calculated, but the average value of one half of the cycle could be positive and the other half negative, giving us an answer of zero. Since the average of each half cycle is the same, except for polarity, an average value without a polarity sign (+ or –) might be used. This average value has been calculated and is equal to 63.7 percent of the peak value. If the peak value is 170 volts, the average value is 108.3 volts. While this value is mathematically correct, it is electrically misleading.

Electrically, the sine wave represents either a voltage variation or a current variation. In using electricity, power is the main consideration. The relationship of both voltage and current to power involves the square of the value, as in $P = I^2R$ and $P = V^2/R$. A measure of the ability of a sine wave to produce power can be arrived at by averaging squared values. To do this, we use the same values along the curve that were used for the simple average, but we multiply each value by itself to square it. We then add the squares and divide by the number of values squared. This gives us an average squared value. The square root of this number is the value that represents the average of the square. This is called the root-mean-square, or rms, value.

The rms value for a sine wave is 70.7 percent of the peak value. Note that the rms value is a bit more than 10 percent greater than the simple average. The rms value is also known as the effective value. The effective value of ac corresponds to the same amount of direct current or voltage in heating power. The rms value is equal to the sine wave value at 45°, 135°, 225°, and 315°.

Whenever ac voltage and current are specified, unless otherwise stated the quantities are rms values. This applies to equipment nameplates, labelling, or hard-ware, wire, and cable values given in reference books and textbooks. The standard ac power used in most of the U.S. is referred to as 120 volts. This is the rms voltage; the peak voltage is 170 volts.

To find the peak value when the rms value is known, multiply the rms value by 1/0.707 or 1.414. To summarize the relationship of peak and effective values:

120 volts (rms value) = 70.7% of 170 (peak)

170 volts (peak) = 141% of 120 (rms)

Frequency

Secondly, how often do the sine wave swings occur? How many times does the cycle repeat in a given period of time? This value is determined by the speed of rotation of the generator armature. In the section **Alternating Current Machines** we saw that a single sine wave of voltage was generated each time an armature loop went through 360° of rotation. The number of cycles is called the *frequency* of the ac and is expressed in units called hertz, abbreviated Hz. This unit is named for a German physicist, Heinrich Rudolph Hertz. The hertz unit is based on a time period of 1 second. The standard ac line frequency is 60 Hz. This means the complete cycle of a sine wave from 0° to 360° occurs 60 times in every second. In years past, frequency was given in cycles per second, abbreviated cps. This abbreviation may still be found on older equipment. Hertz and cps mean exactly the same thing.

For this discussion we have used an rms value of 120 volts, a peak value of 170 volts, and a frequency of 60 Hz. These are the most widely used ac power line values in the U.S. Keep in mind, however, that some other values may be found. For example, rms values may be 110 volts, 115 volts, or 117 volts and the frequency may be 50 Hz. Regardless of these variations, the relationship of effective and peak values remains the same. Frequency affects primarily motors, transformers, and other inductive devices. These items may be usable at only one frequency. The nameplate on the device shows the frequency at which it should be used.

Phase Angles and Reactance

So far we have referred to the sine wave only in connection with voltage and current. In a simple ac circuit, such as a single 100-watt lamp, the source voltage and current flow will vary in accordance with the sine waveform. If we assume a 120-volt ac source, the current will be

$$I = \frac{P}{V} = \frac{100}{120} = 0.83 \text{ ampere}$$

The voltage and current sine waves are shown in Fig. 3-38. Although the two sine waves represent different

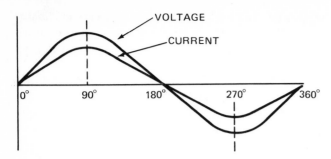

Figure 3-38. Voltage and current in phase.

POWER WAVE FORM

values and different forms of energy, each sine wave has the same form and the same relationship exists between rms and peak values. The peaks and the zero points of both sine waves occur at the same time. This is described as an "in-phase" condition. Voltage and current are in phase when both sine waves are "in step."

Let us now consider how *power* varies in this circuit. Variations in wattage can be calculated by multiplying the values of voltage and current at a number of points. If the power variations in the circuit are plotted, they will look like Fig. 3-39.

We know that the circuit load is 100 watts. This is the product of the rms values of voltage (120) and current (0.83), and is known as true power. We see on the power curve that the peak power is the peak voltage (170) multiplied by the peak current (1.17), or about 200 watts.

The power waveform has two interesting characteristics: first, it consists of *two* sine wave variations, and second, all of the power values are positive. Recall that the position and negative values of voltage and current represent a change in the *direction* of force and flow. Voltage and current produce power without regard to the direction of force and flow. Power, therefore, does just as much work whether voltage and current are positive or negative. The fact that power peaks in a positive direction twice produces the additional sine variation. Power then has twice the frequency of voltage and current. This fact, however, has no special significance in electrical work because power is related only to values of voltage and current, and is not dependent in any way on frequency. For example, power in a dc circuit has zero frequency, but the power is just as real.

We can summarize the relationship of voltage, current, and power as follows.

1. When voltage and current are in phase, all power produced is positive.
2. Positive power is known as true power or real power.

The terms true power and real power refer to power that gets work done. Naturally, we want circuits to

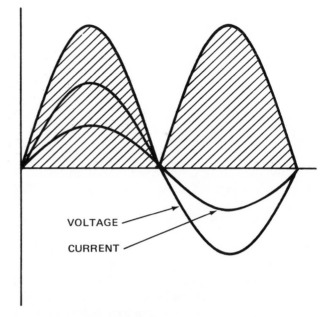

Figure 3-39. Waveform of power.

have as much real power as possible. Inductive and capacitive devices are the enemies of real power. We will see next how this happens.

INDUCTIVE REACTANCE • In the beginning of this chapter we used a simple test circuit (Fig. 3-10) to see how induction works. If necessary, review that portion of the text again.

There are two additional points of interest about voltage and current in the test circuit that we will now consider. We will use the same test circuit values, but we will substitute an ac power source (Fig. 3-40). When the switch is closed, the voltage drop across the coil—according to Kirchhoff's voltage law—will be equal to the generator voltage of 120 volts. The action of the coil in opposing changes in current flow means it requires a brief instant of time for current to flow through the coil. In other words, voltage reaches its peak value before current; or, current lags voltage in inductive devices. When the switch is opened, voltage across the load immediately drops to zero, but the collapsing field causes current to continue to flow. Again, current lags voltage.

The second point of interest about the test circuit is that the expanding and collapsing magnetic field occurs every time there is a change in amplitude or polarity of the power source. With an ac source this means there is constant opposition to current flow. In the test circuit

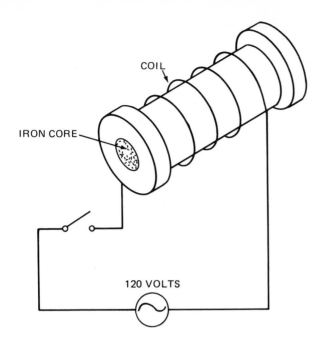

Figure 3-40. Inductive test circuit.

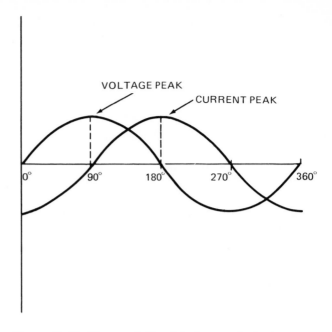

Figure 3-41. Phase relationship in pure inductance.

of Fig. 3-10 a dc source was used. Inductive opposition to current flow occurred only for an instant when the circuit was turned on and off. We could determine current flow (8 amperes) in that circuit from the dc resistance of the coil and the applied voltage.

With an ac source, however, there is another kind of opposition to current flow. This opposition to current flow is called inductive reactance. The presence of inductive reactance means that current flow in the circuit of Fig. 3-40 will be greatly reduced. The symbol for inductive reactance is X_L and it is measured in ohms.

If the inductance, L, in henries is known, the value of X_L can be calculated. The formula is $X_L = 2\pi fL$. The value of π is the ratio of the diameter of a circle to its circumference. This ratio is constant and equals 3.14. Twice the value is 6.28. The symbol f refers to the frequency of the source or 60 Hz. The first part of the formula, then, is a constant value, 6.28 multiplied by 60, or about 377.

When the source frequency is 60 Hz, the inductive reactance of any coil is equal to 377 multiplied by the coil inductance in henries. If the coil in Fig. 3-40 has an inductance of 2 henries, the inductive reactance in the circuit would be 377 multiplied by 2, or 754 ohms. If we assume the ac source is 120 volts, there would be only 0.16 ampere (or 160 milliamperes) of current in the circuit.

INDUCTIVE PHASE SHIFT • An analysis of what happens in an inductive circuit when power is applied has shown that the inductive device opposes changes in current flow. This causes the sine wave of current flow to shift with respect to the sine wave of voltage. The shift is such that the peaks and valleys of the current waveform occur after the peaks and valleys of the voltage waveform. We know that voltage is the force that causes current to flow. If voltage is strongest when current flow is zero, and voltage is zero when current flow is heaviest, little or no work will be done.

If a device could be built that was a pure inductance, the situation noted above could, in fact, happen. A pure inductance would cause a 90° phase shift between voltage and current. The result would be as shown in Fig. 3-41. If we calculate the power throughout a complete cycle we will see that the net result is zero. During the first 90° the voltage force is in a positive direction, but current flow due to reactance is in a negative direction. In other words, the voltage source is opposing the flow of current. We can multiply the values of voltage and current, but with voltage positive and current negative, the product will be negative. Negative power is a theoretical idea for use in circuit analysis, but it has no practical value. As was seen in the in-phase voltage and current relationship, all *real* power is *positive* power. Between 90° and 180° voltage and current are both moving in the same direction, but voltage is decreasing while current is increasing. This produces a positive power peak, but it is equal only to the negative peak, so the net result is zero. The second half cycle produces the same result.

It is electrically impossible to construct a coil having pure inductance. The wire itself has resistance and there is some capacitive effect between any two points of different potential, so a phase shift of 90° does not occur in practical situations. However, *some* power is lost in *any* phase shift. This means that any circuit containing a coil, a transformer, a motor, or any other inductive device will consume more power from the source than it uses to do work.

CAPACITIVE REACTANCE • Capacitors are devices that allow electrical charges to be stored for short periods of time. The plates of the capacitor act like reservoirs for energy. Current can flow onto the plate of an uncharged capacitor with little opposition. With no resistance to current flow, there is initially no voltage drop across the capacitor. This action means that current leads voltage in a capacitive circuit, exactly the reverse of the inductive effect. Capacitive reactance is the opposition to current flow that results from the capacitor's opposition to changes in voltage. Like inductive reactance, capacitive reactance is measured in ohms and it reduces current flow in a circuit. The symbol for capacitive reactance is X_C; the formula for calculating X_C is

$$X_C = \frac{1}{2\pi f C}$$

This formula is similar to the X_L formula, but is a reciprocal value.

As in the formula for X_L the expression $2\pi f$ can be considered a constant in 60-Hz power circuits. The value, again, is approximately 377. X_C, then, is equal to 1 over 377 multiplied by the capacitance in farads. The capacitance of capacitors used in power circuits is generally in millionths of a farad, or microfarads. For example, the value of X_C in a 60-Hz circuit of a 10-microfarad capacitor is

$$X_C = \frac{1}{377 \times 0.000010}$$

$$= \frac{1}{0.00377}$$

$$= 265 \text{ ohms}$$

The same calculation for a 20-microfarad capacitor is

$$X_C = \frac{1}{377 \times 0.000020}$$

$$= \frac{1}{0.00754}$$

$$= 133 \text{ ohms}$$

Because of the reciprocal form of the calculations, the larger the capacitor, the smaller the capacitive reactance.

Note that if we were working with a source voltage of higher frequency—such as 400 Hz—the value of $2\pi f$

would be 2512. If the capacitor value stayed the same, this would result in a decrease in capacitive reactance. In the same way, a lower frequency—for instance 50 Hz—reduces $2\pi f$ to 314 and consequently increases the capacitive reactance.

CAPACITIVE PHASE SHIFT • As in inductance, the phase shift would be 90° if the circuit contained pure capacitance. The direction of the shift would be the reverse of the inductive shift (Fig. 3-42). An analysis of a 90° phase shift in which current leads is essentially the same as the analysis when current lags. The same conditions of out-of-phase voltage and current exist, except in different 90° quadrants. The result is the same as in the pure inductive circuit. That is, zero real power would exist in a pure capacitive circuit.

It is just as impossible to build a pure capacitive device as it is to build a pure inductive one. In practical applications, then, capacitors in a circuit cause some phase shift, and consequently some power loss, but the phase shift is never 90° and power is never completely lost.

Impedance

If an ac circuit contains either inductive or capacitive reactance, this opposition to current flow as well as resistance must be taken into account when calculating current flow in the circuit. Impedance is the name for the total opposition to current flow in an ac circuit. The symbol for impedance is Z. Z represents a combination of resistance and reactance. Either inductive reactance alone, capacitive reactance alone, or both may be present in a circuit.

SERIES IMPEDANCE • When only one type of reactance is present, impedance in a *series* circuit is

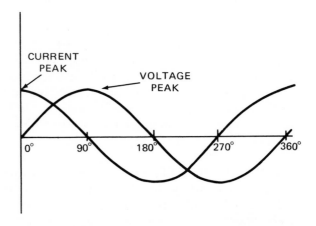

Figure 3-42. Phase relationship in pure capacitance.

calculated by squaring the resistance value in ohms, squaring the reactance value in ohms, adding the two, and finding the square root of that number. The formula is

$$Z = \sqrt{R^2 + X^2}$$

When both inductive and capacitive reactance are present in a circuit, they tend to cancel each other. The X term in the formula above is then the *net* reactance. Capacitive and inductive reactances cancel each other in accordance with their value in ohms. For example, in the circuit of Fig. 3-43, reactance of the capacitor X_C is is 70 ohms. The reactance of the coil X_L is 40 ohms. The net value of reactance X, then, is 30 ohms. This value is combined with the resistance value to calculate Z.

$$
\begin{aligned}
Z &= \sqrt{R^2 + X^2} \\
&= \sqrt{(20)^2 + (30)^2} \\
&= \sqrt{1300} \\
&= 36 \text{ ohms}
\end{aligned}
$$

Current flow is then

$$I = \frac{V}{Z} = \frac{120}{36} = 3.3 \text{ amperes}$$

PARALLEL IMPEDANCE • When resistive and reactive loads are in parallel, the total current flow is found by the formula

$$I_T = \sqrt{I_R{}^2 + I_X{}^2}$$

This formula says that the total current (I_T) is equal to the square root of the sum of current flow through the resistive load squared and current flow through the reactive load squared. Current flow through the resistive load is calculated by Ohm's law. If more than one resistive load is involved, the equivalent parallel resistance must be calculated using the methods described in Chapter 2.

When both inductive and capacitive reactances are in the circuit, the current flow through each must be calculated by Ohm's law and the net current will be determined by subtracting the smaller current from the larger. The calculation for the circuit of Fig. 3-44 is as follows.

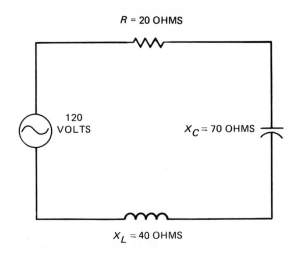

Figure 3-43. Series impedance circuit.

The current in the inductive load is

$$I_L = \frac{V}{X_L} = \frac{120}{40} = 3 \text{ amperes}$$

The current in the capacitive load is

$$I_C = \frac{V}{X_C} = \frac{120}{70} = 1.7 \text{ amperes}$$

The net reactive current is

$$I_X = I_L - I_C = 3 - 1.7 = 1.3 \text{ amperes}$$

The current in the resistive load is

$$I_R = \frac{V}{R} = \frac{120}{20} = 6 \text{ amperes}$$

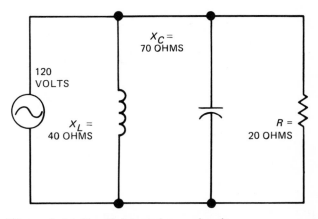

Figure 3-44. Parallel impedance circuit

The total current in the circuit is

$$I_T = \sqrt{I_R{}^2 + I_X{}^2}$$

$$= \sqrt{(6)^2 + (1.3)^2}$$

$$= \sqrt{36 + 1.7}$$

$$= \sqrt{37.7}$$

$$= 6.14 \text{ amperes}$$

SERIES-PARALLEL IMPEDANCE • Current flow in some series-parallel reactive circuits can be calculated by using the series and parallel formulas for each portion of the circuit. These formulas can be used only if the series part of the circuit and the parallel part of the circuit are either wholly resistive or wholly reactive. For example, current flow in a circuit such as that in Fig. 3-45 can be calculated as shown in the figure. Current flow calculations in more complex series-parallel circuits require mathematical skills that are beyond the scope of this book.

Power Measurement

The phase shift that occurs between voltage and current when a circuit has either inductive or capacitive reactance results in a loss of power. How then is power measured in the circuit? Do we measure the power the generator must produce or do we measure the power the circuit uses? The answer to this question is that we need to know both values of power and the ratio between them.

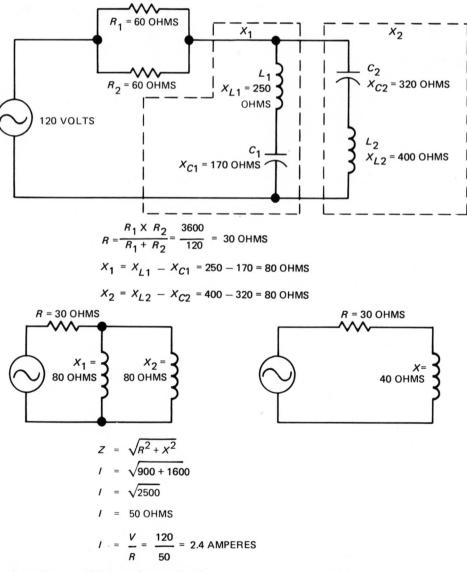

$$R = \frac{R_1 \times R_2}{R_1 + R_2} = \frac{3600}{120} = 30 \text{ OHMS}$$

$$X_1 = X_{L1} - X_{C1} = 250 - 170 = 80 \text{ OHMS}$$

$$X_2 = X_{L2} - X_{C2} = 400 - 320 = 80 \text{ OHMS}$$

$$Z = \sqrt{R^2 + X^2}$$

$$I = \sqrt{900 + 1600}$$

$$I = \sqrt{2500}$$

$$I = 50 \text{ OHMS}$$

$$I = \frac{V}{R} = \frac{120}{50} = 2.4 \text{ AMPERES}$$

Figure 3-45. Simple series-parallel impedance circuit.

Electricians use test instruments to measure voltage, current, and power in circuits. These test instruments are described in Chapter 5. For this discussion of power measurement, we will refer to one test instrument, a volt-amp-watt meter. This instrument contains scales calibrated in volts, amperes, and watts. Terminals on the instrument allow it to be connected to various points in a circuit. Depending on how the instrument is set up, and how the connections are made, the meter needle will move across the scales to show how much voltage is present, how much current is flowing, or how many watts of power are being used.

If this instrument is used to make measurements in a circuit containing both resistance and reactance, it will be found that the wattage in the load is less than the product (multiplication) of the applied voltage and the current. The difference between the values is the amount of power lost because of reactance. Several terms are associated with the subject of ac power measurement. They are defined below.

WATTS • In any electrical circuit watts refers to the real or true power in the load. Watts *always* refers to the actual work being done at any given moment. The terms real power and true power mean watts.

VOLTAMPERES • This term refers to the product of the applied voltage and the current as measured by the meter. In dc circuits when all loads are resistive, volts multiplied by amperes equals watts. In ac circuits if reactance is present, the product of volts and amperes will be greater than the wattage. The difference will be the reactive loss. Voltamperes tell what electrical load the circuit must be able to carry and what the load on the generator will be.

POWER FACTOR • This term, abbreviated *PF*, is the ratio of watts to voltamperes. It is a measure of the efficiency with which electrical power is being used. In a 120-volt circuit, if a current flow of 10 amperes was measured, voltamperes would equal 1200. If the wattage measured in the load is 960, the power factor *PF* would be

$$PF = \frac{watts}{voltamperes} = \frac{960}{1200} = 0.8$$

This would be described as a power factor of 80 percent.

In practical situations, power factor becomes a problem primarily in electrical circuits used in light and heavy industry. Industrial circuits often provide power for several electric motors. This creates a large inductive reactance which in turn causes a low power factor. Low power factor means wasted energy. To get the most work out of the power we generate, power factor must be kept as close to 100 percent as possible.

To review, in an inductive circuit current *lags* voltage and the phase shift between current and voltage causes power to be lost. Consequently the current flowing in the circuit multiplied by the applied voltage is greater than the wattage consumed by the motors. Recall, too, that in a capacitive circuit current *leads* voltage. This provides a method of improving the power factor.

Large capacitors wired in parallel with the motors add capacitive reactance to the circuit. This offsets the inductive reactance and raises the power factor to an acceptable level.

• REVIEW QUESTIONS •

1. The patterns formed by the iron filings demonstrated what two things about magnetic poles?

2. When current flows through a conductor a magnetic field is formed around the conductor. What can be done to the conductor to increase the strength of the field?

3. If a conductor is moved down through a magnetic field in such a way that it cuts the magnetic lines of force, current will flow in the conductor. What happens if the conductor is then moved up through the magnetic field?

4. Mechanical energy is converted to electrical energy in a generator. The electrical energy is taken from the armature of the generator through either sliprings or a commutator. What difference does this make in the form of the electrical energy?

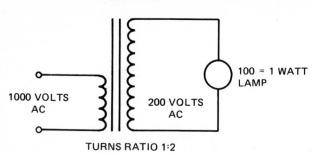

TRANSFORMER A

1000 VOLTS AC

200 VOLTS AC

100 = 1 WATT LAMP

TURNS RATIO 1:2

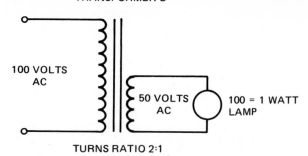

TRANSFORMER B

100 VOLTS AC

50 VOLTS AC

100 = 1 WATT LAMP

TURNS RATIO 2:1

5. Two transformers are shown above. If power is applied, lamps A and B will light. Answer the following for both transformers A and B.
   a. How much current flows in the primary?
   b. How much current flows in the secondary?
   c. What is transformer A called? Transformer B?

6. Two significant points in a sine wave are the peak value and the rms value. Complete these statements. The _____value is 70.7 percent of the _____value. The _____value is 141 percent of the _____value.

7. Frequency is expressed in units called hertz. On what unit of time is the hertz based?

8. In inductive circuits a phase shift occurs between the sine waves of voltage and current. What effect does this have on the efficiency of the circuit?

9. Opposition to current flow caused by inductive or capacitive devices is called reactance. The total opposition to current must include _____as well as reactance. What is the combination called?

10. In a 120-volt circuit, current flow is measured at 6 amperes. A wattmeter measures 576 watts in the load. What is the power factor of the circuit?

4
SAFE WIRING--
SAFETY ON THE JOB

• INTRODUCTION •

When Thomas Edison's incandescent electric light was first introduced to the public, one of its main advantages was thought to be reduced risk of fire. Kerosene lamps, candles, and gas lights were, of course, extremely hazardous. Fires were an ever-present danger and some—such as the Chicago fire of 1871 that destroyed almost one-third of the city—were major disasters. Anything that reduced the risk of fire was sure to be popular. It wasn't long, however, before it became evident that electricity could also be hazardous if not properly handled.

Everyone knew that lightning was a form of electricity and could be fatal. What was not so well known was that the relatively low levels of dc power generated in the early days of electric lighting could also cause fatal shock and could be a serious fire hazard.

The dangers of electricity became known as the use of electricity became more widespread. The need for uniform standards for electrical wiring and safe, reliable electrical devices was clear. This need was met by the introduction of the National Electrical Code (NEC) and the establishment of the Underwriters' Laboratories, Inc. (UL). As time went on, local building codes were expanded to cover electrical installations. Local electrical codes generally were based on the NEC.

This chapter tells you how the NEC, the UL, and local codes affect the electrician's work. After long experience, electricians and inspectors become thoroughly familiar with the NEC and local codes and consult them only when unusual situations occur. For the student, however, the NEC should be thought of as a reference book, like a dictionary. To use it efficiently you must know how it is organized and what kinds of information it contains. This chapter also tells you about the other documents that control electrical work. You will learn how all these things are related and what kinds of information you can expect to get from each.

For an electrician, the rules for safety on the job are much the same as they are for other construction trades:

1. Stay alert.
2. Use tools properly.
3. Wear protective clothing.
4. Follow established procedures.
5. Avoid unnecessary risks.
6. Don't take short cuts.

These rules apply in any area where building construction or modification is taking place. Electricians, of course, must also know how to work safely with, and near, live electrical conductors. Electric shock is a serious matter. It can be fatal.

However, electric shock can be avoided by following a few simple rules. These rules make sense if you understand how electric shock happens and what it is.

• THE PALACE OF ELECTRICITY •

In 1893 the 400th anniversary of the discovery of America was celebrated with the huge World's Columbian Exposition in Chicago. The site of the exposition on the shores of Lake Michigan became known as "White City" because of the construction material used for the buildings. The most spectacular feature of this exposition was the wide use of electric lighting, especially in one building called the Palace of Electricity.

This exposition represented, by far, the greatest use of electric lighting that had ever been attempted. The white construction material that appeared solid was, in fact, only wood framing covered with jute fiber and plaster. Wiring for the lights was strung around and through this material in an unplanned, random way. The result was a series of fires that, luckily, did not seriously damage the exposition, but did impress Chicago insurance executives with the dangers of haphazard electrical wiring and the use of untested electrical devices.

• THE NATIONAL ELECTRICAL CODE •

As early as 1881 an organization known as the National Association of Fire Engineers recognized the need for uniform, country-wide rules and guidance for electrical wiring. Their first meeting—held in Richmond, Virginia—resulted in a proposal that covered basic rules of insulating and grounding for safety.

The lessons learned from the Palace of Electricity made it clear that expert advice must be made available to people doing electrical wiring. In 1895, the National Board of Fire Underwriters (an association of fire insurance companies) published the proposals made at the fire engineers' meeting. This was the first appearance of a nationally recommended electrical code.

In the years since it was introduced the code has grown and changed as technical knowledge has increased and the uses of electricity have increased. Today it is known as the National Electrical Code (NEC). It is printed and distributed by the National Fire Protection Association (NFPA) and is universally accepted as the basis for safe electrical wiring. The

NFPA has established a procedure for periodic review and revision of the Code. Recently, the NEC has been endorsed by the American National Standards Institute (ANSI) and is also known as NFPA 70-1978 (ANSI).

As published by the NFPA and endorsed by the ANSI, the NEC is an advisory document only. The Code is offered for use by lawmakers and regulatory agencies as the basis of local electrical standards and building codes. The NEC becomes "law for electricians" only when it is made a part of local building codes. However, almost all local codes *do* reference the NEC. Statements such as the following are frequently used in local codes. (Material is from the "Code Manual for New York State Building Construction Code" and is quoted with permission.)

> *Electrical Wiring and Equipment*
> *General Requirements*—Electrical wiring and equipment designed and installed in conformity with ANSI, National Electrical Code, shall be deemed to meet the requirements of this code.

In this way the NEC becomes legally enforceable. Local codes generally contain additional requirements and restrictions which must also be followed.

The NEC is concerned only with electrical wiring practices that offer maximum protection against personal injury or death and property loss from shock or fire. Local codes are often concerned also with maintaining community housing standards, specifying height and location of overhead lines, providing adequate electric service, and similar considerations.

Local electrical codes are generally included as part of a complete building code for the area. The local code may be a page or two or many pages containing much detail. It is essential that every working electrician be familiar with the local electrical codes as well as the NEC.

The National Electrical Code is a detailed and comprehensive volume of almost 900 pages. The organization of material is logical, however, and numbering of items and cross-references makes it easy to use with a little study. The Code contains nine chapters, an appendix, a table of contents, and an index. Chapter 1 provides definitions of terms used in the NEC and general requirements for electrical installations. Chapters 2, 3, and 4 provide most of the information needed for wiring in homes and general commercial buildings. Chapters 5, 6, 7, and 8 cover special installations, special conditions and requirements for communications, radio, and TV wiring. Chapter 9 consists of reference tables and examples of how the tables should be used.

Code chapters are divided into articles and articles are divided into sections. Figure 4-1 identifies the main features of a typical page of the NEC.

The NEC is available in two forms. Figure 4-1 is a page from the basic edition, which contains only the text of the Code itself. Another edition, known as "The National Electrical Code Handbook," not only contains the full text of the Code, but includes additional text and illustrations that provide commentary on the provisions of the Code. Different color printing is used to indicate which material is the Code and which is commentary.

There is also a simplified version of the NEC known as "One- and Two-Family Residential Occupancy Electrical Code." This edition contains only the most popular and widely used wiring methods; it is limited to material related to the buildings mentioned in the title.

• THE UNDERWRITERS' LABORATORIES •

The Chicago Board of Fire Underwriters retained an early electrical expert named William Henry Merrill to investigate the fires in the Palace of Electricity. As a result of these investigations, Merrill realized the need for establishing standards for electrical devices. In 1894 he founded a company called the Underwriters' Electrical Bureau. Merrill and a few associates tested electrical devices and issued reports on their performance. The company grew rapidly. In 1901 it became known as Underwriters' Laboratories, Inc. (UL). Today this company is the most widely used testing laboratory in the United States. Its initials and seal should be known to every electrician (Fig. 4-2).

Manufacturers submit products to Underwriters' Laboratories for test. After testing, UL issues a report to the manufacturer indicating its findings. If any product shortcomings are uncovered, they must be corrected by the manufacturer and new samples submitted for test. Products that perform satisfactorily are listed in various UL product directories. The directories of interest to electricians are the Electrical Appliance and Utilization Equipment Directory, Electrical Construction Materials Directory, and Hazardous Location Electrical Equipment Directory.

In addition to the initial test, Underwriters' Laboratories performs follow-up tests on listed products to make sure that quality standards are maintained. Manufacturers whose products are listed in the UL directory are permitted to display the UL symbol on product nameplates or elsewhere on the product.

The NEC does not specify the use of UL-listed products by name, but does specify that equipment used be "tested by an electrical testing laboratory that is nationally recognized." Most local codes interpret this to mean UL-listed products, and most local inspectors look for the UL symbol on all materials used.

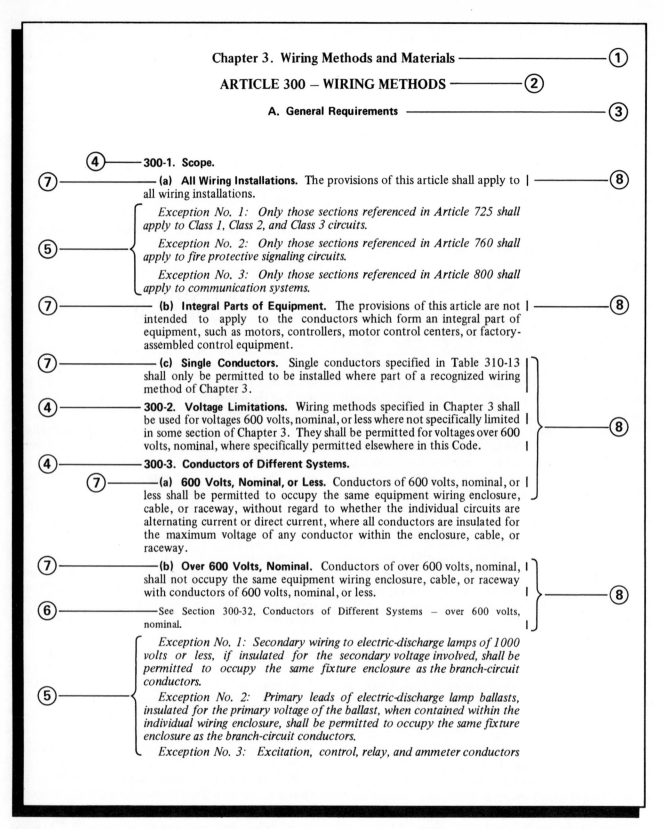

Chapter 3. Wiring Methods and Materials ————————— ①

ARTICLE 300 – WIRING METHODS ————— ②

A. General Requirements ——————————— ③

④ —— **300-1. Scope.**

⑦ ————————— **(a) All Wiring Installations.** The provisions of this article shall apply to | ———— ⑧
all wiring installations.

⑤ {
Exception No. 1: Only those sections referenced in Article 725 shall apply to Class 1, Class 2, and Class 3 circuits.

Exception No. 2: Only those sections referenced in Article 760 shall apply to fire protective signaling circuits.

Exception No. 3: Only those sections referenced in Article 800 shall apply to communication systems.
}

⑦ ————————— **(b) Integral Parts of Equipment.** The provisions of this article are not | ———— ⑧
intended to apply to the conductors which form an integral part of equipment, such as motors, controllers, motor control centers, or factory-assembled control equipment.

⑦ ————————— **(c) Single Conductors.** Single conductors specified in Table 310-13 shall only be permitted to be installed where part of a recognized wiring method of Chapter 3.

④ —— **300-2. Voltage Limitations.** Wiring methods specified in Chapter 3 shall be used for voltages 600 volts, nominal, or less where not specifically limited | in some section of Chapter 3. They shall be permitted for voltages over 600 volts, nominal, where specifically permitted elsewhere in this Code. | ———— ⑧

④ —— **300-3. Conductors of Different Systems.**

⑦ ————— **(a) 600 Volts, Nominal, or Less.** Conductors of 600 volts, nominal, or | less shall be permitted to occupy the same equipment wiring enclosure, cable, or raceway, without regard to whether the individual circuits are alternating current or direct current, where all conductors are insulated for the maximum voltage of any conductor within the enclosure, cable, or raceway.

⑦ ————————— **(b) Over 600 Volts, Nominal.** Conductors of over 600 volts, nominal, | shall not occupy the same equipment wiring enclosure, cable, or raceway with conductors of 600 volts, nominal, or less.

⑥ ————————— See Section 300-32, Conductors of Different Systems – over 600 volts, nominal. | ———— ⑧

⑤ {
Exception No. 1: Secondary wiring to electric-discharge lamps of 1000 volts or less, if insulated for the secondary voltage involved, shall be permitted to occupy the same fixture enclosure as the branch-circuit conductors.

Exception No. 2: Primary leads of electric-discharge lamp ballasts, insulated for the primary voltage of the ballast, when contained within the individual wiring enclosure, shall be permitted to occupy the same fixture enclosure as the branch-circuit conductors.

Exception No. 3: Excitation, control, relay, and ammeter conductors
}

Figure 4-1. How to read the National Electrical Code (Sheet 1 of 2). *(Reproduced by permission from NFPA 70-1981, National Electrical Code, ® Copyright © 1980, National Fire Protection Association, Quincy, MA)*

① Chapter headings appear in the table of contents and in boldface type in the index.

② Article numbers and titles are printed in capital letters. The first numeral in the article identifies the chapter in which it appears.

③ Some articles are divided into parts identified by a capital letter (A, B, C, etc.) and a title.

④ Each boldface heading in an article is called a section. Sections are numbered with a two-part number. The first part identifies the article in which the section appears; the second part includes the order in which the sections appear.

Article numbers and section numbers are not necessarily in sequence. That is, not all numbers are used. For example, the next article after 300 is 305; the last section in Article 300-A is 300-22; the following section in 300-B is 300-31.

⑤ Exceptions to the provisions of each section are in italic type.

⑥ Items in small type are provided to explain provisions of the section, or to provide cross-references to related sections or articles. Read cross-references carefully to make sure you refer to the correct article. Always check listed cross-references. Tables in chapters 1 through 8 have the same number as the section to which they apply, and are located as close as possible to the section. (Tables in chapter 9 are numbered sequentially.) Tables often have notes following that add important information. Read the notes as well as the tables.

⑦ Sections may be further subdivided by (a), (b), (c), etc. In addition, there may be numerical breakdowns (1), (2), (3) under these headings.

⑧ Vertical lines in the outer margin indicate areas in which changes have been made in the current edition.

Figure 4-1. How to read the National Electrical Code (Sheet 2 of 2).

Note that Underwriters' Laboratories lists all acceptable *products* (not manufacturers). The fact that one product has the UL symbol does not mean that all products from the same manufacturer are necessarily acceptable; however, reputable electrical equipment manufacturers distribute only products listed by UL.

It is important to know how the words "listed" and "approved" are used in connection with electrical products and installations. Underwriters' Laboratories lists in their directories all products they have found to be acceptable. This is not the same as approval. Approval can come only from a local electrical or building inspector. Approval means that a listed product is properly installed, mounted, or wired in accordance with the local electrical code.

• ARCHITECTURAL DRAWINGS AND SPECIFICATIONS •

The basic requirements for an electrical installation are established by the architect, often with a staff of experts, who designs a building and prepares the working drawings from which the building will be constructed. These drawings provide information for all the trades involved in construction: plumbers, painters, carpenters, and masons, as well as electricians.

The electrician must be able to read and understand these drawings and determine which items are important to the electrical installation. Chapter 12 covers electrical drawings.

Architect's drawings generally consist of a plot plan, separate drawings for each floor or level of the building, and front and side elevation drawings. The plot plan shows the location of the building on the plot of ground on which it will be built. This drawing usually shows where the electrical service entrance will be located. Floor plans show the location of electrical outlets, switches, fixtures, and appliances. Special features, such as exhaust fans, are shown or covered by notes. Front and side elevation drawings show the layout and dimensions of the sides of the building. In some cases, a separate electrical diagram, such as Fig. 4-3, will be provided.

Because it is not practical to include all necessary information on drawings, the architect prepares another document called a building specification. A part of this specification covers electrical work. The electrical specification details the material to be supplied by the electrical contractor. Information such as type of ceiling and wall fixtures, quality of materials, and colors and finishes of switches and receptacles are covered in the electrical specification. The specification generally states that all work must be done in accordance with local codes, and the National Electrical

Types of labels which appear on products covered by Underwriters Laboratories Inc. Follow-Up Services.

Figure 4-2. Use of the Underwriters Laboratories symbol. (*Underwriters Laboratories Inc.*)

Code. The specification may also state that all material must be UL-listed.

The electrical contractor must be able to review the architectural drawings and specification and prepare an estimate for the work to be performed.

• LOCAL ELECTRICAL CODES •

Local electrical codes are part of local building codes. They may contain many requirements to suit special local conditions. Some examples of local code information follow. (Material is from the "Code Manual for New York State Building Construction Code" and is quoted with permission.)

Electrical Service

Service conductors shall have an insulating covering which will normally withstand exposure to atmospheric and other conditions of use and which will prevent any detrimental leakage of current to adjacent conductors, objects or the ground, except that a grounded service conductor without insulating covering may be used where the voltage between any conductor and ground does not exceed 300 volts.

Underground service conductors shall be installed in duct, conduit, or in cable approved for the purpose. Such conductors installed in duct or conduit shall be TW or of other types approved for the purpose.

Electrical Layout for One-Family Dwelling

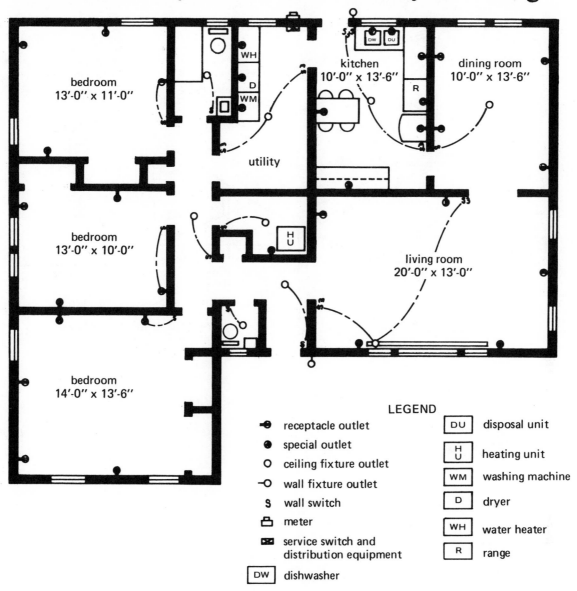

Figure 4-3. Electrical layout for one-family dwelling. (*Reproduced by permission of State of New York Division of Housing and Community Renewal*)

Service conductors within 8 feet of the ground, extending along the exterior of buildings, shall be installed in rigid conduit, electrical metallic tubing, busways or in cables approved for the purpose. Service conductors should not be run within the hollow spaces of frame buildings.

Overhead service conductors shall be installed at sufficient height to maintain the clearances indicated in the illustration entitled, "Overhead Electrical Service," . . .

NOTE: The illustration "Overhead Electrical Service" is Fig. 4-4.

Electrical Loads

In dwelling occupancies other than hotels the load for general lighting shall be determined on a basis of 3 watts per square foot of floor area. The floor area shall be computed from the outside dimensions of the building, apartment or area involved, and the number of floors, but not including open porches, garages or unfinished and unused space unless adaptable for future use.

To the calculated general lighting load there shall be added a load of 4500 watts for each dwelling unit in order to provide for portable appliances used in kitchen, dining room, and laundry.

Overhead Electrical Service

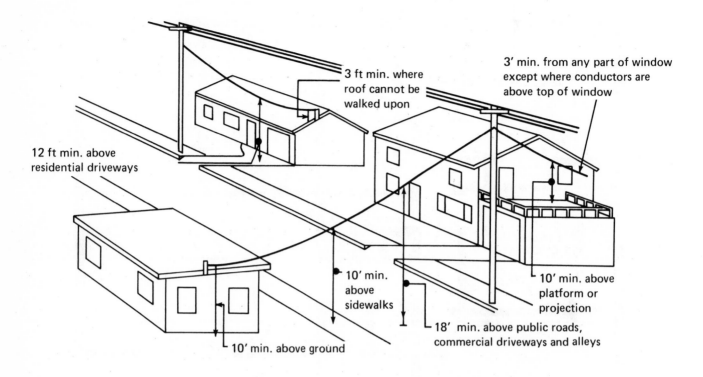

3 ft min. where roof cannot be walked upon

3' min. from any part of window except where conductors are above top of window

12 ft min. above residential driveways

10' min. above sidewalks

10' min. above platform or projection

10' min. above ground

18' min. above public roads, commercial driveways and alleys

MINIMUM CLEARANCES OF OVERHEAD SERVICE CONDUCTORS

Figure 4-4. Minimum clearances of overhead conductors (typical). (*Reproduced by permission of State of New York Division of Housing and Community Renewal*)

A demand factor of 100 per cent shall be applied to the first 3000 watts or less of general lighting load, 35 per cent to the next 117,000 watts, and 25 per cent to any amount over 120,000 watts.

NOTE: The figure of 3 watts per square foot for general lighting is used to estimate what the lighting load will be when the house is occupied. If a house has a floor area of 1500 square feet, it is assumed that the total wattage of all the lamps in the house will not be greater than 4500 watts.

NOTE: The phrase "demand factor" refers to the amount of lighting or other load that would *actually be in use* at any one period of time compared to the maximum load possible. In the example above, the general lighting load of 4500 watts would be reduced as follows:

3000 watts × 100 percent	=	3000 watts
1500 watts × 35 percent	=	525 watts
		3525 watts

The figure of 3525 watts would be used to calculate the electrical load in the house.

Local codes may specify minimum lighting requirements for residences.

Artificial Lighting

General Requirements—Lighting fixtures should be installed in kitchens, bathrooms, laundry rooms, dining rooms, basements, cellars, accessible attics, stairways and hallways, and wherever artificial light is required.

A lighting fixture should be installed so as to illuminate the front of the furnace. A lighting fixture should be installed over laundry equipment in a basement or cellar. A lighting fixture should be installed at the mirror in the bathroom.

Lighting fixtures, where provided in clothes closets, shall be installed on the ceiling or on the wall above the door. Pendant fixtures shall not be installed in clothes closets.

Switches to control the lighting fixtures should be located at the main entrance to each room. It is recommended, where no lighting fixture is provided in a room, that at least one receptacle outlet be controlled by a switch located at the entrance. Switches in bathrooms should not be located within reach of bathtub or shower.

At least one lighting fixture in an accessible attic should be controlled by a switch located at the foot of the stairs. At least one lighting fixture in a basement or cellar should be controlled by a switch located at the head of the stairs. Lighting fixtures illuminating a stairway between stories in a one- or two-family dwelling should be controlled by switches located at the head and foot of the stairs.

Switches controlling required artificial lighting, where accessible to other than authorized persons, should be of the lock type.

Television antennas and lead-in wire must be installed so that contact with power lines is not likely. Local codes may cover this subject.

Television Antenna Installation

Mast and supporting structure of television antenna should be substantially constructed so as to be capable of withstanding the wind and ice loads to which they are subject. It is recommended that masts extending more than 8 feet above the top of their support be guyed. Guys should be of galvanized steel, copper-covered steel, bronze, or other corrosion-resistant material, not less than No. 14 AWG. Where stability of the mast is dependent on guys, there should be at least three, approximately equally spaced about the mast.

Masts may be secured to side walls, to parapet walls, or to chimneys in good structural condition. Attachment to chimneys is not recommended, but where unavoidable, should be made by means of two or more substantial iron straps encircling the chimney. Attachment should not be made to chimneys in poor structural condition or by means of holes drilled into the chimney.

Antenna should not be installed in close proximity to electric service lines. A distance sufficient to prevent contact if the antenna overturns is recommended.

Lead-in conductors attached to buildings should be installed so as to maintain not less than the following clearances from other conductors: 4 inches from conductors of circuits not exceeding 150 volts; 2 feet from conductors of circuits from 151 to 250 volts; 10 feet from conductors of circuits over 250 volts.

Lead-in conductors should be insulated or enclosed in a grounded metallic sheath where they enter the building. Each lead-in conductor should be provided with a lightning arrester, except that where the lead-in conductor is protected by a continuous effectively grounded metal sheath, the lightning arrester may be omitted.

Mast and supporting structure, if of metal, should be effectively grounded. If a system of lightning protection exists, the mast and structure should be bonded to the nearest lightning conductor; where such system does not exist, the mast structure should be grounded as described in the text entitled, "Lightning Protection for Metal on Buildings," below.

Electrical materials and installation should be in conformity with the National Electrical Code.

• ELECTRIC SHOCK •

What It Is

Electric shock occurs when any part of the human body becomes a conductor of electric current. The seriousness of the shock depends on two things: the amount of current that flows through the body and the path that the current takes.

The amount of current that will cause serious injury varies with age and physical condition. In general, the following amounts of current produce the listed effect.

Current Flow (milliamperes)	Effect on the Body
1 or less	Barely noticeable
2–8	Noticeable, but not unpleasant
9–15	Unpleasant, but usually not injurious
15–75	Will cause muscle freeze. May be fatal
More than 75	Almost always fatal. Remember, 75 milliamperes is well below 1/10 of an ampere.

The amount of current that will flow through the body depends on exactly the same factors that determine current flow in any circuit conductor:

The amount of resistance to current flow
The amount of voltage across the body
The number of paths along which current may flow

How to Avoid Shock

Electrically, the general rules for avoiding shock are based on the three factors listed above. First, provide maximum resistance to current flow through the body. Second, avoid placing your body between points of large voltage difference. Third, provide alternate low-resistance paths for current flow. Let's see what each of these rules means in practice.

DON'T PROVIDE AN EASY PATH FOR CUR-RENT FLOW • Maximum resistance to current flow can be provided by proper clothing. Gloves protect hands from accidental contact with live conductors or terminals. Rubber-soled shoes provide good insulation from ground in damp areas. General clothing suitable to construction areas (hard hats, coveralls, etc.) also protect against accidental contact with live wires. Covering conductors and terminals with temporary insulating material before final wiring is done reduces the chances of accidental contact if power is turned on for test purposes.

The voltage difference the human body is likely to "bridge" in residential wiring exists between one live conductor and ground (120 volts). The best way to avoid having this voltage difference across the body is to avoid contact between any part of the body and ground. Particularly on construction sites, many un-expected "grounds" may exist; for example, structural parts of metal buildings, water pipes, ladders and scaffolding, and air conditioning ducts. Direct contact with any grounded item should be avoided.

Providing alternate low-resistance paths for current flow is what grounding is all about. From our study of current flow through parallel resistance paths we know that when two or more paths are available, current will divide, with the heaviest flow following the lowest re-sistance path. This is the reason all metal enclosures that contain electrical conductors or devices must be connected to a ground point. If accidental electrical contact is made between a "hot" conductor and the enclosure, the connection to ground will provide a low-resistance path for current flow. If any part of a human body comes in contact with the enclosure, the current flow through the high resistance of the body will be negligible. Additional information on the subject of grounding is given in the section **Grounding for Safety**.

The current flow path through the body that is most likely to be fatal is across the chest. Current flow through this part of the body can cause muscular freeze of a heart muscle. The most likely way for current to flow across the chest is if both hands come in contact with points having a voltage difference. The use of in-sulating work gloves reduces this risk. In addition, whenever possible, keep one hand in a pocket when working near live circuits. Keep in mind, also, that moisture is a good conductor of electric current. Work gloves may feel uncomfortable in warm locations, but hands moist with perspiration can be dangerous. Keep the gloves on.

Dampness underfoot can also be dangerous. In ad-dition to wearing rubber-soled work shoes, stand on dry boards, a rubber mat, or similar material when working in damp areas.

TEST BEFORE YOU TOUCH • The best safeguard against shock is this simple rule: always test before you touch. Never assume that power is off. Use a meter or voltage tester to check exposed conductors or terminals before working with them or near them

Other General Rules

KEEP WORK AREAS AS NEAT AS POSSI-BLE • Clutter and junk in the work area make it dif-ficult to spot dangers such as dampness or exposed wires.

USE INSULATED TOOLS • But don't rely on the tool insulation alone. Insulation on tools can become broken or coated with conducting material.

DON'T BYPASS SAFETY DEVICES • In later chapters you will learn about interlocks, fuses, circuit breakers, and ground fault circuit interrupters. Under some circumstances you may be tempted to disable or "short out" these devices temporarily. Resist the temptation. These devices protect you not only against shock, but also against fire and the dangers that go with it.

• SAFETY DEVICES •

A primary safety feature in residential and com-mercial electrical systems is automatic power cutoff when current flow exceeds the circuit rating, or when dangerous circuit faults are detected. The safety fea-tures provided by these devices are briefly described below. Chapter 10 covers the installation and use of these items.

Fuses. A special piece of metal within the fuse is in series with the hot side of the power line. When current flow exceeds the fuse rating, the piece of metal gets hot enough to melt and open the circuit.

Circuit Breakers. These are mechanical devices. Excessive current flow heats a special strip of metal, causing it to bend. When the metal bends, it releases a spring-loaded switch that turns off power to the circuit.

Ground Fault Circuit Interrupters. These de-vices monitor the current flow in each conductor on a circuit. If the current is greater in one conductor than the other by a preset amount, power to the circuit is automatically turned off.

Interlocks. Interlocks are switches that auto-matically shut off power in a cabinet or other enclosure when the door is opened or panels are removed.

Fuses and circuit breakers are primarily fire protection devices. They turn off power before the heavy current flow resulting from a short circuit can cause sparking and overheating of switches and other devices.

Ground fault circuit interrupters (GFCI) greatly reduce shock danger. GFCI's are used on circuits to bathrooms, laundry rooms, and outdoor outlets when user contact with ground is likely. From the previous section on electric shock we know that fairly small amounts of current can cause severe, perhaps fatal, shock. GFCI's protect users against accidental contact with any point from which a current flow of as little as 5 milliamperes might be possible.

Interlocks are also primarily shock prevention devices which turn off power in enclosures when doors are opened or panels are removed. It is *sometimes* necessary to bypass interlocks in order to make voltage measurements or do other work in a cabinet containing exposed terminals. Make sure the interlock is properly reset when work is finished.

• GROUNDING FOR SAFETY •

Proper grounding of electrical circuits and devices is an essential part of an electrician's work. It involves numerous wiring procedures, special devices, and NEC requirements. Information on the subject of grounding will be found in all the remaining chapters of this book. This section deals only with the general subject of grounding for safety. First, working definitions of some words and phrases are given below.

Power Ground, Current-carrying Ground, Grounded Conductor, White (or natural gray) Wire. All of these terms refer to one of the wires from the power source. This wire normally carries current and must be connected from the source to the load for the load to operate (Fig. 4-5).

Grounding Wire. Note the "ing" ending. This wire does not normally carry current. It can be a bare wire or it can have green insulation or green with yellow stripes. It provides a second path for current flow as a safety measure. The symbol for this connection is shown in Fig. 4-6.

Neutral Wire. Strictly used, this term refers to the white or natural gray wire in a supply line containing three or more conductors. The term is sometimes used to mean the white or natural gray wire in a two-conductor supply also. This is an incorrect use of the word "neutral," but electrically the wires are identical (Fig. 4-7).

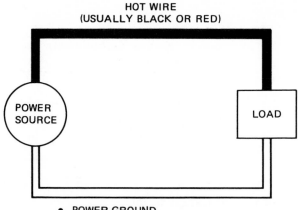

- POWER GROUND
- CURRENT−CARRYING GROUND
- GROUNDED CONDUCTOR

(MUST BE WHITE OR NATURAL GRAY)

Figure 4-5. Current-carrying ground.

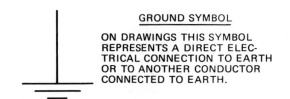

GROUND SYMBOL

ON DRAWINGS THIS SYMBOL REPRESENTS A DIRECT ELECTRICAL CONNECTION TO EARTH OR TO ANOTHER CONDUCTOR CONNECTED TO EARTH.

Figure 4-6. Grounding wire and ground symbol.

System Ground. This refers to the practice of connecting one of the current-carrying wires (the white or natural gray wire) to a cold water pipe or copper rod driven into the ground (Fig. 4-8).

Equipment Grounding. This term refers to the connection of a grounding wire from the non-current-carrying metal parts of a piece of equipment to some ground point (Fig. 4-9).

System Ground

In residential wiring, system ground is connected as shown in Fig. 4-8. The grounded wire (white or natural gray) and all white or natural gray wires connected to it must never be interrupted by a switch, fuse, circuit breaker, or any other device. For maximum safety the ground path must always be continuous. The ground path should also have as little resistance as possible. All ground wire connections must be clean and mechanically solid.

A continuous, low-resistance ground wire increases the safety of an electrical system by improving the performance of circuit protective devices. Fuses blow faster and circuit breakers and GFCI's trip faster when the ground circuit resistance is low. The faster these protective devices work, the sooner current flow is

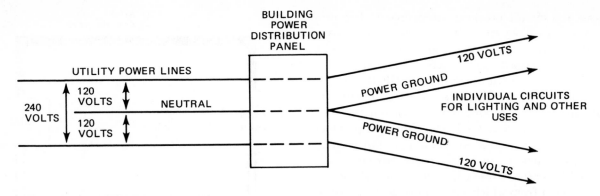

Figure 4-7. Neutral conductor.

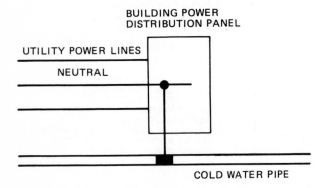

Figure 4-8. System ground.

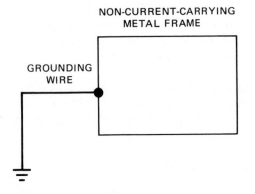

Figure 4-9. Equipment ground.

turned off and consequently the lower the risk of overheating and fire.

Equipment Ground

Equipment grounding is done as shown in Fig. 4-9. The non-current-carrying metal parts of refrigerators, freezers, air-conditioners, clothes washers and dryers, and dishwashers are some of the items that must have grounding wires.

Figure 4-10 represents a typical large appliance. The power line is connected to some voltage source and then is connected within the unit to an electrical load, such

as a motor. One side of the power cord is grounded by being connected to the system ground. However, if a short occurs at point A, because of frayed or worn insulation, there will be no current path to ground. If the user touches the frame and is in contact with any grounded item, such as a water faucet, the user's body will provide the ground path, resulting in severe shock, perhaps death.

When a grounding wire is added, the device is safer in two ways. The low-resistance path to ground results in heavy current flow and a fuse or circuit breaker will cut off power to the appliance. In addition, if for any reason power is not cut off, the grounding wire will still provide a lower resistance path to ground than the user's body so that no significant shock will be felt.

If the appliance is connected by cord and plug, the grounding wire is a third wire in the power cord. This wire connects the third prong (grounding prong) on the plug to the frame of the appliance. When plugged into a receptacle, the grounding prong is connected to a grounding circuit.

• IF AN ACCIDENT HAPPENS •

The next best thing to preventing accidents is knowing what to do if one happens. Injuries can be reduced and lives can be saved by quick, correct first-aid action. The basic rules are summarized below.

Separate the Victim from the Power

Turn off power as quickly as possible. If power cannot be readily turned off, break contact between the victim and the "live" line or terminal. **Do not touch the victim directly.** Be sure you are insulated from contact with the victim's body as well as the exposed conductor. Use dry wood, a blanket, a dry piece of clothing, a dry rope or similar non-conductor to break contact between the victim's body and the source of power. Severe shock causes muscle "freeze." It may

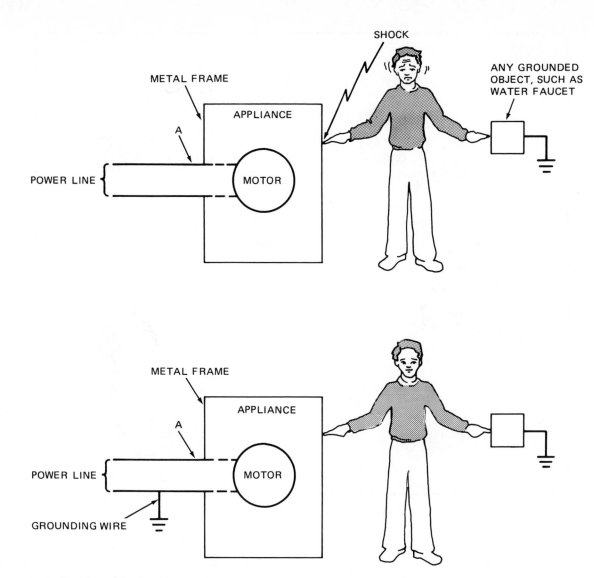

Figure 4-10. Shock protection by equipment ground.

take considerable force to separate a victim from a live line or terminal.

If the victim has stopped breathing, begin mouth-to-mouth resuscitation *immediately*. Even a 10-second delay can be the difference between life and death.

Mouth-to-Mouth Resuscitation

Victims of severe shock often suffer muscle spasms or temporary paralysis that stops breathing. In mouth-to-mouth resuscitation (Fig. 4-11) the rescuer forces breath into the victim's lungs to simulate breathing. This procedure supplies the victim with oxygen to lessen the chances of brain damage. In many cases the procedure restores normal breathing. Act as follows:

Step 1. Place the victim flat in a face up position.

Step 2. If there is any foreign matter (such as chewing gum, food, sand) visible in the mouth, turn the

victim's head to one side (Fig. 4-11a). Wipe the mouth out quickly with your fingers or a handkerchief wrapped around your fingers. Reach far back into the throat, if necessary.

Step 3. Position the victim so that the throat passage is clear. This can be done by putting one hand under the victim's neck and tilting the head back (Fig. 4-11b). Open the victim's mouth by pushing the top of the head back and pulling the lower jaw forward (Fig. 4-11c).

Step 4. Fill your lungs with air. **Open your mouth wide** over the victim's mouth or nose. You may place a handkerchief over the mouth to avoid direct contact. Seal your lips around the victim's mouth. Be sure to keep the victim's mouth open. To prevent leakage of air through the nose, pinch the subject's nostrils with your thumb and finger (Fig. 4-11d).

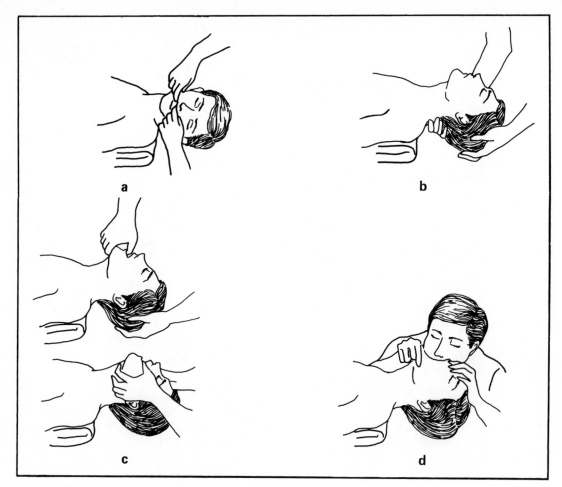

Figure 4-11. Mouth-to-mouth resuscitation.

Step 5.   Blow your breath vigorously into the victim until you see the chest rise. Then remove your mouth to let the breath out.

Step 6.   Take your next breath as you listen to the sound of breath escaping.

Step 7.   Inflate the victim's lungs again as soon as the chest falls back to normal.

Step 8.   Repeat ten to fifteen times a minute (every 5 or 6 seconds). Two or more rescuers can take turns every 5 or 10 minutes.

Step 9.   Gurgling or noisy breathing indicates the need to improve the head backward position or to clear the throat again as in Step 3.

Step 10.   Continue rescue breathing until the victim begins to breathe or until a physician arrives.

AIRWAY (OR RESUSCITATION) TUBE • First-aid kits may contain a tube (Fig. 4-12) which can be used to resuscitate adults and children over 3 years of age. One end is inserted *over the victim's tongue*, the other end serves as a mouthpiece for the rescuer. This enables the rescuer to avoid the direct contact of mouth-to-mouth breathing.

Figure 4-12. Airway (or resuscitation) tube.

Treating Electrical Burns

Electrical burns are treated the same as any burn. First, separate the victim from the source of electricity as described previously. Next, have the victim lie down, and loosen clothing around the neck. Call a doctor. If a first-aid kit is available, follow the instructions for burn treatment in the kit. If no kit or instructions are available, do the following:

Step 1. Place sterile dressings (or the cleanest available cloth material such as part of a shirt or a clean handkerchief) over the burned area to keep the air out.

Step 2. Do not try to clean the burn. Do not disturb the blisters.

Step 3. Keep the victim quiet and comfortably warm until the doctor comes.

Checking Victim's Pulse

A physician reached by telephone may ask the rate and condition of the victim's pulse. Every major artery has a pulse. Many arteries are close enough to the body surface for the pulse to be felt. The easiest place to feel the pulse is at the wrist (Fig. 4-13). Hold the wrist so that the balls of your four fingers are over the pulse and your thumb is against the top of the wrist for support. Exert slight pressure with your four fingers until you feel the pulse most strongly.

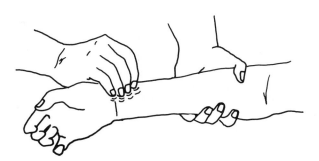

Figure 4-13. How to check pulse.

Once you have felt the pulse, you are ready to count it. You will need a clock or watch with a second hand. Count the pulse beats you feel in a 10-second period and multiply by six to get the pulse rate (per minute). Repeat this at least once to check your accuracy.

Note whether the pulse is strong and easily felt, or is weak and difficult to feel, and whether the pulse is regular or irregular. Irregular pulse will cause marked differences in the 10-second count if repeated three or four times.

The Other Kind of Shock

A victim who has had a severe electric shock or has been badly burned may suffer from another form of shock. This type of shock is due to a collapse of the circulatory function. That is, insufficient blood is being circulated through the victim's body. This form of shock can be serious, and if the typical signs of shock are noticed the condition should be treated. These are the things to look for:

1. Shallow and rapid breathing.
2. Pale face, lips, and fingernails.
3. Perspiration on forehead. Skin cold and moist.
4. Pulse weak and rapid.

This is the simple treatment:

Step 1. Place the victim in the approximate position shown in Fig. 4-14. If this position appears to cause pain or difficult breathing, change the position as necessary until the victim appears comfortable.

Step 2. Make sure the victim's mouth is free of blood or other fluid so that easy breathing is possible.

Step 3. Keep the victim warm with blankets or coats, if necessary.

Step 4. Call a doctor.

Step 5. Do not give the victim any alcoholic drink.

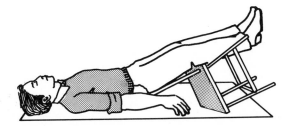

Figure 4-14. Position for circulatory shock victim.

1. What was learned about electric lighting at the 1893 exposition in Chicago?

2. What is the purpose of the National Electrical Code?

3. How does the NEC become "law"?

4. Why do communities and states have local electrical codes?

5. What do the letters UL mean on a product label?

6. What is the meaning of the words "listed" and "approved" when applied to electrical devices?

7. Which is listed by Underwriters' Laboratories: products or manufacturers?

8. What does the term "demand factor" mean?

9. Why do local codes often cover home TV antenna installations?

10. In addition to the NEC and local codes, what other things control electrical installations?

11. What two things determine the seriousness of electric shock?

12. A current flow of a fraction of an ampere can cause a fatal shock. About how small is the fraction?

13. Maximum resistance to current flow through the body can be provided by proper clothing. Name two important items.

14. What is the most dangerous current flow path through the body?

15. A simple safety rule is "Test before you touch." What does this mean?

16. Four safety devices are commonly used to prevent electrical fire and shock. Name the four devices.

17. Two of the four devices in question 16 are primarily for fire prevention and two are for shock prevention. Which devices are in each group?

18. In residential wiring, how is system ground provided?

19. In residential wiring, what is a typical use of equipment grounding?

20. If an accident victim receives a shock or burns from contact with a live conductor, what is the first thing that must be done?

5
TEST EQUIPMENT AND TOOLS

• INTRODUCTION •

Electricians use testing and measuring instruments as safety aids, installation checkers, and troubleshooting tools. This chapter tells you how meters and testers work. If you understand these instruments, you will be able to use them correctly and make the best use of the information they give you. Chapter 14 tells how meters are used to test and troubleshoot circuits.

There are many special handtools designed to make the electrician's work easier, quicker, and neater. This chapter describes some of the popular, widely used tools. Detailed information on the use of these tools is given in subsequent chapters when they are required for wiring procedures.

For much electrical work electricians use the standard tools of other trades. For installations in wood frame buildings, some carpenter's tools are needed. In brick, cinder block, and concrete construction, electricians use some stonemason's tools. When electrical installations require heavy wall conduit (a form of metal pipe), electricians use some of the plumber's tools to do the job. This chapter introduces you to things you will work with every day.

• TESTERS •

The simplest testers are voltage and continuity test lights. The voltage test light (Fig. 5-1) tells you if voltage is present on exposed conductors or conductor terminals. The continuity tester (Fig. 5-2) tells you if there is a continuous path for current flow in a circuit.

Voltage Tester

The voltage tester consists of two insulated test leads, a neon bulb, and a bulb holder. When one test lead is in contact with a live, hot wire and the other test lead is in contact with a neutral line or a ground point, the neon bulb glows (Fig. 5-3). Neon bulbs have high internal resistance and so automatically limit current flow through the tester. The voltage tester can also be used to determine that conductors are *not* live before touching them.

Voltage testers are small and light. They are easy to carry and easy to use. However, care must be taken to avoid contact with exposed live leads or terminals while using the tester. Also, remember to test the instrument from time to time by inserting the test leads into a live socket. A break in a test lead or a defective bulb gives the same indication as no voltage.

Continuity Tester

The continuity tester contains a battery and an incandescent bulb, and also has two test leads. This tester is used *only* on circuits and devices from which all power has been removed by turning off switches or disconnecting wires. When the test leads are placed on any two points in a circuit, the bulb will light if current can flow between the points (Fig. 5-4). The lighted bulb means there is a continuous path for current flow—or continuity—in the circuit under test. Continuity testers can be used to check for breaks in wiring, defects in switches, burned out lamp filaments, and other circuit faults.

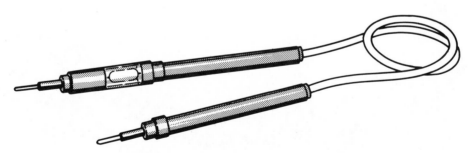

Figure 5-1. Voltage tester.

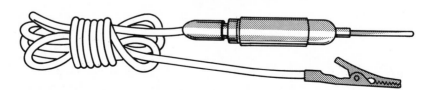

Figure 5-2. Continuity tester.

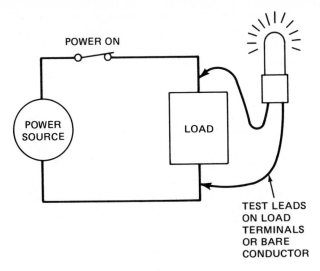

Figure 5-3. Voltage tester in use.

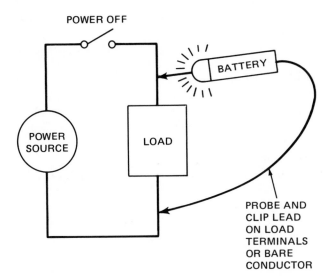

Figure 5-4. Continuity tester in use.

The typical pocket-type continuity tester uses one or two penlight (size AA) batteries, providing only 1.5 or 3 volts at low current drain. When continuity tests must be made on long circuit runs, heavier duty testers must be used. One such circuit tester uses four No. 6 dry cells in series to provide 6 volts and higher current capacity. Procedures for using the tester to trouble-shoot circuits are given in Chapter 14. The low voltage used in continuity testers allows circuits to be tested and faults detected without danger to the electrician or damage to equipment.

Outlet Analyzer

Outlet analyzers are used to check wiring and to locate sources of trouble (Fig. 5-5). When plugged into a live 120-volt outlet, lights on the analyzer indicate if

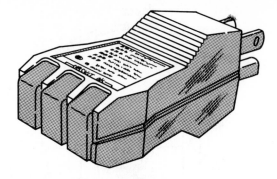

Figure 5-5. Outlet analyzer.

the wiring is good or if a fault condition exists. Wiring defects indicated are hot and neutral reversed, open hot, open neutral, hot and ground reversed, and open ground.

• METERS •

Many times an electrician must know more about a circuit than the "go/no-go" type of indication given by the voltage and continuity testers. An electrician often must know *how much* voltage, current, resistance, or power is present in a circuit. To get this information, meters must be used. The meters electricians use have calibrated dials, moving pointers, switches, and test leads (Fig. 5-6). When the test leads are correctly connected or held in place on terminals or conductors, the meter pointer moves across the dial to a reading that shows the amount of voltage, current, resistance, or power at that point.

Figure 5-6. Typical utility meter.

The force that causes the meter pointer to move is magnetic repulsion; that is, the force that causes like magnetic poles to repel each other. This force is used in

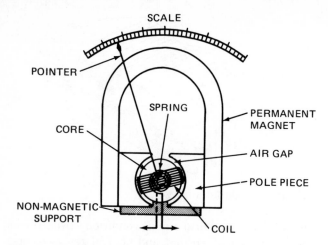

Figure 5-7. Moving coil meter movement.

a moving coil meter movement, known as the d'Arsonval movement, named for its inventor, a French physicist. The basic operation is shown in Fig. 5-7. The movement consists of a permanent magnet and a coil of wire. The coil is mounted on pivots between the poles of the magnet. When current flows through the coil, a magnetic field is induced. The polarity of the field is such that the north pole of the coil field is next to the north pole of the magnet and the south pole of the coil field is next to the south pole of the magnet. The force of repulsion of like poles causes the coil to rotate on its pivot points. The meter needle is attached to the coil and moves with it. If the coil were completely free to move, it would rotate continuously like the armature of a small motor. To prevent this, spiral springs are attached to the coil and the frame of the meter. The coil then moves until the magnetic force can no longer overcome the spring tension. The coil and needle then stop and remain at that position as long as the current flow through the coil is constant.

The magnetic field of the coil, the field of the magnet, and the tension of the spiral springs are all carefully chosen and calibrated so that the amount of movement of the coil and needle is proportional to the amount of current flowing in the coil. The accuracy and reliability of this basic movement have been improved over the years since it was first introduced more than 100 years ago. The permanent magnet is made of special alloys to produce a strong field and to maintain it for a long period of time. The coil rotates around a soft iron core. Soft iron pole pieces have been added to the magnet. The core and pole pieces strengthen and shape the magnetic fields to gain maximum coil movement for the smallest current flow. The coil rotates on jewel bearings to reduce friction to a minimum.

All these improvements have made moving coil meters rugged, dependable, and accurate. When proper wiring and switching are added, this meter movement

can be used to measure either voltage, current, or resistance. Some meters can make all three measurements when a selector switch is properly set. These are known as volt-ohm-milliammeters or VOM's.

Ammeter

Current flow through the meter coil causes the meter needle to deflect. To measure current flowing in a circuit it is necessary only to connect the meter so that circuit current flows through the coil. This requires breaking the circuit at some point and then connecting the meter in such a way that the circuit is completed through it (Fig. 5-8). This puts the meter in series with the circuit load.

All moving coil meters give full-scale deflection for some fixed value of current flow. For example, the coil design and the spring tension may allow full-scale deflection when 1 ampere flows through the coil. How, then, can this meter be used to measure larger amounts of current? The meter range is extended by *shunting* the meter movement with resistors of various sizes.

To shunt simply means to put in parallel. The meter movement has some resistance. If a resistor is placed in parallel with the meter movement, current will divide in accordance with the ratio of the resistance. For example, if the 1-ampere meter resistance is 100 ohms, we can place an 11-ohm resistor in parallel and increase the meter range to 10 amperes (Fig. 5-9). Current flowing into the meter will divide, with the larger amount of current flowing through the smaller resistance. As the shunt resistor is about one-ninth the meter resistance, 90 percent of the current will flow through the shunt and 10 percent through the meter. For example, if this meter and shunt are used to measure current in a circuit in which 5 amperes is flowing, 90 percent of the current or 4.5 amperes will flow through the shunt and 0.5 ampere through the meter. A flow of

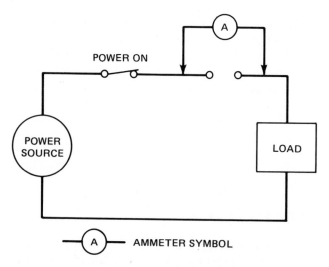

Figure 5-8. Ammeter connections in a circuit.

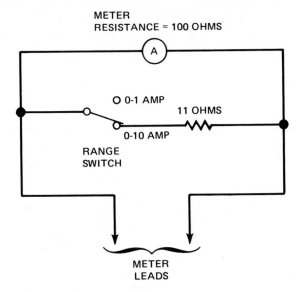

Figure 5-9. Ammeter shunt.

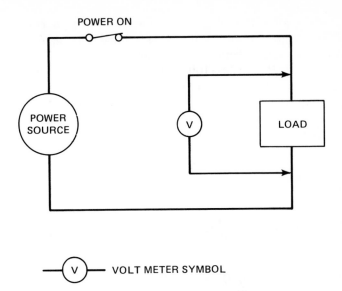

Figure 5-10. Voltmeter connection in a circuit.

0.5 ampere will deflect a 1-ampere movement to midscale. If we multiply that indication by 10, we have the correct current flow in the circuit: 5 amperes. Shunt resistors of other values can be added to provide additional current ranges.

Moving coil ammeters are limited to dc circuits and to low-level current measurement. In an ac circuit the changing polarity of the current flow would result in zero coil movement. A rectifier could be added to limit current flow to one direction, but this would increase the resistance of the meter movement. For current measurement the meter must be in series with the load. Therefore, meter resistance must be kept as low as possible so current flow in the circuit is not significantly reduced by the addition of the meter. A different type of meter—known as a clamp-on meter—operates on a different principle that allows ac current measurement without interrupting the circuit. This type of meter is described later in this section.

Voltmeter

If resistance is known and current is known, voltage can be calculated by $V = I \times R$. The moving coil movement can be used to measure voltage by adding resistance in series with the meter resistance and placing this combined known resistance between two points at different voltage levels (Fig. 5-10). The amount of current that flows through the known resistance is proportional to the voltage between the two points. Assume a meter movement that requires 1 ampere through the meter coil for full-scale deflection and has a total meter and series resistance of 100 ohms. The voltage required for full-scale deflection would then be 100 volts. If only 75 volts is applied, the current flow will be $I = V/R = 75/100 = 0.75$ ampere. The meter needle will deflect three-quarters of full

scale. Voltmeter dials are calibrated directly in volts for easy reading. Various voltage ranges can be obtained by switching additional resistance in series with the meter movement (Fig. 5-11). The higher the series resistance, the greater the voltage must be for full-scale deflection. In the example given above, if the total meter and series resistance is 1000 ohms, it will take 1000 volts for 1 ampere to flow through the meter.

Note that voltage readings are made by placing the meter *in parallel* with the circuit load. From the discussion of parallel resistance in Chapter 2 we know that the effective resistance of two resistors in parallel is less than either resistor and can be calculated as the product divided by the sum, or $R_1 \times R_2/(R_1 + R_2)$. This means that the meter resistance must be *much* higher than the load resistance to avoid having the

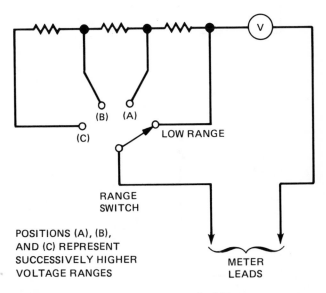

Figure 5-11. Voltmeter range switching.

meter affect the voltage across the load. During normal operation, the voltage across loads A and B in Fig. 5-12 is 60 volts. Total current flow in the circuit is $I = V/R = 120/(100 + 100) = 0.6$ ampere. The voltage across each load is $V = I \times R = 0.6 \times 100 = 60$ volts. If a meter with a total resistance of 100 ohms is placed across load A, the effective resistance at that point drops to 50 ohms.

$$R_E = \frac{R_L \times R_M}{R_L + R_M}$$

$$= \frac{100 \times 100}{200}$$

$$= \frac{10,000}{200}$$

$$= 50 \text{ ohms}$$

Current flow in the circuit then becomes

$$I = \frac{120}{150} = 0.8 \text{ ampere}$$

The voltage across load A is then $V = I \times R = 0.8 \times 50 = 40$ volts as long as the meter is in place. Figure 5-13 shows how a high resistance in series with the meter overcomes this problem.

Moving coil meter movements can be used to measure both dc and ac voltages. For ac measurement, a rectifier is added to the meter circuit to limit current flow to one direction (Fig. 5-14). Because voltmeters are used in parallel with the load and a high meter resistance is necessary for an accurate reading, the resistance added by the rectifier presents no problem. The input current to the meter after the rectifier consists of positive half cycles. For any 50 to 60 Hz or higher frequency, the meter indication will be a steady value. The half cycle pulses occur too rapidly for the meter movement to respond. The meter needle deflection actually represents an average value, but the ac voltage scale is calibrated in rms volts (refer to Chapter 3). If the voltage measured is 120 volts, the meter will indicate 120 volts, unless otherwise marked.

Ohmmeter

Ohm's law can be used to calculate resistance if current and voltage are known. $R = V/I$. For resistance measurement an internal voltage source is added to the meter. Small dry cells are generally used. With all power removed from a circuit, the meter leads are connected across a load of unknown resistance (Fig. 5-15). Current flows from the battery through the load and through the meter coil. With voltage constant, the current in the circuit decreases as the measured resistance increases. The higher the measured resistance, the *less* the meter needle is deflected. The face of the meter is calibrated in ohms; needle deflection represents load resistance. Most VOM's have zero on the ohms scale at the opposite end from zero on the voltage

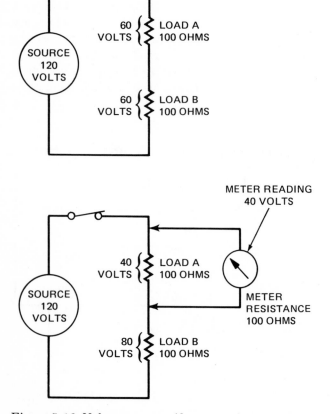

Figure 5-12. Voltmeter error if meter resistance is low.

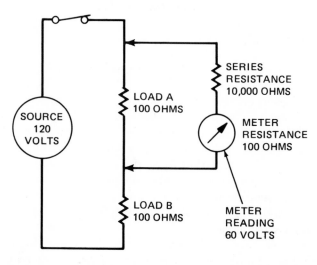

Figure 5-13. Correct voltmeter reading when meter resistance is high.

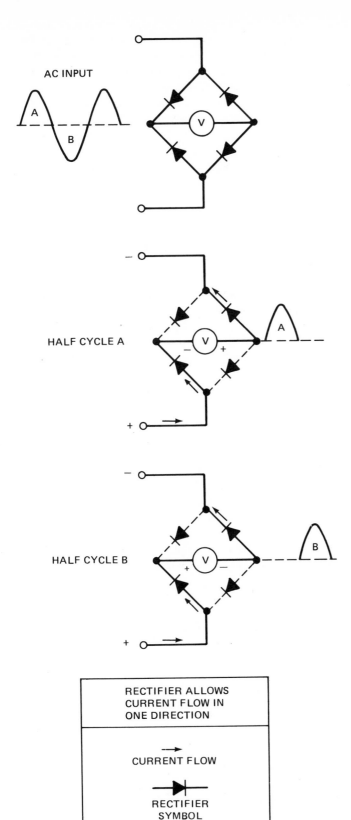

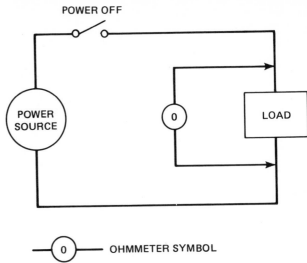

OHMMETER SYMBOL

Figure 5-15. Ohmmeter connection in a circuit.

of the scale. These are called shunt ohmmeters. However, battery life is shorter in shunt ohmmeters because there is a constant current drain even when no measurement is being made. The back-off type of ohmmeter circuit shows current only when current can flow between the test leads. Figure 5-17 shows the basic circuit of each type of ohmmeter. Note that in the back-off meter, even with the meter switch on, there must be continuity between the test leads before current will flow through the meter. In the shunt ohmmeter current flows through the meter continuously when the switch is turned on.

In both types of ohmmeters circuit voltage output will drop as the battery ages. In back-off meters the decreased battery voltage is offset by adjusting a variable resistor in series with the meter. This is called the zero adjustment. Before an ohmmeter is used, the test leads must be shorted together to represent zero resistance and the variable resistor must be adjusted until the meter needle shows zero resistance. Shunt ohmmeters have a similar adjustment, but adjustment

RECTIFIER ALLOWS
CURRENT FLOW IN
ONE DIRECTION

⟶ CURRENT FLOW

▶| RECTIFIER SYMBOL

Figure 5-14. Full-wave rectifier operation.

and current scale. This is called a back-off scale (Fig. 5-16). Some VOM's are wired internally to reverse the meter movement so all scale zeros are on the same end

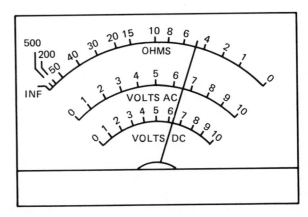

Figure 5-16. Ohmmeter back-off scale.

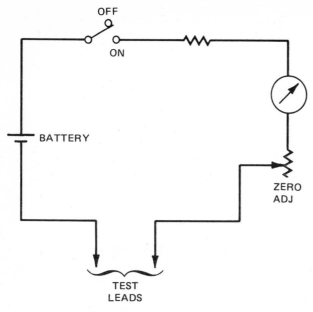

BACK-OFF OHMMETER

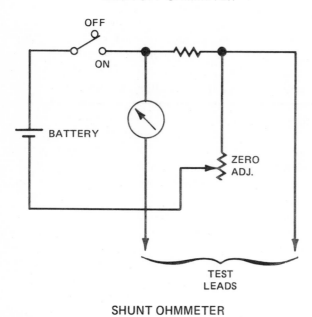

SHUNT OHMMETER

Figure 5-17. Ohmmeter circuits.

is made with the test leads separated to represent the highest reading on the scale. The variable resistor is adjusted until the needle is aligned with the high end of the scale.

Ohmmeter scales are all not equally divided all across the range. For example, if half the scale covers 0 to 10 ohms, from 10 to 20 ohms will cover only about one-third of the remaining half, and so on. The high end of the scale is marked "INF" or "∞" to represent infinite resistance. The uneven (nonlinear) scale results from the fact that current flow through the meter decreases disproportionately as resistance increases. For example, if we have a meter requiring 1 milliampere for full-scale deflection and a 1.5-volt meter battery, 1500

ohms internal resistance would produce full-scale deflection, indicating zero external resistance. If the test leads are put across an external resistance of 1500 ohms, total resistance becomes 3000 ohms and current through the meter is 0.5 milliampere for half-scale deflection. If we measure a much larger resistance, say 15,000 ohms, total resistance becomes 16,500 ohms, current drops to less than 0.1 milliampere, and the meter deflection is only one-tenth of scale. Thus, 50 percent of the meter scale represents an external resistance of 1500 ohms, and 90 percent of the scale represents 15,000 ohms. Because of these uneven scale units (nonlinearity), ohmmeter scales are easier to read and are most accurate from zero to about midscale.

Ohmmeter ranges are increased by adding shunt resistors in parallel with the meter and the series resistor. The shunt resistance is small compared to the meter and series resistance, so most battery current flows through the shunt resistor. This produces midscale meter deflection for higher external resistances. Figure 5-18 shows how shunt resistance changes an ohmmeter scale.

With the meter ON and the HIGH-LOW switch in the HIGH position the meter, series, and zero-adjust resistance is 1500 ohms. With the test leads shorted, 1 milliamp will flow through the meter to produce full-scale (0) resistance. If the test leads are placed across a 1500-ohm resistance, total resistance is 3000 ohms, current flow is 0.5 milliampere, and the meter will deflect to midscale. A midscale reading in the HIGH position represents 1500 ohms measured resistance.

With the HIGH-LOW switch in the LOW position, the 167-ohm resistor is placed in parallel with the meter, series, and zero-adjust resistances. This results

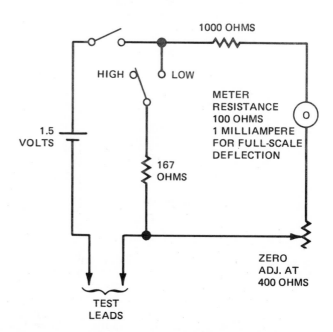

Figure 5-18. Ohmmeter range switching.

in an effective resistance of about 150 ohms. With the test leads shorted, 10 milliamperes will flow in the circuit. Current will divide with 90 percent or 9 milliamperes flowing through the shunt resistor and 1 milliampere flowing through the meter to produce full-scale deflection. If the test leads are placed across a resistor of 150 ohms, total resistance is 300 ohms, and total current flow is 5 milliamperes. Current will divide with 4.5 milliamperes flowing through the shunt and 0.5 milliampere through the meter to produce midscale deflection. A midscale reading in the LOW position, then, represents 150 ohms measured resistance.

Ohmmeters that have several ranges have selector switches marked with multipliers. For example, the meter scale may be simply marked from 1 to 100 and a bit past the 100 point, with INF or ∞. The selector switch might be marked $R \times 1$, $R \times 10$, $R \times 100$, or more. In the $R \times 1$ position the numbers on the dial indicate the measured resistance directly. In the next position, the measured resistance is ten times the dial reading. If the pointer stops at 65, the resistance being measured is 650 ohms. In the next switch position the same point on the dial would represent 6500 ohms.

As noted previously, ohmmeter scales are easier to read and are most accurate between zero and midscale. If the approximate resistance to be measured is known, select a range multiplier that will give a reading as near as possible to midscale. If the approximate resistance is unknown, start with the highest multiplier setting and work down until a reading near midscale is obtained.

Wattmeter

Real watts of power in an ac circuit can be measured directly by a special combination of ammeter and voltmeter called a wattmeter (Fig. 5-19). The wattmeter has two low-resistance fixed coils to measure current, and a movable coil and resistor to measure voltage (Fig. 5-20). The meter action is similar to the moving coil meter except that the current coils replace the permanent magnet. The strength and polarity of the magnetic field is then controlled by current flow in the circuit being measured. The current-measuring coils are connected in series with this circuit load. The movable coil and series resistor provide a high resistance for voltage measurement. The voltage-measuring leads are connected in parallel with the load. The deflection of the meter needle represents the magnitude of both current and voltage. Current determines the field strength and polarity. Voltage determines current flow in the coil and, therefore, the coil's magnetic field and polarity.

The action of the meter is such that either dc or ac power can be measured. No rectifier is required for ac

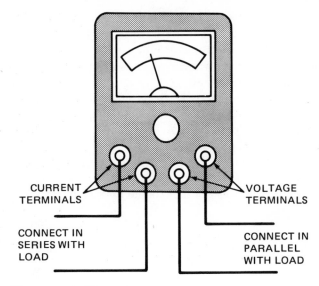

Figure 5-19. Wattmeter.

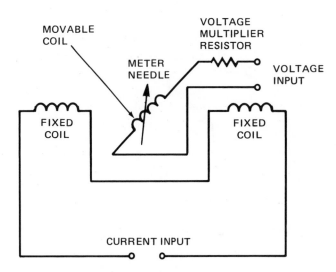

Figure 5-20. Wattmeter circuit.

measurement because polarity changes occur simultaneously in both the current coils and the movable coil. Magnetic repulsion remains constant for any input values of current and voltage. Note, however, that the force of magnetic repulsion will be affected by the time (phase) relationship of voltage and current. Meter needle deflection will be greatest when voltage and current peaks occur at the same time to produce the strongest magnetic repulsion. At the other extreme, if voltage and current peaks are 90° apart, the voltage coil will have zero magnetic field when the current coil field is strongest, or vice versa. This condition would produce no needle deflection. This sensitivity to the phase relationship of voltage and current is the reason that the wattmeter measures real watts of ac power. This measurement can be compared to the product of voltmeter and ammeter readings (VA) to determine power factor.

Clamp-on Ammeter

The clamp-on ammeter provides a quick and easy way to measure current flow in an ac line without interrupting the circuit even momentarily. A typical clamp-on ammeter (Fig. 5-21) consists of a trigger-actuated set of jaws on a body containing a meter scale. The trigger opens the jaws so that they can be closed around a conductor. The meter then indicates the current flow in the conductor. The clamp-on ammeter can be thought of as a highly sensitive transformer secondary (Fig. 5-22). The conductor is equivalent to a single-wire transformer primary. Current flow in the conductor creates a magnetic field around the conductor. This field induces a voltage in windings within the ammeter jaws. The resulting current flow is rectified and applied to a moving coil indicator. Meter shunts provide various meter ranges.

Many clamp-on ammeters have test leads and can also be used for voltage and resistance measurements. Clamp-on ammeters are available with digital readouts and "freeze" controls. Freeze controls allow readings that are taken in dimly lighted locations to be held as long as required for reading elsewhere.

CLAMP-ON VOLT/AMP/OHMMETER HAS ROTARY SCALE WHICH PERMITS ONLY THE OHMMETER SCALE AND THE SELECTED CURRENT OR VOLTAGE SCALE TO SHOW.

CLAMP-ON VOLT/AMP/OHMMETER HAS DIGITAL READOUT WHICH JUST ABOUT ELIMINATES READING ERRORS. DIGITAL INSTRUMENTS GENERALLY HAVE BETTER ACCURACY THAN ANALOG INSTRUMENTS.

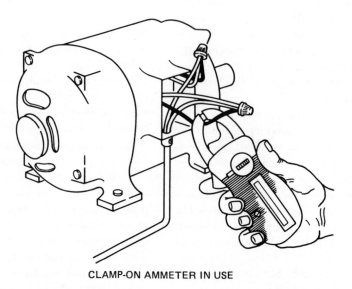

CLAMP-ON AMMETER IN USE

Figure 5-21. Clamp-on volt/amp/ohmmeters.

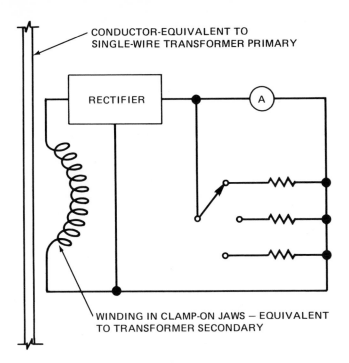

Figure 5-22. Clamp-on ammeter circuit.

Megohmmeters

As the name suggests, these meters measure high resistances (up to 500 megohms, full scale). They are used mainly to measure insulation resistance where even small current leakage can be dangerous to personnel and can cause equipment breakdowns. Megohmmeters—often called "meggers"—are basically the same as standard ohmmeters. The principal difference is that a higher voltage source is required in order to produce sufficient current flow for meter movement. The voltage source is generally a constant-voltage, hand-cranked generator that produces 500 to 1000 volts. Clip-type test leads and test probes are provided to connect the meter to the resistance to be measured. For example, one clip can be attached to a conductor, and a test probe can be held in contact with the outer insulation of a cable. When the generator is cranked, the meter will show the insulation resistance.

Voltage Level Test Instruments

These testers show measured voltage level by lighted indicators or a moving bar indicator, rather than by calibrated dials and pointers (Fig. 5-23). The testers are designed so that the probes may be set into holders on the instrument housing. In this position the probes are spaced to fit directly into the slots of wall receptacles. The probes may be removed from the instrument housing to make measurements on widely separated points. Various models can check as many as ten ac/dc voltage levels.

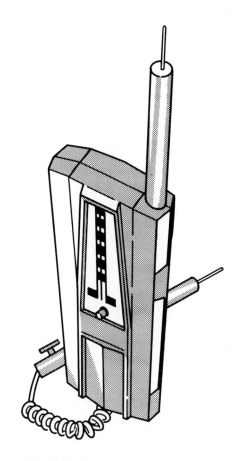

LIGHTWEIGHT, POCKET SIZE VOLTAGE TESTER INDICATES VOLTAGES "THERMOMETER STYLE" USING A VERTICAL COLUMN OF NEON LAMPS.

Figure 5-23. Voltage level test instrument.

• POWER TOOLS •

The power tool electricians used the most often is a drill. Power drills are used to drill holes in studs and joists for cable runs. With special bits, power drills can be used to install various kinds of mounting and anchoring devices on masonry walls. An extension for overhead drilling is useful (Fig. 5-24), as is a right-angle attachment. For general use in wood frame construction a single spur ship auger power bit is used. For masonry, carbide-tipped bits are used with a low-speed low rpm) drill, or the low setting on a variable-speed drill. Figure 5-25 illustrates some of the bits used in wiring. Protective goggles or an eye shield should always be worn when working with a power drill.

Many other jobs can be done with special-purpose power tools. On large projects these special power tools may be used to reduce the time spent on manual operations. For example, when large amounts of conduit must be installed, a power bender may be used in place of the conduit hickeys, described in the next section. Power saws may also be used on some jobs. The manufacturers of power equipment supply information on

Figure 5-24. Power drill with extension.

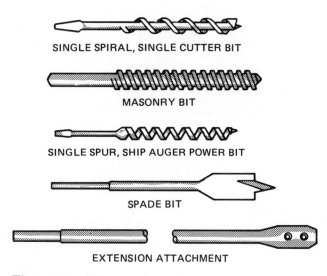

SINGLE SPIRAL, SINGLE CUTTER BIT

MASONRY BIT

SINGLE SPUR, SHIP AUGER POWER BIT

SPADE BIT

EXTENSION ATTACHMENT

Figure 5-25. Bits used in wiring.

the proper use of their tools. Read and follow these instructions. Special information and precautions are often stencilled or printed directly on some part of the power tool. For safe and efficient use, take the time to read and follow these instructions.

• ELECTRICIAN'S HANDTOOLS •

Fuse Pullers

Fuse pullers are used to remove cartridge fuses from service panels (Fig. 5-26). Various sized gripping jaws

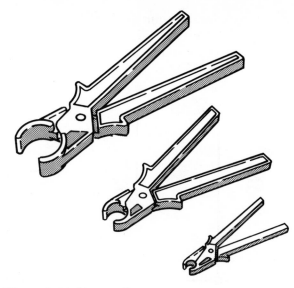

Figure 5-26. Fuse pullers.

are available to fit different sized fuses. The tool is made of non-conducting material. It is used by closing the jaws of the puller over the fuse and pulling straight out to remove the fuse.

Wire Cutters

These are diagonal-cutting tools designed to give maximum leverage for cutting through conductors up to no. 2 in size. Larger sizes of stranded cables can be cut a few strands at a time. Wire cutters have short cutting jaws and long handles for easy cutting.

Wire Strippers

Several types are available, but all do about the same job (Fig. 5-27). These tools have pliers-like jaws with various sized cutouts to fit different sized conductors. When the stripper is closed over the proper sized wire, the jaws cut through the layer of insulation, but do not cut into or nick the conductor. A few quarter-turns of the stripper make a clean cut through the insulation. With the jaws still closed, pulling toward the cut end of the wire causes the insulation to slide off.

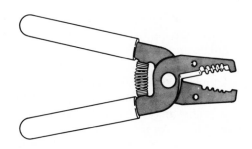

Figure 5-27. Wire stripper.

Cable Strippers

These tools simplify removal of the outer covering from non-metallic cable (Fig. 5-28). The stripper (sometimes called a ripper) is slipped over the end of the cable and pressed closed on the flat side of the cable. A knife edge inside the stripper penetrates the cable covering. The stripper is then pulled toward the cut end of the cable. The knife edge cuts cleanly through the outer covering, and the tool keeps the cut centered on the flat side between the inner conductors of two-conductor cable. Care must be taken when using strippers on three-conductor cable to avoid cutting into conductor insulation.

Electrician's Knife

This is a folding pocketknife which contains several specially shaped blades that can be used to strip insulation when working in tight places (Fig. 5-29). The knife has a safety lock to prevent the blades from closing in use.

Fish Tape

This tool consists of a coiled length of flexible steel tape in a holder with a special grip built into the body (Fig. 5-30). The end of the tape is hooked. The tape is fed through a hollow wall or conduit. Cable is fastened to the hooked end. The tape is then pulled or reeled in to pull the cable through the wall or conduit.

Pipe Cutter

Pipe cutters are used to cut heavy wall conduit (Fig. 5-31). To make a cut the conduit is clamped in place under two carbide wheels. The tool is then rotated to

Figure 5-29. Electrician's knife.

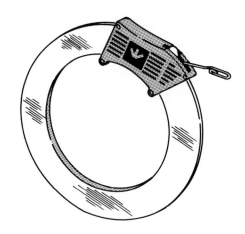

Figure 5-30. Fish tape.

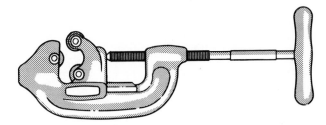

Figure 5-31. Pipe cutter.

cut a groove in the conduit. With each revolution the handle is tightened to cut the groove deeper until the conduit is cut. Pipe cutters tend to leave a sharp internal edge on cut conduit. This must be smoothed with a conduit reamer before the conduit is installed to prevent damage to the conductor insulation.

Conduit Hickeys

Hickeys are tools used to bend conduit. Two types are available; one for rigid or intermediate conduit and another for thin-wall conduit or EMT (electrical metal tubing) (Fig. 5-32). Hickeys hold the conduit securely and provide the leverage needed to make a smooth bend. Hickeys for rigid conduit are used to make a series of small bends along 4 to 8 inches of conduit. The series of small bends adds up to a smooth large bend without distorting the conduit. Hickeys for thin-wall conduit (EMT) have high supporting side walls to prevent buckling or kinking of the conduit. The EMT hickey has a long arc that makes a 90° bend in a single "bite."

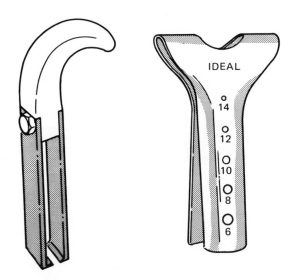

Figure 5-28. Cable strippers.

FOR RIGID CONDUIT

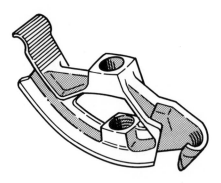

FOR THIN WALLED CONDUIT

Figure 5-32. Conduit hickeys.

Conduit Reamer

Whenever conduit is cut, burrs and sharp edges are left on the inner edge of the cut. If not removed, these sharp points dig into insulation when wires are pulled through the conduit. A reamer inserted in a brace and given a few turns smooths the inner surface (Fig. 5-33).

Conduit Indenting Tool

This tool is used to join EMT conduit. A coupling—similar to the conduit, but slightly larger in diameter—is slipped over the ends of two pieces of conduit. The coupling is 3 or 4 inches long and is positioned to cover the joint. The indenting tool is then placed over one end of the coupling. When pressure is applied to the tool handles, two "beads" or indentations are pressed into opposite sides of the coupling and the conduit (Fig. 5-34). This process is repeated to make four beads around each end of the coupling.

Electrician's Hammer

The head of an electrician's hammer is longer from striking surface to claw than a carpenter's hammer. The extra length makes it easier to drive nails into studs at the back of electrical boxes.

Electrician's Pliers

Electricians should have several types of pliers. Side-cutting pliers can be used to cut wires and to strip insulation. Diagonal pliers give extra leverage for cutting large-sized wires and are handy for cutting wires in tight places. Long-nose pliers are useful in bending conductors for connection to screw-type terminals. Slipjoint pliers grip and hold locknuts so they can be tightened on cable clamps and box connectors. Crimping pliers have various sized cutouts in the jaws to crimp solderless terminal lugs to secure them to conductors (Fig. 5-35).

• COMMON HANDTOOLS •

Soldering Tools

Soldering jobs can be done with an electric soldering iron or soldering gun, or a gas-fueled torch (Fig. 5-36). The tool used depends on the amount of soldering to be done and the size of the conductors to be joined.

Soldering irons are slow, but may be satisfactory for small amounts of soldering on no. 14 or smaller conductors. Larger conductors absorb heat and carry it away from the joint. This makes soldering slow, and insulation may be damaged by overheated conductors.

Soldering guns heat up much faster than irons, and many have variable heat settings. For joining electrical conductors the highest available heat level should be used.

Torches fueled with methane or propane are best for large conductors. The extremely high heat of the blue flame allows the conductor to be quickly heated to the

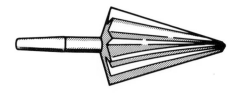

Figure 5-33. Conduit reamer.

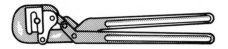

Figure 5-34. Conduit indenting tool.

Figure 5-35. Lug-crimping tool.

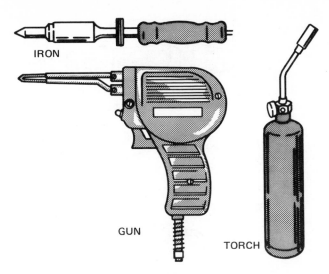

IRON

GUN

TORCH

Figure 5-36. Soldering tools.

melting point of the solder before the heat can be carried away by the conductors.

SOLDERING HINTS • Clean joints are the secret of good soldering. Carefully scrape conductors clean before making the splice. Cover the solder joint with a thin layer of resin-base, non-corrosive flux. This makes the solder flow better. Apply heat to the conductors until they are hot enough to melt the solder. Allow solder to flow into the joint for a second or two before removing heat. A good solder joint is smooth and has a slight sheen. Dull or "grainy" joints should be reheated until the right appearance is obtained.

Powder Cartridge Fasteners

These tools use a powder cartridge in a gunlike device to set fasteners in concrete or steel (Fig. 5-37). When the trigger is pulled, the powder cartridge fires with the force necessary to drive the fastener into the masonry or metal. Various sizes of cartridges are used for different thicknesses and hardnesses of material. These devices are efficient and can set mounting fasteners rapidly. Like firearms, however, they are extremely dangerous and must be used only in complete accordance with the manufacturer's safety rules. Special permission may be required for use of powder cartridge fasteners in some areas. Their use may be prohibited in other areas. Check local codes and regulations before using any powder-driven tool.

Hammer Drill

When electricity for a power drill is not available, these drills can be used to drill holes in masonry (Fig. 5-38). Carbide-tipped bits are set in a handle, then driven into the wall by hitting the handle with a heavy hammer and rotating the handle slightly between hits.

Wear safety glasses or an eye shield when using a hammer drill. Hammer drills can make large-sized holes to match the lead shields used to mount heavy items, such as large panel boxes.

Holes can also be drilled in masonry by using a star drill and hammer (Fig. 5-39). The star drill is so named because when viewed from the end it looks like a four-point star. The star point is placed against the surface to be drilled and the drill is hit with a hammer. After each hit, the drill is rotated a bit. The process of

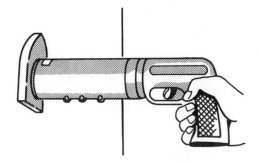

Figure 5-37. Powder cartridge fastener.

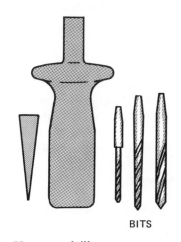

BITS

Figure 5-38. Hammer drill.

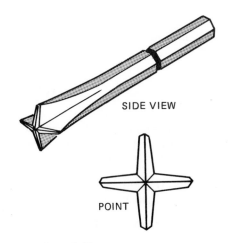

SIDE VIEW

POINT

Figure 5-39. Star drill.

hit and turn is repeated until the required hole depth is reached.

Drive Pin Set

This tool can be used to mount lighter objects, such as cable or conduit clamps (Fig. 5-40). Steel pins fit into the barrel of the tool and are then driven into the wall by hitting the drive pin with a hammer. The steel pins come in various lengths from 1/2 to 1-1/2 inches. There are straight pins with nail heads and pins with internal and external threaded studs.

Brace and Bit

A brace and bit can be used to bore holes for cable runs when a power drill cannot be used (Fig. 5-41). Using a conduit reamer in the brace, the inner surface of conduit can be smoothed after cutting.

Folding Rule

A wooden folding rule is best for most measurements. The folding rule can easily be used with one hand. Wood protects against shock if a live conductor is accidentally touched.

Screwdrivers

A good selection of sizes and types is essential. Both slot-head and Phillips-head screwdrivers are useful (Fig. 5-42). If several sizes are available, you can use the correct size for each screw. This makes for easier work and prevents "chewing up" the screw head.

Keyhole Saw

This saw is useful for notching studs for cable runs and making cutouts for switch and outlet boxes (Fig. 5-43).

Hacksaw

Hacksaws can be used to cut conduit and to remove cable armor.

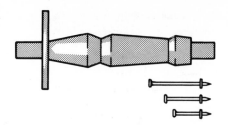

Figure 5-40. Drive pin set.

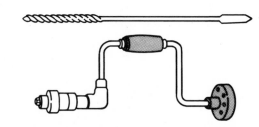

Figure 5-41. Brace and bit.

Figure 5-42. Slot-head and Phillips-head screwdrivers.

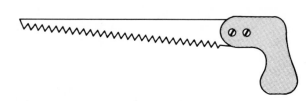

Figure 5-43. Keyhole saw.

• REVIEW QUESTIONS •

1. Voltage testers should be tested from time to time by inserting the test leads in a live socket. Why is this necessary?

2. Which tester, the voltage tester or the continuity tester, uses penlight batteries for power? Why?

3. What magnetic force causes the moving coil in a meter to move?

4. How must an ammeter be connected to measure current in a circuit?

5. The internal resistance of an ammeter should be as low as possible. Why?

6. How must a voltmeter be connected to measure voltage across a load?

7. The internal resistance of a voltmeter should be as high as possible. Why?

8. Ohmmeters must only be used to measure resistance when all circuit power is turned off or disconnected. What power source causes the meter to move?

9. If an ohmmeter needle is pointing to 50 on the meter dial and the range switch is set at $R \times 100$, what is the measured resistance?

10. Why is it desirable to select an ohmmeter range that gives a reading at the low end of the meter scale?

11. Why is a wattmeter useful in calculating power factor?

12. What is the main advantage of a clamp-on ammeter?

13. For what measurement are megohmmeters often used?

14. What are hickeys used for?

15. Why is a gas-fueled torch better than an electric gun for soldering large-sized conductors?

6
WORKING WITH WIRE AND CABLE

• INTRODUCTION •

An electrical circuit has four essential parts: a source of power, conductors to carry power where it is needed, control devices to turn power on and off, and a load to use the power. This chapter tells you how to work with the second circuit part, conductors.

Conductors are the pathways along which the electrical energy moves. The word conductor describes any material which allows electrical current to flow easily. Since electrical current *is* electron flow, materials that allow electrons to move freely are good conductors (Fig. 6-1).

Insulators may be thought of as the opposite of conductors. Insulating materials such as plastic, rubber, and paper allow just about zero electron flow (Fig. 6-2). Most conductors are covered by one or more layers of insulating material.

To get the most from this chapter you should know what current flow is, what an electrical circuit is, what kinds of things are covered by the National Electrical Code (NEC), how to work safely with electricity, and the kinds of tools electricians use. These subjects are covered in Chapters 1 to 5. Now you will learn how to work with the kinds of wire and cable that are most widely used, how to select the right size and type for a circuit, and how to remove the cable coverings and conductor insulation so that good electrical connections can be made.

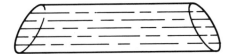

Figure 6-1. Electron flow in a conductor.

Figure 6-2. Electron flow in an insulator.

• WIRE AND CABLE •

In the electrical industry the words *wire* and *cable* have specific meanings. You must know these meanings to understand the NEC, to order material from a catalog, or to talk with electrical suppliers and contractors.

The word wire applies to a single electrical conductor. The conductor may be one strand of material or several strands twisted together. Conductors are usually covered with insulating material (Fig. 6-3). Wires

used only for grounding circuits may or may not have insulation.

Wiring in residential buildings and stores requires two or more conductors for each circuit. So, for easier installation and greater conductor protection, various outer coverings are manufactured to enclose these groups of insulated conductors. The NEC requires grounding wires on each circuit for added safety. For this purpose, an additional conductor, which may or may not be insulated, is included with the conductors. This combination of insulated conductors and grounding wire enclosed in an outer covering is an electrical cable (Fig. 6-4). Wiring may also be done by enclosing insulated conductors in metal tubing, called conduit. This is described in Chapter 7.

The electrical industry produces a great variety of wires and cables to fill all wiring needs. To understand why this range of types and sizes is needed, you must be familiar with the subjects discussed in the sections that follow. To begin with, let us look at the system of wire sizes.

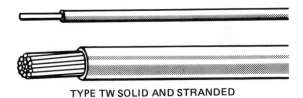

TYPE TW SOLID AND STRANDED

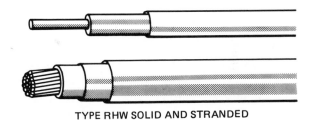

TYPE RHW SOLID AND STRANDED

Figure 6-3. Typical insulated conductors.

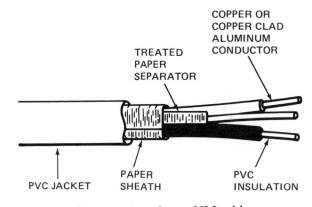

Figure 6-4. Construction of type NM cable.

• WIRE SIZES •

Cross-Sectional Areas

The system of electrical wire sizes is based on the cross-sectional areas of the conductors. Because wires are round, the end view of a straight cut through a wire is a circle, so the size system is based on the area of a circle. A unit of measurement is needed to express wire sizes. Because wire cross sections are rather small, it is convenient to use a small unit of measurement. The unit used is one-thousandth of an inch; this is known as a mil, an abbreviation of milli-inch.

Mils are units of length and they can be used to describe the diameter of wire. However, a more important consideration in conductor use is the cross-sectional *area* of the conductor. The area of a circle is proportional to the square of its diameter. The area of a circle whose diameter is 1 mil has been chosen as the unit to express area. The unit is called a circular mil, abbreviated cmil (Fig. 6-5). The use of this unit gives a clearer picture of the true relationship of wire sizes.

Wire Numbers

A group of wire sizes has been established to cover all wiring needs. For easy reference, the various sizes are identified by number. The larger the number, the smaller the cross-sectional area of the wire. The numbers range from 50 down to 0, then 00, 000, 0000. Sizes larger than 0000 are identified directly by their area in circular mils. Wire sizes 8 and smaller have either solid or stranded conductors. Wire sizes 6 and larger are stranded.

Wire size can be determined by using a standard wire gauge (Fig. 6-6). The NEC requires that wire size be marked on the insulation of all wires larger than no. 16. The wire gauge, then, is ordinarily used to size only small or bare wires. Large-size wires with unmarked insulation should not be used. They may not be safe and probably will not be passed by an inspector.

Common Wire Sizes

Although all sizes are manufactured for various uses, only the even numbers are used in electrical wiring. Many reference tables in the NEC list more than 30 wire sizes, but most wiring in residential buildings and stores is done with about six sizes. These are numbers 14, 12, 10, 8, 6, and 4 (Fig. 6-7).

A comparison of two of these commonly used sizes shows how the units used for wire sizes actually work (Fig. 6-8). To simplify the example, some numbers are rounded off. No. 14 wire is 0.064 inch (64 mils, 1/16 inch) in diameter and its area is 4100 cmil. No. 8 wire

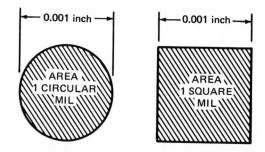

Figure 6-5. Conductor cross-sectional areas.

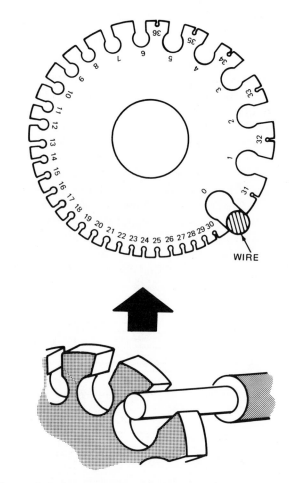

Figure 6-6. American standard wire gauge.

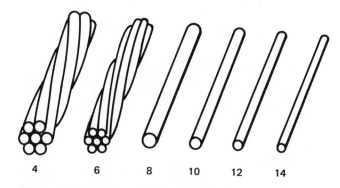

Figure 6-7. Commonly used wire sizes.

DIAMETER
0.064 INCH
(64 MILS, 1/16 INCH)

NO. 14 WIRE
AREA 4100 CMIL
(APPROX)

DIAMETER
0.128 INCH
(128 MILS, 1/8 INCH)

NO. 8 WIRE
AREA 16,500 CMIL
(APPROX)

Figure 6-8. Comparison of no. 14 and no. 8 wire areas.

is 0.128 inch (128 mils, 1/8 inch) in diameter and its area is 16,500 cmil. This tells us that no. 8 wire is twice as thick as no. 14, but has four times as much cross-sectional area.

The cross-sectional area of stranded wire is the sum of the areas of each individual strand. Because there is some space between the strands even though they are twisted together, stranded wires have slightly larger actual diameters than solid conductors of the same area. Both no. 4 and no. 6 wires consist of seven individual strands: one in the center and six surrounding it. Each strand of no. 4 wire, for example, is 77 mils in diameter. The diameter of the bundle of twisted strands is 230 mils, or just under 1/4 inch.

Special Sizes

In addition to the six common sizes, some larger and some smaller wires are used in special locations. Larger sizes are used at the service entrance. This is the place where power company lines enter the building. Wires used here must be capable of handling the total electrical load of all the circuits in the building. This wire must be capable of carrying at least 100 amperes. Smaller-size wires are used for electrical fixtures and low voltage circuits. These wires are described in the section on **Cable Types**.

• CONDUCTOR MATERIALS •

From the point of view of electron theory, the ideal electrical conductor is the material that will carry the greatest amount of current with the least amount of temperature rise. In practice, cost, strength, and ease of use must be taken into account, as well. The materials that satisfy all these conditions best are: copper, aluminum, and copper-clad aluminum.

Copper

The most widely used and, generally, the preferred conductor material is copper. Copper has been in use since prehistoric times and may well be the first metal

used to make useful objects. In the modern world copper is especially well suited to use as an electrical conductor. The atomic structure of copper makes it an excellent conductor. It has high tensile strength; that is, it resists breakage. Copper is highly resistant to corrosion. Copper wiring can be safely used in all circuits and can be connected to all electrical devices. Other conductor materials have some restrictions in these areas.

Aluminum and Copper-Clad Aluminum

In recent years the cost of copper wire has steadily increased to the point where manufacturers have introduced other materials. Aluminum seemed best suited to substitute for copper. It is a good conductor, light in weight, and reasonably strong.

The use of aluminum wire has, however, caused some unexpected problems. Because it is not as good a conductor as copper, aluminum wire must be larger than copper in cross-sectional area to handle the same current flow. It was also soon evident that at the point where aluminum was connected to a terminal made of some other metal—brass, for example—a chemical action took place that caused an increase in resistance to current flow at that point, and thus greater heat. The additional heat caused the metals to expand. When current flow was turned off, the metals cooled and contracted. This repeated expansion and contraction loosened the connection. This in turn further increased the resistance and more heat was generated. Several serious fires were traced to the use of aluminum wire.

As a result of this experience, the NEC was revised to limit the use of aluminum wire to specially made electrical devices. Manufacturers of switches and outlets, for instance, changed the materials used for terminals and added markings to the devices to show which ones could be used with aluminum wire. These markings are described in Chapter 9.

Another type of conductor was devised when a way was found to bond copper to aluminum. This conductor consists of aluminum wire covered with a thin film of copper and it is called copper-clad aluminum.

All these developments have made it safe to use aluminum wire *when the special rules for aluminum wire are followed*. Because of their importance to safe wiring and to compliance with the NEC, the rules for using aluminum and copper-clad aluminum wire are emphasized whenever they apply to the information given in this book. So that you can become familiar with the areas covered by these rules, they are briefly listed below:

1. For the same current flow, aluminum wire must have a larger cross-sectional area than copper. A rule often used is that aluminum wire must be

two sizes larger than the copper wire that would be right for the job. For example, if a circuit requires no. 14 copper wire, no. 12 aluminum or copper-clad aluminum should be used. This rule of thumb is a guide only. Refer to tables in the NEC for exact sizes to be used under various conditions.

2. Aluminum and copper-clad aluminum can be connected only to switches, outlets, and fixtures especially marked for that use. This is covered in detail in Chapter 9.

Conductor Ampacity

The NEC has established the maximum safe current flow for each size of wire and type of insulation. This maximum current flow is the ampere capacity of the wire. The NEC refers to this value as the conductor *ampacity* (Fig. 6-9). A partial listing of ampacities for usual residential wiring is given in Table 6-1. Note that conductor ampacity must always include, in addition to conductor size, conductor material, temperature of the surrounding air, and whether the conductor is in a cable or in free air.

Conductor Temperature and Air Temperature

The NEC refers to two temperatures when specifying ampacity. One is *ambient temperature*; the other is *conductor temperature rating*. The ambient temperature is the normal air temperature in the area in which the conductor will be installed. The conductor

AMPERE + CAPACITY AMPACITY

Figure 6-9. Ampacity.

temperature is the maximum temperature of the wire itself when carrying full rated current.

The NEC ampacities are based on a maximum surrounding air temperature of 80°F (30°C). If wiring is to be installed where air temperatures are usually or frequently higher than the specified maximum, the ampacity must be reduced by a correction factor. For example, no. 12 copper conductor with plastic insulation is rated for 20 amperes where temperatures do not exceed 86°F (30°C). If this cable is to be installed in an area where the usual temperature is 110°F (43°C), the allowable ampacity must be reduced to 58 percent of 20 amperes, or 11.6 amperes.

• INSULATION MATERIALS •

The insulating material that covers copper or aluminum conductors prevents the bare current-carrying wires from touching other conductors or metal objects

TABLE 6-1. ALLOWABLE AMPACITIES OF SOME INSULATED CONDUCTORS IN CABLE. AMBIENT TEMPERATURE OF 86°F (30°C)

Size	Copper		Aluminum or Copper-Clad Aluminum	
	A	B	A	B
14	15	15	-	-
12	20	20	15	15
10	30	30	25	25
8	40	45	30	40
6	55	65	40	50
4	70	85	55	65
2	95	115	75	90
0	125	150	100	120

Column A—This column applies to conductors such as NEC types TW and UF. These conductors have plastic insulation and may be used in wet or dry locations.

Column B—This column applies to conductors such as NEC types TWH, RH, and RHW. These conductors have plastic, rubber, or synthetic rubber insulation that will withstand more heat than those in column A. These conductors may also be used in wet or dry locations.

NOTE: This table applies only to cables having a maximum of three conductors.

and causing short circuits. The insulation also prevents death or injury to humans from accidental contact with "live" wires.

The material most often used for insulation is referred to as *thermoplastic*. This word describes a plastic material that can withstand heat. When copper or aluminum conductors are carrying the maximum rated current flow, they become hot. Insulation must be able to withstand this heat and yet be flexible and easy to remove when cold. All plastic insulation approved by the NEC has these characteristics. Insulation is made in several colors to help you install wiring correctly. The colors used, the meaning of the colors, and how you use them is described later in this section.

Except when using conduit (Chapter 7), electricians usually work with cables rather than individual insulated conductors because they are easier to install and can be routed through walls and ceilings. Cables are made by enclosing two or more insulated conductors and a grounding wire in another layer of protective material. Cables for 120-volt circuits are made of two conductors and a grounding wire; 240-volt circuits require three conductors and a grounding wire.

There are two general types of covering used to make cables; one is heavy plastic, the other is flexible, spiral-wound metal. At the points where connection must be made the outer cable covering, whether plastic or metal, is removed, the conductor insulation is removed, and the bare conductor is used to make the electrical connection. Instructions for doing this are provided in the section **Removing Conductor Insulation**.

You will often hear the plastic-covered cable referred to as "Romex" and the metal-covered as "BX." These are trade names which actually apply only to the cables made by one manufacturer. In this book all plastic-covered cables are called "nonmetallic" and all metal-covered cables are called "armored." These names are the basis of the abbreviations that the NEC uses for cable types.

Conductor and Cable Markings

The NEC requires that all conductors (except those that are too small, no. 16 and smaller) and cables be marked to show important electrical characteristics.

Conductor markings show the wire size, insulation type, and maximum voltage. The Underwriters' Laboratories symbol (UL) is also included if the conductor has been tested and listed. Typical conductor markings are shown in Fig. 6-10.

Cable markings show wire size, number of conductors, cable type, and voltage rating. Typical nonmetallic cable markings are shown in Fig. 6-11. Armored cable cannot be marked because of the uneven surface; tags are attached to provide the required information.

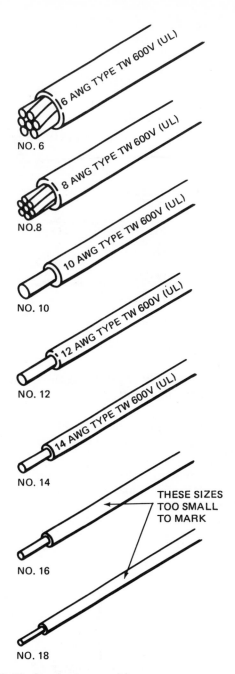

Figure 6-10. Conductor marking.

Wet and Dry Locations

Because water is a good conductor of electricity, moisture on conductors can cause power loss or short circuits. Protection against moisture is an important part of the design and use of cables. In describing where and how cables can be used, the NEC refers to dry, damp, and wet locations. This is what these words mean in general use; for full definitions consult the NEC.

Dry. Fully protected against dampness and water under all normal conditions

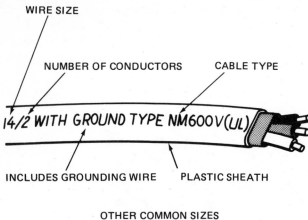

WIRE SIZE

NUMBER OF CONDUCTORS   CABLE TYPE

14/2 WITH GROUND TYPE NM600V(UL)

INCLUDES GROUNDING WIRE   PLASTIC SHEATH

OTHER COMMON SIZES

14 - 2G

14 - 3G

12 - 2G

12 - 3G

Figure 6-11. Cable marking.

Damp. Generally or frequently subjected to heavy moisture or humidity

Wet. Submerged in or enclosed in wet or damp material, as in concrete or below ground.

All of these conditions apply to finished buildings. The fact that an area may be exposed to the weather during construction but not when the building is finished, does not mean that weatherproof cable must be

used. The moisture conditions apply to normal conditions in finished buildings.

Insulation Color Code

The insulation on conductors is colored to make the electrician's job easier. If you understand and follow the color code system, joining and connecting wires is easier and quicker, and wiring errors are much less likely to happen. The insulation colors generally used in the trade are as follows (Fig. 6-12).

1. Two-conductor cable contains one white wire, one black wire, and a grounding wire.
2. Three-conductor cable contains one white wire, one black wire, one red wire, and a grounding wire.
3. Four- and five-conductor cables are available, but are rarely used in home wiring. In these cables the fourth conductor is blue and the fifth is yellow.

The grounding wire in all of these cables is usually a bare wire, but it may have either green or green-with-yellow-stripe insulation.

Chapter 4 described the difference between power ground (or neutral) and grounding wires. If the purpose and use of these wires is not clear, review that portion of Chapter 4 before continuing with this section.

The NEC specifies that the white (or natural grey) wire in all cables shall be the power ground or neutral conductor. Switches or other devices should *never* be

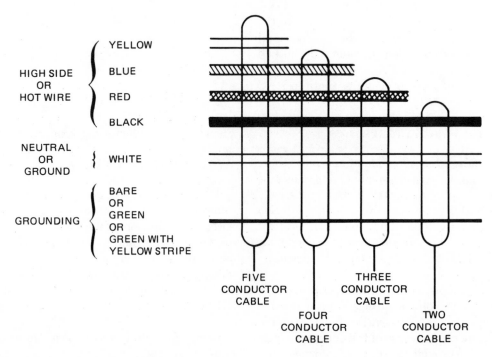

Figure 6-12. Insulation color codes in general use.

wired into the white line. The white line must always be electrically continuous from the power source to the end of the circuit. The grounding wire must also be continuous from the ground point (a cold water pipe or a rod driven into the ground) to all points in the circuit.

Remember: the white wire is the normal return path for current flow from the load back to the source. The grounding wire is a safety path for current flow to prevent shock in the event of failure of a part of the electrical circuit or failure of something connected to it.

• CABLE TYPES •

The most widely used types of nonmetallic (plastic-covered) and armored (metal-covered) cable are described below.

Nonmetallic Cable

Three types of nonmetallic cable are in general use for home wiring. All three are light in weight, easy to work with, and relatively inexpensive. All are rated for 600 volts or less. The current-carrying capacity depends on the conductor material and size. Most cables contain a grounding wire, usually a bare wire wrapped in paper.

Type *NM* consists of insulated conductors wrapped in paper with a fiber filler added, which are enclosed in a plastic cover or sheath. The cable is available in sizes from no. 14 to no. 2 with copper conductors, and no. 12 to no. 2 with aluminum conductors. Most types contain a grounding wire. Because the filler material can act as a wick and absorb moisture and corrosive materials, type NM may be used only in *dry locations.*

Type *NMC* is similar to NM, but is modified to withstand some moisture (Fig. 6-13). In NMC cable the insulated conductors are enclosed in solid plastic. NMC is available with either copper or aluminum conductors in the same range of conductor sizes as type NM. Type NMC may be used in *damp locations*, but must not be used in concrete or masonry or underground.

Type *UF* (underground feeder) cable is similar in appearance to NMC, but its outer cover is tougher and less porous so that it can be used in all *wet locations* and underground.

Armored Cable

Armored cable does the same job electrically as nonmetallic cable. It is available in the same range of sizes in two- and three-conductor cable. The main advantage of armored cable is the protection provided by the metal covering. The main disadvantage is that the

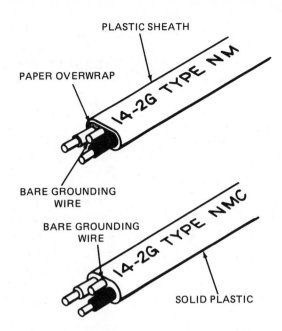

Figure 6-13. Construction of nonmetallic cable.

metal is subject to rust and corrosion in damp locations.

The NEC requires a grounding wire in armored cable, even though the armor itself is a conductor. The grounding wire assures a continuous ground connection throughout the circuit. Armored cable is approved only for use in *dry locations* (except type ACL noted in the following paragraph). With this limitation it is useful in areas where wiring must be protected from accidental damage (Fig 6-14).

There are two common NEC designations for armored cable: *AC* and *ACT*. Type AC has rubber conductor insulation; type ACT has plastic conductor insulation. There is, also, a type *ACL* armored cable which contains a lead sheath over the wires. This type of cable may be used in *wet locations.*

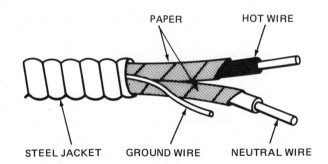

Figure 6-14. Construction of armored cable.

Fixture Wire

A special kind of wire is approved for use within electrical fixtures. This is a stranded, flexible wire with various kinds of insulation (Fig. 6-15), designed for and

INSULATION MAY BE COTTON,
ASBESTOS, OR SILICONE

Figure 6-15. Construction of fixture wire.

used only for internal fixture wiring and for connecting the fixture to the power wiring.

This wire is made with cotton, asbestos, or silicone insulation. The types are *CF*, *AF*, and *SF*. The choice of insulation depends on the heat generated inside the fixture when it is turned on. Cotton can be used for temperatures up to 140°F (60°C), asbestos for temperatures to 194°F (90°C), silicone up to 312°F (156°C).

Low-Voltage Wire

For some special household circuits, like doorbells and control circuits for heating and air conditioning, operating at 24 volts or less, low-voltage wiring may be used. This wire is available in nos. 18 and 16 sizes, may be either solid or stranded, and is covered with a thin plastic insulation.

• SELECTING THE RIGHT CONDUCTOR SIZE AND CABLE COVERING •

Conductor Check List

Let us review briefly the things to be considered when selecting the right conductor size.

Conductor Materials. Copper is the standard. For aluminum or copper-clad aluminum the rule of thumb is to use wire two sizes larger than specified for copper, but NEC requirements should be checked.

Maximum Expected Current Load. Conductor size (based on ampacity) must be sufficient to carry the maximum expected load safely.

Wet or Dry Location. Conductor insulation must be suitable to moisture conditions.

Area Air Temperature. Areas where air temperatures are usually above 86°F (30°C) require larger conductors and special insulation.

Cable Check List

To select the right cable, these additional factors must be considered.

Wet or Dry Location. Cable type must be suitable for moisture conditions.

Protection from Damage. Cable outer covering must be stronger if cable damage can occur.

Corrosive Materials. If corrosive materials are present, a cable must be selected that can withstand them.

In addition, some circuits have special problems and must be specially treated. A common problem of this sort is caused by power loss in long cable runs. Power loss of this kind is caused by a line voltage drop.

Line Voltage Drop

Because all conductors offer some resistance to current flow, some power is always lost between the source and the load. Ohm's law can be used to determine what the voltage drop is in any circuit, if the resistance of the conductor and the current flow are known. The formula is $V = I \times R$.

In this calculation, I is the current flow in amperes in the circuit, R is the conductor resistance in ohms. When they are multiplied, the result is the voltage drop in volts that is due to conductor resistance. In Chapter 2 a basic electrical law was explained that applies to this subject. The law, called Kirchhoff's law, states that the sum of all the voltage drops in a circuit is exactly equal to the applied voltage (Fig. 6-16).

In a 120-volt circuit, then, the sum of the voltage drops must be 120 volts. If we have, for example, a two-wire circuit to a single load, these will be two voltage drops; one is the voltage drop across the load, the other is the voltage drop in the conductors (Fig. 6-17).

Operating the load, a motor, for example, is the reason for the circuit; there should be, therefore, as much voltage as possible applied to the load and as little as possible wasted in the line. Motors and other devices operate most efficiently when the applied voltage is close to the rated voltage.

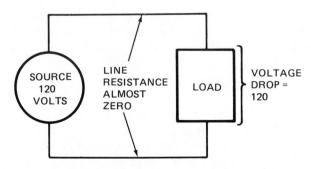

Figure 6-16. Kirchhoff's law.

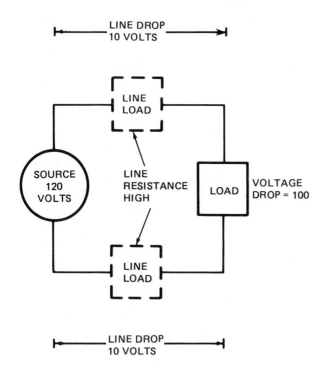

Figure 6-17. Conductor voltage drop.

The voltage wasted in the line is what electricians call "line load," "line drop," or just "voltage drop." There is, of course, a voltage drop across every load. What the electrician must do is select wiring that will make as much voltage as possible available to the working loads (motors, heaters, lights, etc.) and waste as little voltage as possible in the line.

The conductors used for most of the wiring in homes are nos. 12 and 14 for copper. The resistance of these wires is quite small for the short runs that are usually installed within the building. In these cases line voltage drop is so small it can be ignored. Every additional foot of wire increases the line resistance by a fixed amount. When lines become long, as they sometimes do in outdoor wiring, and especially if there is heavy current flow, line voltage drop can cause a significant loss. As a general rule, electricians and circuit designers try to keep line losses to about 2 percent. In a 120-volt circuit this would mean 2.4 volts.

HOW TO DETERMINE CONDUCTOR RESIST- ANCE • Tables in the NEC give the resistance at various temperatures for all types of conductors. A portion of this information for the most common sizes and types of conductors at average temperatures is shown in Table 6-2. Note that the resistance is given for 1000 feet of wire and that the resistance decreases as the wire size increases. Remember that 1000 feet of wire means a 500-foot-long, two-wire circuit. For longer wire runs the resistance value in the table must be increased and for shorter runs it must be decreased.

TABLE 6-2. CONDUCTOR RESISTANCE

Wire Size	Resistance in Ohms per 1000 Feet at 77°F (25°C)	
	Bare Copper	Aluminum
18	6.51	10.7
16	4.10	6.72
14	2.57	4.22
12	1.62	2.66
10	1.018	1.67
8	0.6404	1.05
6	0.410	0.674

HOW TO CALCULATE LINE VOLTAGE DROP • An example will show how line voltage drop is calculated and how the choice of wire size can keep the line voltage drop to a low value.

First, consider a fairly long outdoor line. There are four 150-watt lamps on a line 350 feet long. This might be entrance lights on a long driveway, or outdoor floodlights to illuminate a large area (Fig. 6-18). For an outdoor circuit, type UF cable would be a good choice. For a circuit of this type, the NEC requires at least no. 14 wire. The question is: Would a larger and more expensive conductor size improve the circuit ef- ficiency enough to justify the cost?

The following steps show how to calculate the line drop for various sizes of conductor.

Step 1. Find the total working load on the line. Multiply the number of loads (in this case, lamps) by the wattage of each.

$$4 \times 150 = 600$$

The total working load is 600 watts.

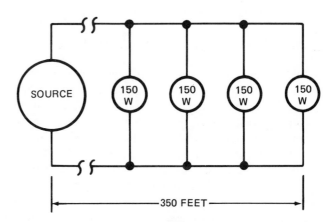

Figure 6-18. Calculating voltage drop.

Step 2. Find the current flow in the circuit. Divide the wattage by the applied voltage (120 volts).

$$\frac{600}{120} = 5$$

The current in the circuit is 5 amperes.

Step 3. Find the resistance of the conductor. Use the value in the table for 1000 feet of line and adjust for the length of line in the actual circuit. From the table, the resistance for no. 14 copper wire is 2.57 ohms per 1000 feet. The actual circuit line is 350 feet long, so conductor length is:

$$350 \times 2 = 700$$

Total wire length is 700 feet. The resistance of 700 feet of wire is 700 divided by 1000 multiplied by the 1000-foot resistance.

$$\frac{700}{1000} = 0.7$$

$$0.7 \times 2.57 = 1.799$$

The line resistance (rounded off) is 1.8 ohms.

Step 4. Find the line voltage drop. Multiply the line resistance by the current flow.

$$1.8 \times 5 = 9$$

The line voltage drop is 9 volts. This represents 7.5 percent of the applied voltage. This is too high. Recalculate, using the resistance for no. 12 wire.

If the same calculation is made using the resistance of no. 12 wire which is 1.62 ohms, the line voltage drop is 5.7 volts. This might be acceptable if the circuit is to be turned on for short periods of time. If usage is greater, no. 10 wire should be used. Line voltage drop using no. 10 wire would be about 3.5 volts.

Other Considerations in Cable Selection

While line voltage drop is one important consideration in choosing a conductor size, other things must also be considered. First, in some communities, local electrical codes specify what conductor sizes must be used for certain types of circuits. It is always acceptable to do more than the Code requires, but (unless an exception is made) never less. Begin your calculation with the minimum Code requirement and then consider whether or not it is necessary to exceed the Code requirement in order to have the circuit do the job it is supposed to do. As noted in the preceding example of

voltage drop calculation, the use of the circuit must be considered in determining acceptable voltage drop. Circuits that are used only infrequently or for short periods of time can have line voltage drops of 5 or 6 percent and still be satisfactory. In addition to calculating the present load on a circuit or the present wire length, consideration must also be given to future growth or changes. If there is reason to believe that circuit additions will be needed, it might be wise to use a larger conductor size to avoid complete rewiring when additions are made.

In all of these decisions cost is an important factor. Larger conductors cost more and in some cases the extra cost may not justify installing a larger size conductor that the NEC or local code requires. On the other hand, when circuits have motor loads, line voltage drop can cause a large drop in motor efficiency. For example, if a motor is operated at 10 percent below its rated voltage, its power output drops 19 percent.

• REMOVING CABLE COVERINGS •

Cable coverings and conductor insulation must be removed before electrical connections can be made between cables or before cables can be connected to electrical devices.

The NEC requires that all electrical connections be made in metal or plastic enclosures known as electrical boxes (Fig. 6-19). The sides of the electrical boxes are scored so that sections can be knocked out as needed to provide openings to bring in cables. All cables that enter boxes must be mechanically fastened to the box by one of two methods.

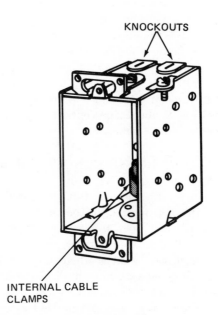

Figure 6-19. Typical electrical box.

One method is the use of a separate cable connector consisting of threaded collar and locknut, and a split-ring clamp (Fig. 6-20). The split-ring clamp fastens the collar to the cable. The threaded part of the collar is put through the box opening. The locknut is then put on the threaded end of the collar, inside the box, and tightened to secure the cable.

Some types of boxes contain built-in cable clamps (second method) that can be adapted for use with either armored or nonmetallic cable (Fig. 6-21). There are two clamps provided in each box. Each clamp can hold two cables.

This section tells how to prepare the cable ends to make electrical connections and how to secure the cable in the electrical boxes. Details on the construction and use of the most common types of boxes are given in Chapter 8.

Nonmetallic Cable

The outer covering on nonmetallic cable can be removed by cutting into the cable with a cable stripper or a utility knife and then peeling the outer cover away from the conductors and cutting it off. This general procedure can be used whether the outer covering is plastic, rubber, synthetic rubber, or a combination of fabric and plastic or rubber.

Cables such as NM, NMC, and UF have two flat surfaces. The cut should be made as nearly as possible straight down the center of a flat side. The easiest way to make a clean, straight cut is to use a stripper for nonmetallic cable (Fig. 6-22). Slip the cutter end of the stripper over the cable first. Set the stripper so that the cutter is about 10 inches from the end of the cable and centered over a flat side. Press the halves of the stripper together and pull the stripper toward the end of the cable. Peel off the outside covering and trim it off.

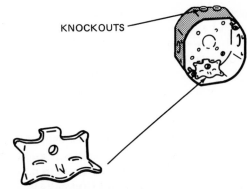

AN ENLARGED VIEW OF A CONNECTOR FOR NONMETALLIC SHEATHED CABLE

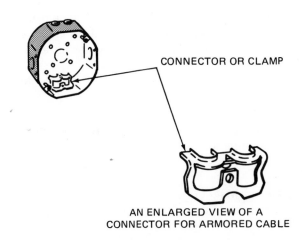

AN ENLARGED VIEW OF A CONNECTOR FOR ARMORED CABLE

Figure 6-21. Internal cable clamps.

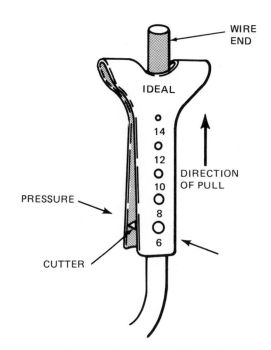

Figure 6-22. Stripper for nonmetallic cable.

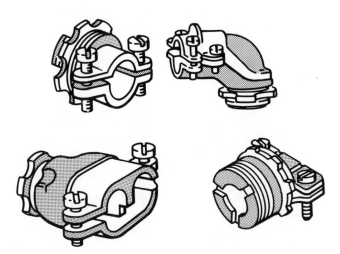

Figure 6-20. Cable connectors.

To use a utility knife to make the cut, you must cut carefully to avoid cutting into the conductors (Fig. 6-23). Each type of outer cable covering has a different "feel." The best way to learn the right amount of pressure to use is to practice with scrap pieces of cable. Place the cable flat on a solid surface to make the cut. After the cut has been made, peel off the outside covering and trim it.

If the cable has filler material, this should be pulled free of the conductors and grounding wire and trimmed away. If the grounding wire is enclosed in paper sleeving, trim the sleeving away also (Fig. 6-24).

The cable end is now ready for installation of a connector. The connector used may be the type that has openings around it so that an electrical inspector can check installation, or it may be the type made expressly for nonmetallic cable which does not have the inspection openings. The installation for either is the same.

To install a connector on nonmetallic cable do the following:

Step 1. Slip the connector over the conductors and the grounding wire and over the end of the outer covering.

Step 2. Make sure the outer cable covering is fully inserted in the connector.

Step 3. Tighten the clamping screws to grip the cable firmly. The cable can now be secured to an electrical box.

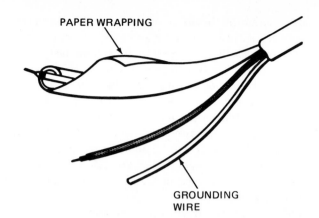

Figure 6-24. Internal wrapping in nonmetallic cable.

Step 4. Remove the locknut from the connector. Insert the conductors and the threaded end of the connector through the box opening. Thread the locknut on the connector and tighten it as much as possible by hand. Use a screwdriver and hammer to tighten the locknut so that the points on the locknut "ears" make solid contact with the wall of the box.

Note that the grounding wire extends into the box with the conductors (Fig. 6-25). When other cables are brought into the box, the grounding wires will be joined. If metal electrical boxes are used, the joined wires will be connected to the box.

Armored Cable

The spiral steel covering on armored cable provides good protection against accidental damage to the insulated conductors. This steel covering, however, causes some problems when working with armored cable. The steel is not easily cut, yet it must be cut carefully to avoid damage to the conductors and grounding strip. After the steel has been cut and removed, a protective bushing must be inserted under the steel to prevent the sharp edges of the armor from cutting into the conductor insulation.

Figure 6-23. Stripping nonmetallic cable with a utility knife.

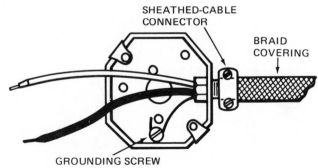

Figure 6-25. Nonmetallic cable connected to an electrical box.

Next, connectors must be fastened on the end of the collar over the armor. These connectors are used to secure the cables to the electrical boxes when connections are made. The connectors also maintain a good ground connection between the steel armor, the grounding strip in the cable, and the electrical box.

Various tools are available for working with armored cable. You may use an ordinary hacksaw or tools especially designed to cut armored cable. The procedures for each method follow.

HACKSAW METHOD • (Figs. 6-26 through 6-30):

Step 1. To expose enough of the conductors for easy connection to other conductors or devices, about 10 inches of the armor must be removed from the end of the cable. Use a firm, flat work surface.

Step 2. Hold the cable and hacksaw so that you can cut across the spiral. This means that the saw will be about a 45° angle to the line of the cable.

Step 3. Saw until you have cut all or most of the way through one section of the spiral. Do not cut deeply enough to damage the insulation or the conductors.

Step 4. Grasp the armor on each side of the cut and twist. Slide the short section of armor off the cable.

Step 5. If necessary, use pliers to trim off sharp pieces of metal at the cut edge. Bend back or break off points that could cut into the conductor insulation.

Step 6. Bend the bare ground strip back along the armor. Unwind the heavy paper wrapped around the conductor. You will need some clearance (about 2 inches) under the end of the armor to install a bushing (step 7). To gain this clearance, remove some paper under the end of the armor. After you have unwound

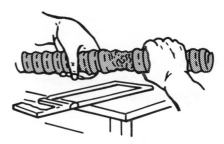

TWIST THE ARMOR TO BREAK IT
AND REMOVE

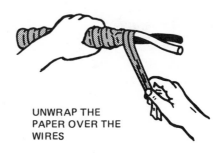

UNWRAP THE
PAPER OVER THE
WIRES

INSERT A FIBER
BUSHING BETWEEN
THE ARMOR AND THE
WIRES

Figure 6-27. Breaking and preparing the armor at the cut.

the paper from the exposed conductors, pull on this paper to unwind more paper under the armor. Tear or cut off the paper.

Step 7. The paper was removed in step 6 to make room inside the armor for a fiber bushing. These bushings are made of red material so that electrical inspectors can spot them, even after other fittings have been added to the cable end. The bushings protect the conductors from insulation breakage and contact with the armor. Slide the bushing over the conductors and under the armor as far as possible.

Step 8. The cable end is now ready for installation of a connector. Loosen the clamp on the connector by loosening both screws, slide the connector over the end of the armor, and tighten the clamp. The grounding strip must be positioned under the clamp to make good contact with the connector when the clamp is tightened. The red bushing must be visible through the openings in the clamp.

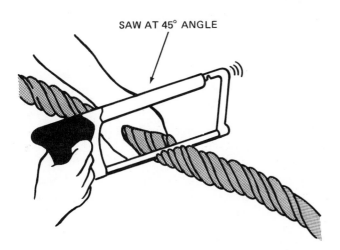

SAW AT 45° ANGLE

Figure 6-26. Cutting armored cable with a hacksaw.

Working With Wire and Cable 97

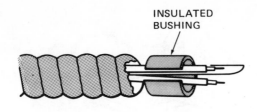

INSULATED
BUSHING

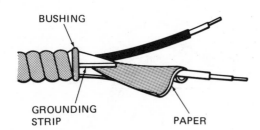

BUSHING

GROUNDING
STRIP

PAPER

Figure 6-28. Fiber bushing in place on cable.

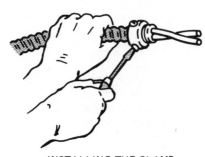

INSTALLING THE CLAMP

SECURING THE CLAMP

Figure 6-29. Installing and securing a clamp on armored cable.

Step 9. The cable can now be secured to an electrical box. Remove the locknut from the connector. Insert the conductors and the threaded end of the connector through the box opening. Thread the locknut on the connector and tighten it as much as possible by hand. Use a screwdriver and hammer to tighten the locknut so that the points in the locknut ears make solid contact with the wall of the box.

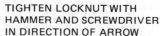

TIGHTEN LOCKNUT WITH
HAMMER AND SCREWDRIVER
IN DIRECTION OF ARROW

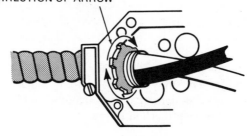

Figure 6-30. Connecting armored cable to an electrical box.

SPECIAL ARMOR CUTTING TOOLS • (Fig. 6-31). Special tools are available for cutting armored cable. These tools are designed to cut the armor cleanly and to limit the cut to the armor so that damage to conductors is almost impossible. The procedures for using these tools vary according to the tool design. Follow the instructions supplied with the cutting tool. After the armor is cut and removed, follow steps 6 through 9 of the hacksaw procedure to complete the job. (Step 5 should not be required if the tool is properly used.)

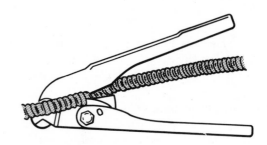

Figure 6-31. Armor cutting tool.

• REMOVING CONDUCTOR INSULATION •

Before conductors can be joined or connected to some electrical device, the conductor insulation must be removed. About 4 inches of exposed conductor will be adequate for making connections. There are three methods of removing insulation.

Wire Stripper

The quickest and easiest way to remove insulation is to use a wire stripper (Fig. 6-32). This is a special tool that is designed to cut through insulation without damaging the conductor. The tool has openings for various wire sizes. Be sure to use the right opening for the wire size you are working with.

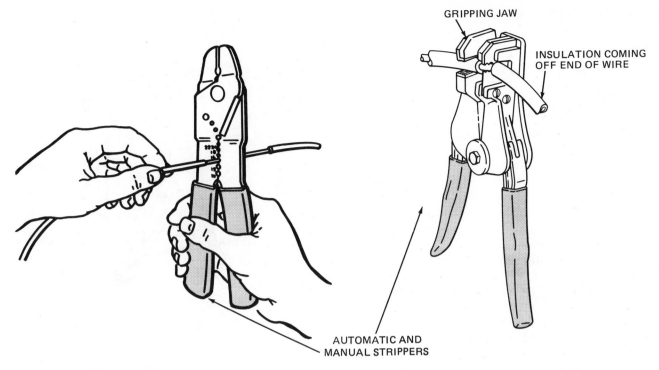

Figure 6-32. Using a wire stripper.

Step 1. Position the stripper about 4 inches from the end of the conductor. This provides enough bare conductor to work with. The excess can be trimmed off when connections are made. Squeeze the stripper to close the jaws.

Step 2. Rotate the stripper back and forth about one-quarter turn several times. This helps to make a clean cut through the insulation.

Step 3. Hold the conductor firmly and pull the stripper toward the end of the conductor to slide the insulation off the end.

Pliers

Ordinary lineman's pliers may be used to remove insulation, but special care must be taken to avoid nicking the conductor and thus weakening the conductor material. The conductor can easily break when bent to make connections or can break shortly after installation, if subjected to any vibration.

Step 1. Before cutting into the insulation, squeeze it a bit between the jaws of the pliers to soften the material.

Step 2. Place the conductor between the cutting edges and close the pliers enough to cut into the insulation, but not the conductor. Rotate the pliers a quarter turn several times.

Step 3. Pull the pliers toward the end of the conductor and slide off the insulation.

Knife

Any sharp knife can be used to remove insulation, but the knife must be used carefully to avoid cutting into the conductor. The best way to avoid conductor damage is to cut at a shallow angle so that the edge of the blade will slide along the metal rather than dig into it (Fig. 6-33).

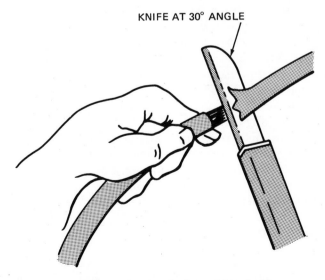

Figure 6-33. Removing insulation with a knife.

Working With Wire and Cable **99**

• JOINING CONDUCTORS •

Connections between conductors must always be made in electrical boxes. A good conductor connection must do three things.

1. Make good electrical contact
2. Have enough mechanical strength to take normal handling without breaking or loosening the electrical connection
3. Provide insulation to prevent shorts and shocks

There are a number of devices manufactured to make connections that meet all of these requirements.

Solderless Connectors

Solderless connectors are made in various sizes to join two, three, or four conductors. They fit wire sizes no. 8 and smaller. You must use the right size to be sure of a good connection. The approved usage is marked on the container the connectors are packed in. If necessary, you can compare loose connectors with those in a marked container to determine the size.

The most widely used device is the solderless connector, sometimes called a *wire nut*. Wire nuts are plastic caps that contain a tapered and threaded metal insert. Push two or more bare conductors into the cap and rotate the cap. This forces the tapered insert over the conductors, making solid contact (Fig. 6-34).

Some solderless connectors are all plastic, with no metal insert (Fig. 6-35). The connectors are tapered and a spiral design is molded into the inner surface. When using these connectors, first twist the conductors together to ensure good electrical contact. Screw the connector over the twisted ends to hold the conductors together.

Screw-on connectors are made of ceramic material for use in high temperature locations. They are used just like the plastic type.

Some connectors have removable metal inserts with setscrews to hold the conductors. Secure the conductors with the setscrew, then screw a plastic cap onto the metal insert.

For wires larger than no. 8, pressure type connectors are generally used (Fig. 6-36). These are clamping devices that hold conductors by tightening a nut or screw to grip the bare wire. Many connectors of this type are not insulated. These connectors must be wrapped with plastic electrician's tape to a thickness equal to the conductor insulation after the connection is made.

Splicing Wires

The NEC requires that all cable connections, including splices, be made in electrical boxes. The best

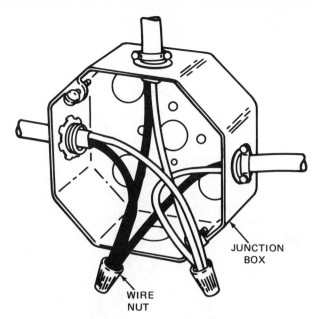

HOW SOLDERLESS CONNECTORS ARE USED

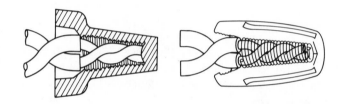

TWO TYPES OF SOLDERLESS CONNECTORS

Figure 6-34. Cross section of solderless connectors.

WIRE-NUT CONNECTORS

Figure 6-35. Common types of solderless connectors.

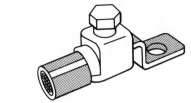

CONNECTING A LARGE WIRE TO AN APPLIANCE

JOINING LARGE WIRES

Figure 6-36. Connectors for large-size wire.

1. CUT AWAY DAMAGED WIRE

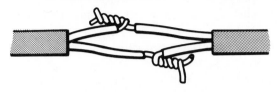

2. MAKE PIGTAIL SPLICE IN EACH WIRE

3. WRAP EACH WIRE WITH TAPE

THREE-STEP REPAIR OF DAMAGED
CORD USING PIGTAIL SPLICES

1ST STEP

2ND STEP

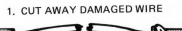

3RD STEP

COMPLETE TAPING JOB

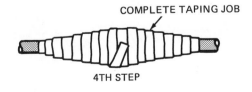

4TH STEP

(A SIMILAR PROCEDURE SHOULD BE USED
FOR THREE-CONDUCTOR CABLES)

Figure 6-37. Splicing two-conductor cables.

way to make most electrical connections is by using the connectors previously described. Always consult your local codes before proceeding with splicing. Splicing can be dangerous and counterproductive. However, when necessary and permitted by law, splices may be used.

When splicing is done correctly, it provides a good electrical connection, good mechanical strength, and insulation equivalent to the original cable. The two most common methods of wire splicing are described below.

TWO- OR THREE-CONDUCTOR SPLICE • (Fig. 6-37). This kind of splice can be used for types NM or NMC cable.

Remember: the splice must be made only in approved electrical boxes or fittings!

Step 1. Trim the conductors unevenly so that the finished splices will not be directly opposite each other. This prevents the splice from being too bulky.

Step 2. Twist together the conductors to be joined.

Step 3. Bend the twisted pair parallel with the wire and solder the joints. Use electrical tape to tape each splice separately. Start the tape on the insulation at one end and wrap it as smoothly as possible.

Step 4. Continue taping over the bare conductors to overlap the insulation at the other side of the splice. The layer of tape should be about the same thickness as the original insulation on the wire.

Step 5. After each splice has been taped, tape the whole area of the connection.

TAP SPLICE • (Fig. 6-38). Tap splices are made when a T-joint is needed, that is, when a conductor must be joined to another conductor without breaking the line. These splices, too, must be made in electrical boxes.

Step 1. Remove a section of insulation from the line and from the end of the conductor to be joined.

Step 2. Wrap the free end of the conductor tightly around the continuous conductor. Solder the joint.

Step 3. Tape the entire splice.

To make this type of splice in large-size wires (no. 8 or larger), use a split-bolt connector (Fig. 6-39).

Step 1. Fit the split-bolt over the continuous conductor and the end of the conductor to be joined.

Step 2. Thread a nut onto the split-bolt and tighten to secure the splice.

Step 3. Cover the splice. Some connectors have clamp-on insulating covers. Others must be covered by taping.

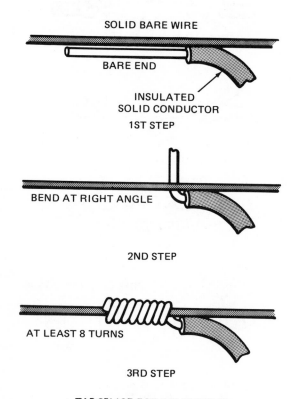

SOLID BARE WIRE

BARE END

INSULATED
SOLID CONDUCTOR
1ST STEP

BEND AT RIGHT ANGLE

2ND STEP

AT LEAST 8 TURNS

3RD STEP

TAP SPLICE FOR SOLID WIRES

Figure 6-38. Tap splice of small-size wire.

Figure 6-39. Split-bolt tap splice for large-size wire.

Soldering Splices

To maintain a good electrical connection in a splice, the conductor joint should be soldered. For a good solder joint, the conductor must be heated until it is hot enough to melt the solder. A soldering iron can do the job on small size wires, such as no. 18 and smaller. For larger wires use a gas-fueled torch. The hot conductor can damage insulation. A torch will heat the joint rapidly so that solder will flow before the conductor carries away too much heat. Whether using a torch or an iron, the conductor must be heated enough to melt the solder, so that hot solder flows onto a hot conductor. A good solder joint will appear smooth and have a slight sheen when cool.

If hot solder flows onto a cool conductor, a cold solder joint will result. This must be avoided because cold solder joints increase electrical resistance at the joint, causing the conductor to overheat. Cold solder joints appear dull and grainy; if one is suspected, apply the torch or iron to the joint and reheat it thoroughly until the solder flows freely.

The general procedure for soldering a splice is as follows:

Step 1. Make certain the conductor surface is clean and bright.

Step 2. Use a rosin-core, wire-type solder or apply nonacid (rosin) flux to the joint separately.

Step 3. Heat the joined conductors in the blue flame of the torch or hold the iron under them. Hold the solder spool so that the end of the solder is touching the joined wires.

Step 4. In a few moments the solder will melt and flow into the conductor joint. Keep the joint hot for a second or two longer to allow the solder to flow freely. Remove heat and allow the joint to cool. Inspect as noted above to make certain the joint is properly soldered.

• CONNECTING WIRES TO TERMINALS •

Screw Terminals

Screw-type terminals are used for electrical connections on many devices for wire sizes up to no. 10. The basic rules for connection to screw terminals follow.

1. Only one conductor may be connected to a terminal.
2. The end of the conductor must be bent into a hook that fits over the screw.
3. Not more than 1/8 inch of bare conductor should be exposed after the connection is made.

The NEC specifies only one conductor per terminal to reduce the possibility of a poor connection and the resulting overheating at the terminal. If it is necessary to connect two or more conductors to a terminal, a jumper must be used.

Step 1. Make a 4-inch jumper by removing insulation from each end of a short piece of conductor.

Step 2. Join one end of the jumper and the other conductors with a solderless connector.

Step 3. Loosen the terminal screw and connect the other end of the jumper to the screw terminal (Fig. 6-40). Tighten the screw.

For proper connection to a screw terminal, bend the end of the conductor into a hook so the screwhead will press down evenly on the wire for good electrical contact (Fig. 6-41). The hook must curve in the direction in which the screw is turned to tighten it (clockwise). The hook should go about three-quarters of the way around the screw. Never make the hook so long that

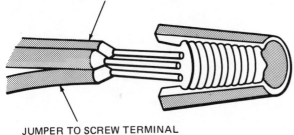

CONDUCTORS TO BE CONNECTED TO TERMINAL

JUMPER TO SCREW TERMINAL

Figure 6-40. How to connect several wires to a single screw terminal.

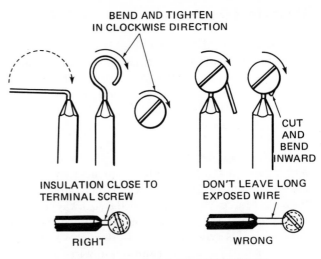

BEND AND TIGHTEN IN CLOCKWISE DIRECTION

CUT AND BEND INWARD

INSULATION CLOSE TO TERMINAL SCREW

DON'T LEAVE LONG EXPOSED WIRE

RIGHT

WRONG

SCREW TERMINALS ARE COLOR CODED TO INDICATE HOW CONDUCTORS SHOULD BE CONNECTED. THIS SUBJECT IS COVERED IN CHAPTER 9

Figure 6-41. How to connect wires to screw terminals.

the conductor can overlap. Keep the length of exposed conductor to 1/8 inch to prevent accidental contact with other wires.

Straight-Wire Terminals

There is a type of screw terminal (Fig. 6-42) that can be used without bending the conductor. To use this type of terminal, place the wire in a grooved slot next to the screw. Tighten the screw so that the screwhead holds the wire in the grooved slot.

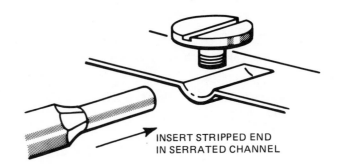

INSERT STRIPPED END IN SERRATED CHANNEL

TIGHTENING SCREW LOCKS WIRE IN PLACE

Figure 6-42. Straight-wire screw terminal.

Push-In Terminals

Another type of terminal (Fig. 6-43) requires only that the bare end of the conductor be inserted in an opening in the back of the switch or outlet. The electrical connection is made by a spring-loaded strip of metal within the device that presses against the conductor. A strip gauge is provided on the back of the device. The gauge shows how much bare conductor is needed for good contact. To release the conductor, insert a small screwdriver into a slot next to the conductor opening. When the screwdriver is pressed into the slot, it forces the spring-loaded strip away from the conductor and the conductor can be pulled free. Devices with this type of terminal are approved for use with copper or copper-clad aluminum wires only.

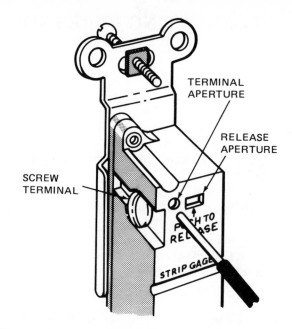

Figure 6-43. Push-in type wire connections.

• REVIEW QUESTIONS •

1. As used in the electrical industry, the words *wire* and *cable* have different meanings. What are the differences?

2. Wires are identified by the cross-sectional area of the conductor. Why is this measure important?

3. Wire sizes are identified by numbers. Which wire is larger, no. 14 or no. 8?

4. Large-size wires consist of several individual strands twisted together. How is the cross-sectional area of these wires determined?

5. Two metals are commonly used as conductor material. What are they? Which is the better conductor?

6. The NEC uses one word to mean the maximum current capacity of a wire. What is it?

7. If a cable is marked "14/3 G TYPE NMC 600V (UL)," what does each item mean?

8. Some cables contain four wires: one with red insulation, one with black insulation, one with white insulation, and one bare conductor. What do the colors mean? What does the bare wire do?

9. All conductors have some resistance to current flow. Tables in the NEC give the resistance in ohms for 1000 feet of various conductor sizes. For what calculation do we use this conductor resistance?

10. What are the three essentials of a good electrical connection?

11. What article and tables in the NEC cover maximum allowable ampacities of conductors?

7
WORKING WITH CONDUIT

• INTRODUCTION •

Electricity can be routed to various parts of a building by installing pipes between the source of power and the places where power is needed, and then running conductors through the pipes to carry the current. The pipes used for this purpose are called conduit. Conduit electrical systems are more expensive to install than systems which use armored or nonmetallic cable. Good planning and careful work are necessary to keep costs as low as possible.

Working with conduit is an important part of an electrician's job. Special skills are needed to cut, bend, join, and secure conduit. The information in this chapter gives you the background you need for conduit work.

Conduit must be installed so that wires can be pulled through the piping without damage to conductors or insulation. Provision should be made for future additions to the system and for repair or replacement of conductors.

This chapter tells you how to work with the types of conduit that are most widely used. You will learn the advantages and disadvantages of each type. The special tools used for conduit work are described and procedures are given for using them.

• TYPES AND USES •

The main advantage of conduit over nonmetallic cable is the mechanical protection it provides for conductors. In any location where accidental damage to conductors can occur, conduit is a good investment. In residential wiring, conduit may be used throughout the house or it may be limited to certain areas where wiring is exposed, such as workshops, storage rooms, and utility rooms. Limited use of conduit keeps costs down, yet provides wiring protection where it is needed. It is often used at the service entrance, the place where utility company power lines enter the building. At this point conduit provides protection against the weather. In some installations it may also provide mechanical strength to support the incoming power lines.

Conduit is desirable; properly installed it will give many years of service. It may be made of steel, aluminum, or plastic. Surface coverings can be added to improve resistance to corrosion. Conduit is especially well suited to installations where explosive gases may be present. It can be combined with explosion-proof fittings which prevent gas from seeping into the conduit, thus preventing explosions caused by electrical sparking.

Another important feature of rigid metal conduit is the excellent grounding conductor it provides. You will

recall (Chapter 6) that armored and nonmetallic cables usually contain a bare (or green-insulated) wire for equipment grounding. In conduit installations, the NEC approves the use of most types of conduit as the grounding conductor. This, of course, requires that there be a continuous electrical path throughout the conduit installation. All fittings, boxes, connectors, and joints must be made so the continuous ground path is not broken. When conduit is combined with cabling, care must be taken in joining the grounding conductor in the cabling to the sections of conduit to maintain a good equipment grounding circuit.

Because conduit completely surrounds the current-carrying conductors, an exposed live conductor cannot be touched accidentally. It is possible that an exposed live conductor will touch some part of the inner wall of the conduit. This will cause a short circuit to ground, and an automatic circuit protective device will turn off power.

Many types and styles of conduit are manufactured, but four types are the most widely used.

Rigid Conduit

This conduit (Fig. 7-1) closely resembles piping used for water or gas. The sidewall thickness is about the same; it is available in the same sizes; it can be cut, threaded, and joined the same way; similar fittings are available. In spite of all these similarities, however, there are two important differences. The inner surface of rigid conduit is specially smoothed so that conductors can be pulled through readily. The pipe itself is made of a special alloy that makes cutting and bending easier. Rigid conduit provides the greatest protection to conductors and is the most desirable. However, it is expensive, heavy, and difficult to work with.

Figure 7-1. Rigid or intermediate conduit.

Intermediate Conduit

This is essentially the same as rigid conduit, but the sidewall is slightly thinner. This lowers the cost and makes cutting and bending somewhat easier. Fittings and connectors used with intermediate conduit are similar to those available for rigid conduit. Intermediate conduit has the same characteristics as rigid conduit, but is somewhat lighter and less expensive.

Electrical Metal Tubing (EMT)

This is also known as thin-wall conduit (Fig. 7-2). Its sidewall thickness is only about 40 percent that of rigid

Figure 7-2. Electrical metal tubing (EMT).

conduit. This makes for easier cutting and bending. The walls are too thin, however, for threaded fittings. Compression or setscrew type fittings are used to make secure joints and tight connections to electrical boxes. EMT is relatively inexpensive and easy to work with. The thin walls provide adequate protection for most uses.

Flexible Conduit

Flexible conduit (Fig. 7-3) looks like oversized armored cable. It is often referred to in the trade as "Greenfield," the name of one manufacturer. It is well suited to cable runs where frequent bends cannot be avoided and to connections (to motors, for example) where vibration would damage rigid conduit. Flexible conduit is cut like armored cable. Unlike the other forms of conduit, the NEC does not allow flexible conduit to serve as a grounding conductor. A grounding wire must be included with the current-carrying conductors when flexible conduit is used.

The NEC allows ordinary flexible conduit to be used in dry locations only. A special type of flexible conduit is made for use in damp or wet locations. This is known as liquidtight flexible conduit. It consists of regular flexible conduit with an outer liquidtight plastic covering. Liquidtight fittings are available for use with this conduit. Because liquidtight flexible conduit is often used in outdoor locations, the outer covering is designed to resist damage from exposure to sunlight.

Flexible conduit provides good protection to conductors and is easy to work with. The relatively high resistance of the spiral construction makes it unsuitable for equipment grounding.

Figure 7-3. Flexible conduit.

• SIZES AND CONDUCTOR CAPACITY •

Rigid and intermediate conduit and EMT are sold in 10-foot lengths. Flexible conduit is sold in coils of 50 to 250 feet, depending on conduit size. The coils can be cut to any convenient length. The available cross-sectional sizes are the same as the trade sizes for plumbing pipes. The trade sizes range from 1/2 inch to 6 inches for rigid conduit and from 1/2 inch to 4 inches

for intermediate, EMT, and flexible conduit. The *actual* inside diameter of conduit is 1/16 to 1/8 inch larger than the trade size. For example, 1/2-inch trade size conduit actually has a 5/8-inch inside diameter and 2-inch trade size is actually 2-1/16 inches.

The minimum size permitted by the NEC for any job is determined by the number and size of conductors that must be enclosed by the conduit. Tables 2 and 3 in Chapter 9 of the NEC list the maximum number of conductors of any *one* size and type that can be installed in each trade size of conduit. For example, 1/2-inch trade size conduit may contain nine no. 14 TW-type conductors or seven no. 12 TW-type conductors, or five no. 10 TW-type conductors. When conductors of different sizes are to be installed in conduit, the maximum allowable number must be calculated by figuring the total cross-sectional area of the conductors and comparing this number with the NEC-permitted fill area for various sizes of conduit. Tables 4 and 5 in Chapter 9 of the NEC provide the size information needed. The following example shows how these tables are used.

A conduit is required for six no. 12 TW conductors and three no. 8 TW conductors. From Table 5 we find that the area in square inches for no. 12 TW is 0.0172, and for no. 8 TW it is 0.0471. Six no. 12 TW conductors, then, have a total area of 0.1032. Three no. 8 TW conductors have a total area of 0.1413. The total conductor cross-sectional area is the sum of these numbers, 0.2445. Next, we refer to Table 4. TW conductors have thermoplastic insulation, so we use the column headed "Not Lead Covered." We have a total of nine conductors, and the column headed "Over 2 Cond. 40°" must be used. This column lists 0.21 square inch maximum conductor area for 3/4-inch trade size conduit. This is less than the calculated area of 0.2445, so the next larger size must be used. Trade size 1-inch conduit can have up to 0.34 square inch conductor area. Trade size 1-inch conduit is then the minimum size that the NEC permits for six no. 12 TW conductors and three no. 8 TW conductors.

In practice, two other considerations play a part in selecting the size of conduit to be used. First, ease of installation of the conductors must be considered. If a conduit run is long and has several bends, it is easier to pull conductors through the conduit if there is room to spare. A size larger than the minimum permitted by the Code might well save time on the job. A second point to be considered is future needs. Conduit installations are more expensive than cabling. The difference in cost between trade sizes, however, is a small percentage of the total cost. In the previous example, the installation of 1-inch conduit would allow either five no. 12 TW conductors or two no. 8 TW conductors or any other combination of conductors not exceeding 0.0955 square inch area to be added if needed.

• RIGID AND INTERMEDIATE CONDUIT •

Rigid and intermediate conduit are basically the same except for the thickness of the sidewalls. The trade sizes are the same (except that the largest intermediate size is 4 inches), and similar fittings are available. These two conduit types may be mixed in a single installation, sections may be interchanged, or one type may be substituted for the other.

Rigid conduit and intermediate conduit are sold in 10-foot lengths threaded at both ends. A coupling—a threaded sleeve that is used to join two pieces of conduit—is screwed on one end of each length of conduit. The threads on the other end are protected by a plastic cover. Shorter threaded lengths of conduit, known as nipples, are available in lengths from 4 inches to 3 feet.

Fittings

Several types of fittings are available for making bends and turning corners.

CONDULETS • Right-angle bends above ground can be made with condulets. These are rectangular, galvanized steel boxes with two or three threaded openings (Fig. 7-4). One opening is usually located on a short side of the rectangle. The other opening can be on either long side or the back of the box. There may be an opening on each long side of the box, as well as on the short side.

One side of the box is covered by a removable hatch. Conduit can be screwed into the condulet openings. The hatch is removed to pull cables through and to make the sharp right-angle bend. When conductors are in place, the hatch is closed and secured with a screw.

Condulets are used whenever a sharp right-angle turn must be made, or when conductors must be spliced or joined. The NEC specifies that conductor splices in conduit must be made where they are permanently accessible. The condulet hatch serves this purpose. Condulets with three openings are useful at T junctions when switches or outlets must be located above or below the main run of conduit.

ELBOWS • Elbow fittings (Fig. 7-5) can be used for simple right-angle or 45° bends. Elbows are joined to straight conduit with standard couplings, providing a smooth bend through which conductors are pulled. Elbows can be used below ground and are often used to join underground conduit to an aboveground run.

OFFSETS • When slight jogs are required in a run, offset adapters can be used (Fig. 7-6). This situation can occur when conduit must be shifted to avoid water pipes, other conduit, or variations in floors or walls.

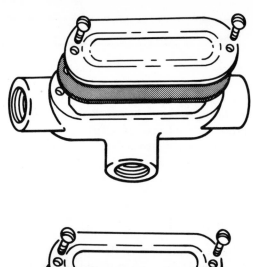
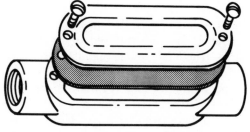

Figure 7-4. Typical condulets.

Figure 7-5. Elbow fitting.

Figure 7-6. Offset adapter.

The offset adapter shifts the line of the run a few inches left or right, up or down. Both lengths of conduit joined to the offset adapter are parallel.

Nipples and fittings can be used in many combinations to make a great variety of turns and bends. Always keep in mind that conductors must be pulled into the conduit. Fittings used too close together or in too many combinations may make conductor installation difficult. When standard fittings cannot do the job—or simply for economical use of material—conduit can be bent on the job to almost any desired shape.

Bending

Two styles of handtools are used to bend conduit. One type—sometimes called a *hickey*—has a head that grips a small section of conduit. Another length of conduit or a special handle fits into an opening on the hickey. The handle provides the leverage needed to bend conduit. With this tool large bends (more than 45°) must be made by making a series of small bends to avoid damage to the conduit.

Another type of handtool—called a *bender*—has a curved head that covers a large section of conduit and provides sidewall support so that bends of 90° can be made in one motion. This type of bender also has a long handle, but in addition has a tread projection on the head. By pressing down on the tread with one foot while pulling back on the handle, enough leverage is provided to bend heavy-wall conduit.

Either a hickey or a bender can be used on any type of rigid or intermediate conduit or AMT. In practice, however, hickeys are used for bending large-diameter rigid conduit while benders are used for small-diameter rigid conduit and for all sizes of intermediate conduit and EMT.

On large jobs where many bends of the same type must be made, bench-type benders or power benders may be used. But for small jobs the conduit hickey or bender is the tool to use.

The procedure for using a hickey or bender varies with the design of the tool and the size of the conduit to be bent. The skill needed to make smooth and accurate bends can be gained only by actual experience with the tool and conduit. You will gain the skill faster if you understand the basic procedure for making standard conduit bends. The right-angle bending procedure for hickeys or benders is different. Each procedure is given below. For other bends, both tools are used about the same way.

RIGHT-ANGLE BEND USING A HICKEY • The walls of rigid conduit are too heavy for a full 90° bend to be made at one point and in one motion when a hickey is used. The bend must be made in three stages or "bites" to get a smooth curve. If conduit bends are

made too sharp, the conduit may be weakened or the inside diameter may be reduced.

The sharpness of bends is defined by the "bend radius." This measurement is the radius of the circle that would be formed if the curve of the bend were continued (Fig. 7-7). The sharpness of conduit bends is limited by the NEC. Table 346-10 gives the minimum bend radius for each trade size. As a guide for bends made by hand, the bend radius will meet Code requirements if it is at least eight times the trade size of the conduit. For example, 1-inch conduit with a bend radius of 8 inches is well within the NEC.

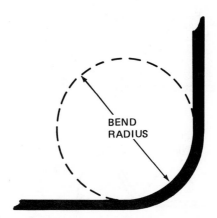

Figure 7-7. Measuring bend radius.

Step 1. Mark on the conduit the point where the bend is to end (Fig. 7-8a).

Step 2. Measure and mark a second point 4 to 8 inches ahead of the end point.

The second point is the point where the first bend will be made. The exact location of the second point depends on the size of the conduit and the skill of the electrician. Four inches is about right for small sizes. For large conduit the first bend point can be as much as 8 inches ahead of the point where the bend is to end. Experience will help in determining the bend point.

Step 3. To make the bend, place the conduit on a flat, solid section of floor.

It is helpful to brace the end of the conduit opposite the bend against a wall. This prevents the conduit from slipping backward when the bend is started.

Step 4. Place the hickey over the end of the conduit and line it up with the bend point. Hold the conduit by placing one foot on it just behind the hickey.

Step 5. Pull the hickey back to make a bend of about 45° (Fig. 7-8b).

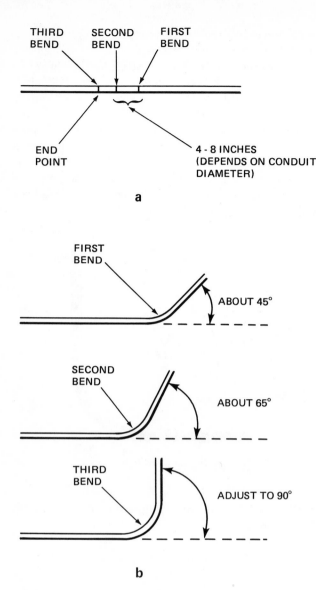

Figure 7-8. Right-angle bend using hickey. (*a*) Mark end point. (*b*) Make bends.

Step 6. Move the hickey back a few inches and make another bend about half of the first bend.

Step 7. Move the hickey back a few inches and make the final bend.

The bend should now be about 90°. If an exact angle is needed, the angle can be checked using a blade and beam square or a bubble level (Fig. 7-9). Adjustments can be made by putting the hickey in the position of the last bend and adjusting the angle as necessary.

RIGHT-ANGLE BEND USING A CONDUIT BENDER •

Step 1. Mark on the conduit the point where the bend is to start.

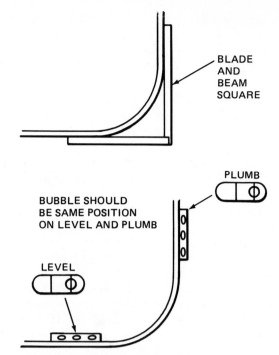

Figure 7-9. Checking right-angle bends.

Step 2. Place the conduit on a solid, flat surface braced against a wall.

Step 3. Slip the bender over the conduit and line up the start mark on the bender with the point marked on the conduit. (The start mark may be an arrow, a number, or some symbol. Follow the instructions supplied with the bender.)

To start, the bender handle is at an angle of about 45°

Step 4. With one foot on the bender tread, pull up and back on the handle to make the bend (Fig. 7-10). When the bender handle is vertical, the bend is 45°.

Step 5. Remove your foot from the tread and put it on the conduit to hold it in place.

Step 6. Complete the bend by pulling back and down on the bender handle until the handle is at an angle of about 45°.

The conduit now has a bend of about 90°. Check the angle and use the bender to make adjustments, if necessary.

If the top of the bend must be an exact distance above the horizontal run of the conduit, more care must be taken and a few measurements must be made. Of course, making the top of the bend longer than necessary, cutting off the excess conduit, and threading the

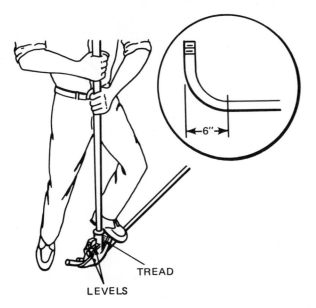

Figure 7-10. Right-angle bend using a bender.

end will get the same result. If many such bends are needed, however, this procedure wastes a considerable amount of expensive material. Accurate bends should be made instead. Usually there is some adjustment possible in locating the conduit and in making a connection at the top end. However, reasonable accuracy is necessary for a satisfactory job. If the distance from a horizontal run to an outlet box, for example, is 12 inches, add about 3/8 inch to this for the projection of the threaded end of the conduit into the box. This point (12-3/8 inches) is then the start point of the bend. If a hickey is being used, a second point must be located, as previously explained. If a bender is used, only the start point is needed.

OFFSET BEND • When conduit is attached to metal boxes, the conduit must enter the box in a straight line (Fig. 7-11). This is not done for appearance alone. Because conduit serves as the grounding conductor, every connection in a conduit run must be mechanically solid and must not offer any significant resistance to current flow. When conduit is fastened to a box, it is secured with one or two locknuts and a

bushing. If the conduit enters or leaves the box at an angle, the locknuts and the bushing cannot be properly seated against the wall of the box. If the locknuts and bushing loosen, the grounding circuit no longer provides a low-resistance path to ground.

When both the conduit and the box it must enter or leave are mounted against a flat surface, a double bend must be made in the conduit to bring the conduit straight into the box. This double bend is called an offset. Offset bends require a bit more practice than right-angle bends because the bender is used a different way.

Step 1. To make the first bend, insert the conduit into the bender, hold the conduit, and let the handle of the bender rest on the floor so that the bender is upside down (Fig. 7-12).

Step 2. Bracing the end of the handle with one foot to hold it in place, press down on the conduit to form a 45° bend.

Step 3. Rotate the conduit one-half turn, and move the bender so that it just clears the first bend.

Step 4. With the bender still resting on the floor, make another 45° bend.

The conduit before the offset should be parallel with the conduit past the offset (Fig. 7-13). If the conduit still does not enter the box straight, the angles of the offset must be adjusted. If the offset projects out too far, reverse the bending process and make each bend shallower. If the offset does not project out far enough, repeat the process and make each angle more than 45°. The amount of the bend will vary with different sizes of conduit and various styles of boxes. Once you have mastered the upside-down use of the bender, with a little practice you will be able to make offset bends in any size conduit for any style box.

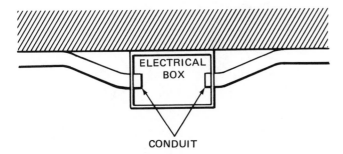

Figure 7-11. Correct conduit connection to a box.

Figure 7-12. Making an offset bend.

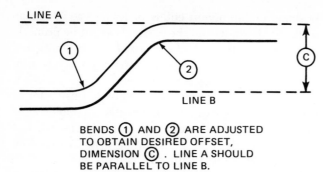

BENDS ① AND ② ARE ADJUSTED
TO OBTAIN DESIRED OFFSET,
DIMENSION ©. LINE A SHOULD
BE PARALLEL TO LINE B.

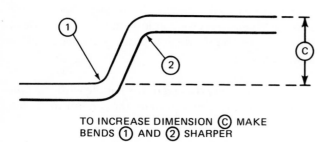

TO INCREASE DIMENSION © MAKE
BENDS ① AND ② SHARPER

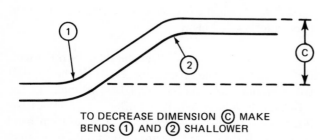

TO DECREASE DIMENSION © MAKE
BENDS ① AND ② SHALLOWER

Figure 7-13. Correct offset angle.

SADDLE BEND • When conduit must bypass objects such as large pipes, beams, or ductwork, a double offset bend—called a saddle—must be made (Fig. 7-14). The bending procedure is similar to the offset procedure, but the distance between bends is greater.

Step 1. Determine the distance necessary to clear the obstruction, and make the first bend using the upside-down bender procedure.

Step 2. Rotate the conduit one-half turn and move the bender to a point 1 to 3 inches farther than the clearance distance.

Step 3. For bend no. 2 measure the distance between the parallel lengths of conduit. If the distance is enough to clear the obstruction, continue with the bending. If the distance is too large or too small, adjust the angles as previously explained for offset bends.

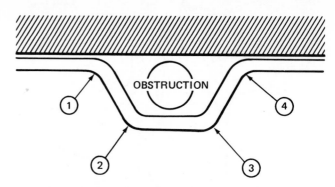

Figure 7-14. Saddle bend.

Step 4. To make the second offset to complete the saddle, keep the conduit and bender in the same position, but slide the conduit through the bender far enough to clear the obstruction. Make bend no. 3.

Step 5. Rotate the conduit one-half turn, and move the conduit the same distance as it was moved to make the second bend. Make bend no. 4.

It is important in this procedure that the conduit be rotated exactly one-half turn before making bends nos. 2 and 4. The best way to do this is by marking guide lines on each side of the conduit before you start.

Cutting

Rigid and intermediate conduit may be cut with a hacksaw (Fig. 7-15) or a pipe cutter (Fig. 7-16). With either method the conduit must be locked in a pipe vise before making the cut. Put the conduit in the vise so that the vise grips the conduit 2 or 3 inches from the point where the cut will be made. This prevents the grip of the pipe vise from damaging the surface of the conduit that must be threaded.

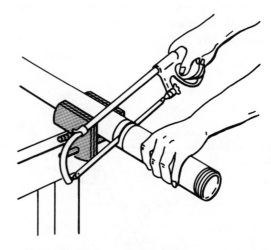

Figure 7-15. Cutting conduit using a hacksaw.

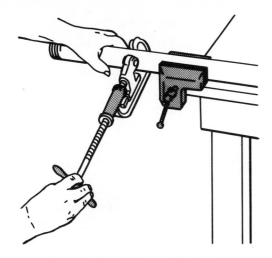

Figure 7-16. Cutting conduit using a pipe cutter.

If a hacksaw is used, use 18- or 24-teeth-per-inch blades. Be sure to install the blade so the cut is made on the forward stroke.

After cutting by any method the inside edge of the conduit must be smoothed with a half-round file (Fig. 7-17) or a pipe reamer mounted in a brace. Be particularly careful with conduit cut with a pipe cutting tool. This tool tends to leave a sharp ridge on the inner edge of the cut. Be sure this ridge is removed and the conduit is smooth before installing a coupling or any fitting.

Threading

Conduit is threaded just as plumbing pipes are by using dies and a die stock. Apply cutting oil to the end of the conduit and cut threads exactly the length of the die. Cutting threads longer than the length of the die will leave exposed threads that are not protected from corrosion.

Joining

Lengths of conduit, or conduit and nipples, are joined by couplings. One coupling is supplied with each length of conduit. Couplings are made of the same material as the conduit, are galvanized, and have internal threads. The coupling is simply threaded onto the conduit and tightened. Ridges on the outside of the coupling make tightening easier.

When conduit is joined to a metal electrical box, it is secured with a grounding locknut and a bushing (Fig. 7-18). The locknut is threaded onto the conduit first, then the conduit is inserted into the box opening. The bushing goes on the threads projecting into the box. The bushing is tightened as much as possible and then the locknut is tightened to secure the conduit to the box. The bushing secures the conduit and also protects the conductors from abrasion by covering the edges of the conduit with a smooth plastic insert. Conduit

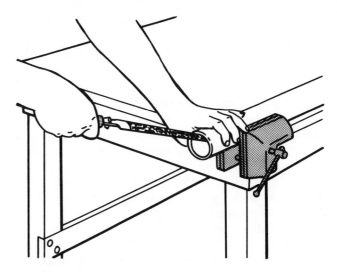

Figure 7-17. Smoothing conduit using a file.

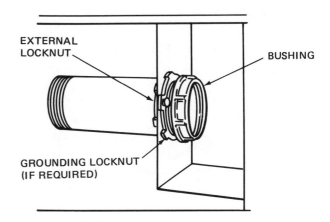

EXTERNAL LOCKNUT

BUSHING

GROUNDING LOCKNUT (IF REQUIRED)

Figure 7-18. Joining conduit to an electrical box.

locknuts may also have setscrews that must be tightened to assure that the locknut cannot work loose.

If any conductor to be installed in the conduit will carry more than 250 volts (measured to ground), the NEC requires additional bonding, such as a second locknut installed on the inside of the box before the bushing is added. Local codes may require two locknuts for lower voltages.

• ELECTRICAL METAL TUBING (EMT) •

Electrical metal tubing (EMT)—like rigid and intermediate conduit—is sold in 10-foot lengths. Trade sizes range from 1/2 inch to 4 inches. Because of the thin walls, EMT is not threaded. Connections are made by compression fittings, setscrew couplings, or couplings secured by indenting the metal. EMT is also used as the grounding conductor.

EMT is made of galvanized steel or aluminum. The materials used are designed to bend smoothly, without kinks or distortion, if proper methods are used. The EMT bender is different from the one used for rigid and intermediate conduit. It is designed to support the sidewalls of the tubing during bending; a foot tread projection is not required.

To make a bend, place the tubing on solid flooring and slip the bender over the tubing to the point where the bend is to be made. The thinner sidewalls allow a full 90° bend to be made easily in a single bite (Fig. 7-19). As with the rigid conduit bender, when the handle is vertical, a 45° bend has been made. The bending procedures given previously for rigid and intermediate conduit bending apply to EMT, as well. The EMT bender is used upside down, just as the rigid bender is, when necessary.

The procedure for making offsets and saddles is similar to the procedure for rigid and intermediate conduit, the principal difference being that bends can be made more easily.

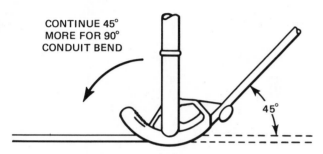

Figure 7-19. Using EMT bender.

Cutting

EMT can be cut readily with a hacksaw. A blade having 24 or 32 teeth per inch should be used. The tubing can be braced by hand, but should be supported on some solid surface so that a clean, straight cut can be made.

Fittings

Fittings for EMT are of two general types. There are watertight fittings that may be used outdoors or in any location, and there are fittings that provide strong mechanical and electrical connections but may be used only in dry locations.

WATERTIGHT FITTINGS • The watertight fittings join sections of tubing by means of a five-piece compression fitting (Fig. 7-20).

Step 1. Place a gland nut and compression ring over the end of each piece of tubing.

Step 2. Slip a double-threaded ring—called the body—over the end of each section.

COMPRESSION-TYPE CONNECTOR

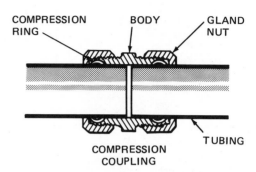

Figure 7-20. EMT watertight fittings.

Step 3. Screw the gland nuts onto the body and tighten them to squeeze the compression ring. The ring forms a watertight seal.

A similar fitting having only three pieces is used to make a watertight joint to metal boxes (Fig. 7-21).

Step 1. Place the large nut and compression ring on the end of the EMT.

Step 2. Place the double-threaded body over the end.

Step 3. Screw the nut onto the body, squeezing the compression ring and making a watertight seal.

Step 4. Use the exposed threads on the body to secure the EMT to a weatherproof box using a locknut and bushing.

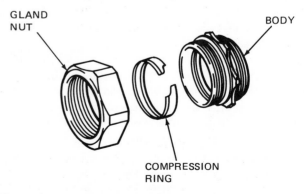

Figure 7-21. EMT watertight box fitting.

FITTINGS FOR DRY LOCATIONS • Fittings for use in dry locations are simpler to use and less expensive. One type consists of a sleeve and two or four setscrews (Fig. 7-22).

Step 1. Slip the ends of the EMT into the sleeve.

Step 2. Tighten the setscrews to make the coupling.

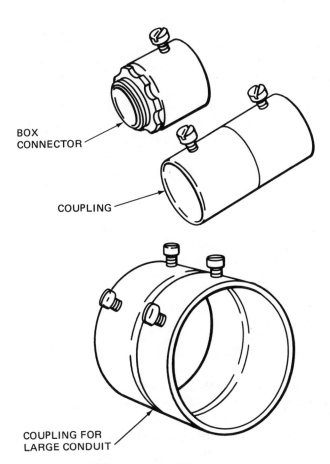

Figure 7-22. EMT setscrew fittings.

Another form of coupling is made by using a plain sleeve and an indenting tool (Fig. 7-23).

Step 1. Place the sleeve over the ends to be joined.

Step 2. Use the indenting tool to make indents in the couping and the tubing to secure the joint. The tool makes two indents at once on either side of the coupling.

Step 3. Use the tool twice, 1/4 turn apart, on each end of the coupling to make a total of eight indents at the joint.

Both setscrew- and indent-type fittings are available with threaded ends. These fittings are secured to the end of the EMT by setscrews or indents. The threaded end of the fitting can then be fastened to an electrical box.

INDENTING-TYPE
CONNECTOR

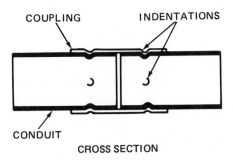

Figure 7-23. Coupling EMT by indenting.

• FLEXIBLE CONDUIT •

Flexible conduit is similar in appearance and construction to the special armor on armored cable. It is available in coiled lengths that can be cut to order. Trade sizes range from 1/2 inch to 4 inches inside diameter.

The spiral construction of flexible conduit causes it to have higher resistance per foot than solid metal conduit. For this reason flexible conduit should not be used as a grounding conductor. An additional bare or green-insulated grounding conductor should be included with the current-carrying conductors in flexible conduit installations. The grounding conductor must be joined to boxes and electrical devices just as is the grounding conductor in armored cable. The wire size to be used for the grounding conductor depends upon the maximum current rating of the circuit in which it is used. The following may be used as a guide.

Maximum Circuit Current (Amperes)	Copper Wire Size
15	No. 14
20	No. 12
30-60	No. 10
100	No. 8

Flexible conduit, of course, presents no bending problems. However, care must be taken to comply with the minimum bend radius requirements of the NEC. These requirements are covered by NEC Table 349-20

(b) or (c). In addition, complex bends that would cause difficulty in installing conductors should be avoided.

Cutting

Flexible conduit can be cut with a hacksaw, as armored cable is cut (Fig. 7-24). Saw through a single ribbon of metal. Break the metal loose by a counterclockwise twist. Use pliers or metal cutters to trim off sharp edges and to make the end square.

Fittings

Fittings for flexible conduit are either internally or externally attached to the conduit. The internal type is designed to screw into the spiral of the conduit. This type of connector covers the end of the conduit completely, protecting the conductors from contact with the cut edge of the conduit. Externally attached connectors are secured to the conduit with clamping screws (Fig. 7-25). When using these connectors, make

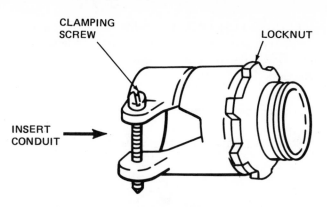

Figure 7-25. Flexible conduit fitting.

sure the cut end of the conduit is pushed as far as possible into the connector, covering the cut end and protecting the conductors from damage.

• NONMETALLIC CONDUIT •

Nonmetallic conduit is used primarily in underground or permanently wet locations. Nonmetallic conduit must have a separate equipment grounding conductor installed. Allowance must be made for this conductor when the maximum number of conductors permitted by the NEC is calculated.

Joints are made by cementing the pieces together. Most nonmetallic conduit is made of rigid polyvinyl chloride (PVC) plastic. The cement used is actually a solvent that softens the plastic at the joint and allows the softened areas to flow together to form a "weld." The resulting joint is watertight and strong. PVC conduit can be cut readily with any fine-tooth saw.

Elbow and offset fittings are available for standard bends. For other bends, a special device called a *hotbox* must be used. The hotbox heats the PVC electrically and softens it so that it can be bent to the desired shape (Fig. 7-26). Before heating the PVC (especially sizes 2 inches and larger), both ends of the section should be plugged. This traps air in the conduit. The air—heated in the hotbox—expands to prevent kinks or dislocation of the conduit when it is bent.

Nonmetallic conduit is durable, easy to work with, and moderate in cost. It is particularly well suited to areas where resistance to moisture and corrosion are essential. The main disadvantage of nonmetallic conduit is that joints cannot be taken apart after they are cemented.

• INSTALLING CONDUIT IN BUILDINGS •

Conduit installations must be made with three main considerations always in mind.

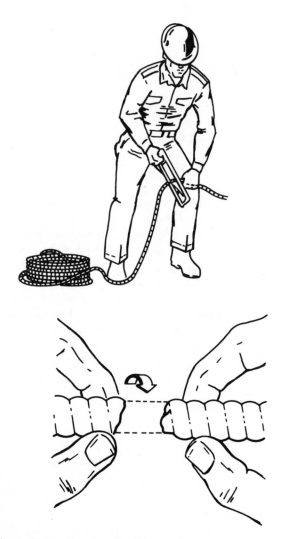

Figure 7-24. Cutting flexible conduit.

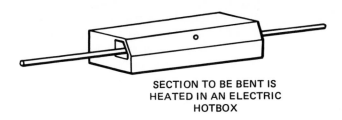

SECTION TO BE BENT IS
HEATED IN AN ELECTRIC
HOTBOX

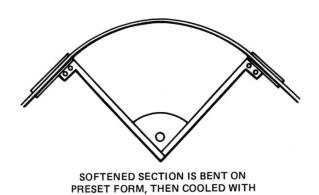

SOFTENED SECTION IS BENT ON
PRESET FORM, THEN COOLED WITH
A DAMP CLOTH

Figure 7-26. Bending nonmetallic conduit.

1. The conduit must be installed so that conductors can be pulled into place with as little difficulty as possible.
2. Conduit installation in wood frame construction requires boring and notching of studs and joists. This must be done in such a way that the structure is not weakened.
3. Conduit—particularly rigid conduit—is much heavier than cable and must be adequately supported so that its own weight does not cause damage.

The plans for any building specify where switches, receptacles, and fixtures must be installed. The usual procedure is first to locate and install the electrical boxes for these switches, receptacles, and fixtures. The conduit is then installed to join those boxes that will be on the same circuit. The types and styles of electrical boxes and the considerations in choosing the right box for a particular location are covered in Chapter 8. This paragraph covers the materials and techniques that are used to install the conduit itself.

Conduit installations must be planned so conductors can be installed with minimum difficulty. After conduit is installed, each electrical box becomes a *pull box*; that is, a point at which fish tape can be inserted and pushed through to the next box and then used to pull the conductors back through the conduit.

The NEC specifies that no more than four 90° bends—the equivalent of a complete circle—can be

made between *pull boxes*. If any run cannot be made without making more than four 90° bends, an additional box must be installed in the run. If a box is to serve only as a pull box and not as a mounting point for a switch, receptacle, or fixture, it represents an increase in material and labor cost that should, if possible, be avoided. Careful planning can generally turn up a way to avoid installing boxes only to serve as pull boxes. For example, a pull box can do double duty as a right-angle elbow.

Rigid conduit of all sizes must be supported within 3 feet of each electrical box and every 10 feet between these supports. The NEC allows longer unsupported straight runs in some larger sizes. Keep in mind that these long runs of large-size conduit will be heavy, and supports used must be strong and securely mounted. Intermediate conduit and EMT must also be supported within 3 feet of electrical boxes and every 10 feet between. No exceptions are made for intermediate conduit or EMT.

Flexible conduit must be supported within 1 foot of every electrical box and every 4-1/2 feet between. The NEC makes some exceptions for connections to motors and recessed lighting fixtures. Refer to Article 350-4. Support requirements for nonmetallic conduit vary with size. NEC Table 347-8 covers support spacing.

Whenever possible, conduit runs should be planned so that when conduit enters and leaves an electrical box the two connections are on opposite sides of the box. This saves time when conductors are being installed. Of course, conduit runs must be planned to keep materials required to a minimum. Runs should be kept as short as possible, and special fittings should be used only where necessary to save space or comply with the NEC. Conduit runs should follow standard trade practices as described in the following paragraphs.

Wood Frame Construction

Horizontal conduit runs in wood frame buildings are made by boring holes in studs or joists and running conduit through the holes. Vertical runs are made by securing conduit to studs and boring top plates for passage between floors. To avoid weakening the studs and joists, holes should be bored approximately in the center. This means that nonflexible conduit must be bowed to thread it through the holes. Smaller-diameter lengths of rigid and intermediate conduit and EMT can be bowed without difficulty. Long lengths of rigid conduit can be bowed slightly. Care must be taken to avoid giving the conduit a permanent bend or "set" that would make connections difficult. Shorter lengths of conduit are easier to thread into bored holes. Short lengths require the use of additional connectors, however, so this should be done only when no other installation scheme is practical.

Conduit runs can be made by notching studs and joists, rather than boring. This makes it easier to install long lengths of conduit without bowing or to install short lengths that cannot be bowed. Notching, however, must be done carefully to avoid weakening the structure. A good rule to follow is that notches should be no more than one-fifth the width of the stud or joist. This means that a 2-by-4 (which is actually about 1-1/2 × 3-1/2 inches) should not be notched deeper than about 11/16 inch. This limits notching in 2-by-4 construction to 1/2 inch trade-size conduit. The NEC specifies that a steel plate at least 1/16 inch thick be installed to protect conductors passing through wood framing members within 1-1/2 inch of the face (Fig. 7-27). The plate is required for exposed and concealed work and must be large enough to cover the areas where nails or screws might penetrate the installed cable.

Three types of supports are used for conduit runs in wood frame buildings (Fig. 7-28). They are the U-shaped two-hole strap, the one-hole strap, and the nail strap. The two-hole strap is a bit more work to install, but may be required to support larger sizes of conduit

indoors or conduit exposed to wind and weather outdoors. Whatever type of strap is used, the size must be suitable for the outside diameter of the conduit.

Conduit supports can be mounted on hollow walls—such as plasterboard or lath and plaster—with expansion fasteners or toggle bolts (Fig. 7-29).

Masonry Construction

Many types of devices are available for mounting conduit on masonry. New products are continually being introduced to make the job easier, quicker, or less expensive. Any current building supply catalog will list the latest products.

Most masonry fastening devices require drilling. This is done using a low-speed power drill and carbide-tipped masonry drill bits (Fig. 7-30). Eye protection must be worn when drilling in masonry. If

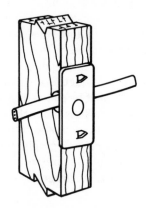

Figure 7-27. Cable protector plate.

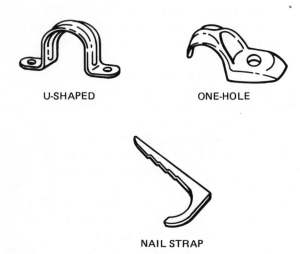

U-SHAPED   ONE-HOLE

NAIL STRAP

Figure 7-28. Straps for conduit.

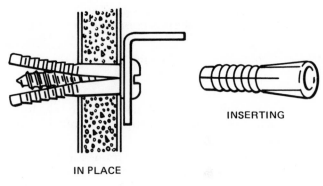

IN PLACE   INSERTING

EXPANSION FASTENER

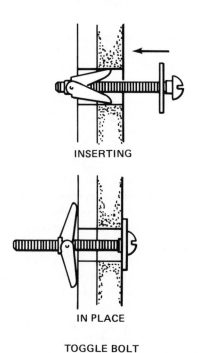

INSERTING

IN PLACE

TOGGLE BOLT

Figure 7-29. Fasteners for hollow wall construction.

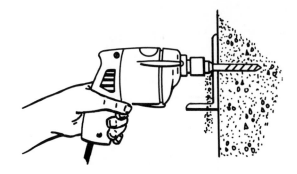

Figure 7-30. Drilling masonry.

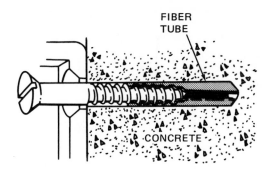

Figure 7-31. Expansion tube.

power is not available at the work site, hammer-powered drills can be used. A few of the available masonry fasteners are briefly described below.

EXPANSION TUBES • These fasteners consist simply of a fiber or plastic tube (Fig. 7-31). A hole is drilled large enough to accommodate the tube and the tube is inserted in the hole. The conduit strap is secured by putting a wood screw through the strap and into the tube. The screw expands the tube so it grips the masonry hole. Expansion tubes are best for light-to-medium loads.

DRIVE PLUGS • Drive plugs consist of a sleeve and special plug (Fig. 7-32). After drilling, the sleeve is inserted in the hole. The plug is then put through the conduit strap and hammered into the sleeve. Ridges on the plug expand the sleeve to secure the strap. These, too, are best for light-to-medium loads.

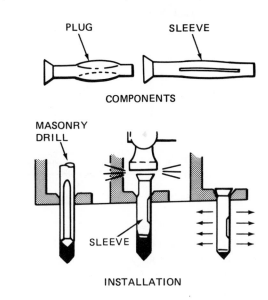

Figure 7-32. Drive plug.

LEAD ANCHORS • These devices (Fig. 7-33) secure a threaded insert in the wall. They are especially useful in any location where removal of the conduit strap may be necessary from time to time. The lead anchor is ridged and has a tapered insert at the base. The anchor is inserted in the hole. A special punch is used to drive in the anchor so that the tapered insert is forced into the base of the anchor to expand it. These fasteners can hold heavy loads.

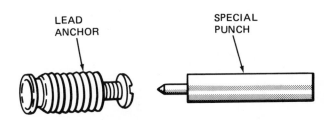

Figure 7-33. Lead anchor and special punch.

• INSTALLING CONDUCTORS IN CONDUIT •

The general procedure for installing conductors in conduit is the same for all types of conduit. Conductors are installed by pulling them through the conduit. The pulling is done with a special tool called a fish tape, which is described in Chapter 5.

The fish tape is fed through the conduit from its storage reel. Usually the tape is fed in at a box installed for a switch or receptacle. The tape is pulled out of the next opening in the line. Conductors are fastened to the end of the tape. The tape is then pulled or reeled in to draw the conductors through the conduit (Fig. 7-34). If the run is long, two people are needed for this job. One feeds the conductors in at one end as the other reels in the fish tape.

In most cases there will be more than one conductor being fed into the conduit. It is important to keep the conductors smooth and free of kinks. Set up the conductor spools so that they unwind freely and can be kept free of bends and crossovers (Fig. 7-35). If the conductors become twisted, they are difficult to pull around bends. For particularly long runs or where there are many bends, wires can be coated with a lubricating compound. Noncorrosive lubricating compounds are available in dry powder form and in paste

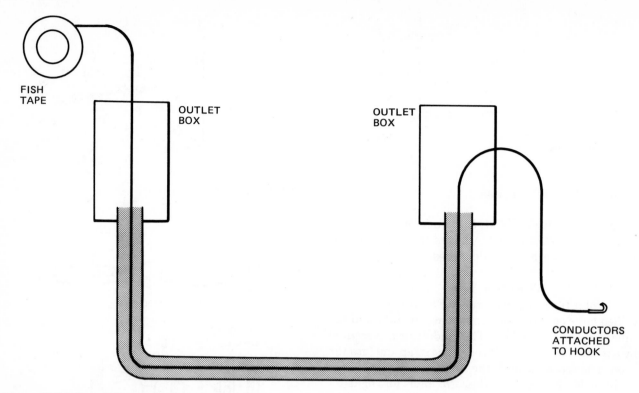

Figure 7-34. Installing conductors in conduit.

form. When more than one conductor is to be pulled, the connection to the fish tape should be staggered to avoid a bulky connection that would make pulling difficult (Fig. 7-36).

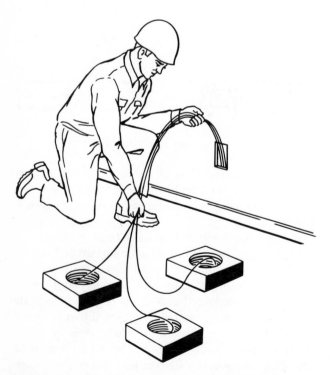

Figure 7-35. Feeding conductors into conduit.

Sometimes nonmetallic cable—rather than individual conductors—is installed in conduit. This is done in locations where special protection is needed, such as below grade in residences. The extra stiffness and larger diameter of cable require that special care be taken when feeding the cable into the conduit to avoid damage to the cable.

The procedure of pulling conductors between switch or receptacle boxes is continued until the complete system is wired. Sometimes if no switch or receptacle box is called for on a long run, a box or condulet must be installed near the middle of the run to make conductor installation easier. Condulets can be used as pull boxes by opening the hatch. Be sure to leave at least 6 inches of wire at each box to make connections to the switch, receptacle, or fixture that will be installed later.

Whenever possible, run conductors from box to box without a break (Fig. 7-37). In particular, the white- or gray-insulated (neutral) wire should be continued unbroken. When the fish tape has been reeled in, hold the red or black conductors and pull out enough white wire to reach the next box or the end of the run. At each box where a connection must be made, leave a loop of white wire. The wire can be connected by removing a section of insulation, without cutting the conductor. In many cases, the red and black wires can be continued in this manner, too. Continuing conductors in this way reduces the number of connections that must be made in electrical boxes. This speeds up work and keeps boxes uncrowded.

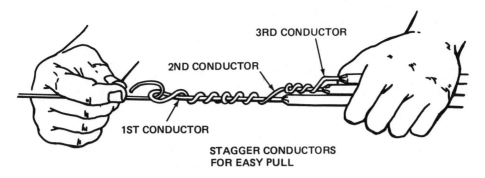

STAGGER CONDUCTORS
FOR EASY PULL

Figure 7-36. Connecting conductors to fish tape.

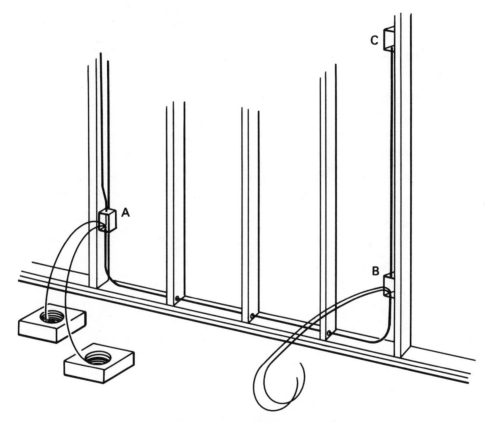

- CONNECTIONS AND SPLICES CAN OFTEN BE AVOIDED,
  IF CONDUCTOR RUNS ARE CONTINUOUS WHENEVER POSSIBLE.
- PULL OUT ENOUGH WIRE AT BOX B TO REACH BOX C,
  ALLOW 6 INCHES EXTRA AT EACH END FOR CONNECTIONS.
- DO NOT CUT CONDUCTORS UNTIL FINAL WIRING IS DONE.

Figure 7-37. Continuing conductors box to box.

• SUPPORTING CONDUCTORS IN CONDUIT •

If an installation of no. 8 to no. 18 conductors includes a vertical conduit run of 25 feet or more, the NEC requires one cable support at the top or near the top of the run. (Longer vertical runs require more supports. Refer to Article 300-19.)

The support is needed to carry the weight of the conductors so that undue strain is not placed on terminals or connections at the top of a run. Special fittings are made to provide this support. The fitting consists of a tapered collar with a wedge-shaped insert. The collar is designed to be installed on the end of the conduit. After the conductors are fed through the conduit and the collar, the wedge is pressed into place between the conductors to provide the required support.

• REVIEW QUESTIONS •

1. Name two advantages of rigid and intermediate conduit and EMT over nonmetallic cable.

2. What are the two principal disadvantages of conduit compared to nonmetallic cable?

3. Rigid and intermediate conduit are similar to water pipe, but there are two important differences. What are they?

4. All but one type of metal conduit can serve as a grounding conductor. Which one cannot? Why?

5. What determines how many conductors can be installed in conduit?

6. What two common fittings can be used to make a right-angle bend?

7. Why does the NEC specify minimum bend radius?

8. Why must solid metal conduit enter and leave electrical boxes in a straight line?

9. What type of bend must be made to bypass large obstructions?

10. What must always be done to conduit after cutting it?

11. Because of the thin wall, EMT fittings cannot be secured by threading. Name at least two ways they are secured.

12. Nonmetallic conduit is especially well suited to one type of installation. What is it?

13. Conduit runs in wood frame buildings can be made by boring holes or cutting notches in studs and joists. What must be kept in mind when this is done?

14. What is the maximum number of right-angle bends that are permitted by the NEC between pull boxes?

15. Use Tables 4 and 5 in Chapter 9 of the NEC to calculate what size conduit should be used to hold three no. 12 TW conductors and six no. 8 TW conductors.

16. What article and table in the NEC cover how bends should be made in EMT?

8
WORKING WITH
ELECTRICAL BOXES

• INTRODUCTION •

Whenever cable or conduit is cut and insulation is removed from conductors, the bare conductors must be enclosed in a metallic or nonmetallic (plastic) box designed for that purpose. These boxes provide protection from both fire and shock. Electrical failures that cause overheating and arcing are much more likely to occur at points where insulation is removed and connections are made. Enclosing the conductors in boxes greatly reduces the danger of fire.

Both metal and plastic boxes provide a means of maintaining continuity in the equipment grounding circuit. They also provide a grounding connection for any device installed in the box. Metal electrical boxes may be used with any type of cable or conduit. Nonmetallic (plastic) boxes, however, must be used only with nonmetallic cable (types NM, NMC, or UF) or rigid nonmetallic conduit.

Boxes are mounted to structural parts of wood frame and masonry buildings in such a way that they provide mechanical support to fixtures and other devices that must be mounted on ceilings and walls. Electricians must know how to install the various types of boxes in buildings, and how to connect cable or conduit to the boxes.

Chapters 6 and 7 cover the general procedures for cable and conduit connections. This chapter presents specific installation procedures for the types of boxes you are most likely to work with, and tells you how to select the right size and type of box for a particular location. The NEC requirements for the use of boxes are also covered. The correct installation and use of electrical boxes is a key factor in a good electrical system and, therefore, is an important part of the knowledge an electrician must have.

Figure 8-1. Typical ceiling boxes.

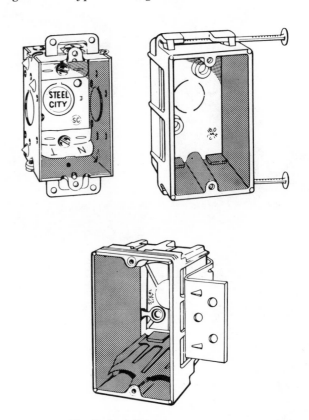

Figure 8-2. Typical wall boxes.

• TYPES AND USES •

There are two general types of electrical boxes. One type, designed for ceiling fixtures (Fig. 8-1), is usually mounted on or between beams or joists. The other type, designed to mount switches and receptacles (Fig. 8-2), is usually mounted on studs in wood frame buildings or is set into masonry walls. Ceiling boxes are square, octagonal, or round. Wall boxes are usually rectangular.

Electrical boxes are also used as junction boxes and pull boxes. Junction boxes are installed whenever conductor splices must be made in a location not suitable for a switch, receptacle, or fixture. The NEC requires that conductor splices be enclosed in a box. In most cases the wiring can be planned so that all conductor joining can be done in boxes that will also mount some electrical device. It is less expensive to install a larger box that can accommodate the extra wiring than to install a separate box only for making connections. However, situations can occur when a box must be installed only as a junction point. For example, the usual method of wiring recessed fixtures to comply with the NEC is to install a junction box not less than 1 foot from the fixture and run the fixture wire to the box through flexible metal conduit 4 to 6 feet long (Fig. 8-3). Any standard rectangular box can be used for this purpose and other connections can be made in it, if the box has sufficient conductor capacity. Electrical boxes used as junction boxes must be covered by a solid plate of the same material as the box. The NEC requires that junction boxes be accessible. They must not be covered by wall, ceiling, or floor material. Pull boxes are used in conduit wiring. Their use is covered in Chapter 7.

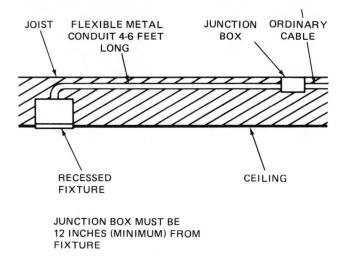

Figure 8-3. Junction box with recessed fixture.

• STANDARDIZATION •

As in most electrical hardware, there is a great degree of standardization in electrical boxes. Box dimensions are essentially the same among all manufacturers for each trade size and type. Spacing and size of mounting holes are the same, and threaded holes fit uniform screw sizes. This standardization has existed for many years, so that additions or replacements in older buildings are usually easy to do with presently available hardware. Some standardized features of electrical boxes are described below.

Knockouts and Pryouts

All boxes are made with some sort of easily removable circular sections called knockouts and pryouts (Fig. 8-4). Knockouts are removed to provide openings for cable or conduit connections. Some ceiling boxes also contain a knockout that is removed to mount the box to a hanger. Two styles of knockouts are used by box manufacturers. One style is made by scoring the metal in such a way that only a thin layer remains. If this knockout is hit sharply using a small hammer and the handle of a screwdriver, the scored metal will break and the knockout will fall out, or it can be twisted out with pliers.

In another type, the knockout section is cut through, part way around, with solid metal tabs left at two points. A rectangular slot is cut in the center of the knockout. To remove the knockout, insert a screwdriver in the slot and twist to break the solid tabs and free the knockout. These are sometimes called pryouts.

Knockouts must be removed only when the opening will be used for a cable, conduit, or other fitting and should not be left open. Metal discs are available to cover unused knockouts. One type snaps into the opening; another type consists of two plates joined in

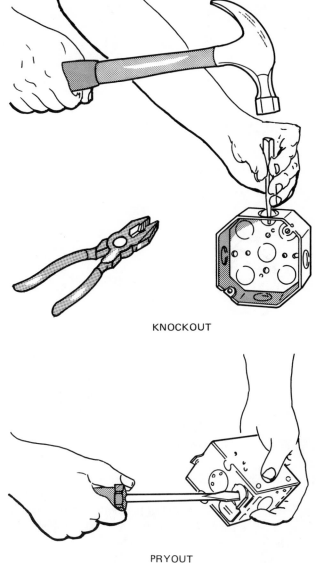

KNOCKOUT

PRYOUT

Figure 8-4. Knockouts and pryouts.

the center by a screw. The second type covers several different sizes of knockout.

Cable Clamps

Most wall boxes and some ceiling boxes contain internal cable clamps to secure nonmetallic or armored cable. A typical box contains two double clamps to secure a total of four cables (Fig. 8-5). One style of internal clamp is designed to be used with either type of cable. The clamp has an outer ring which covers the end of armored cable to protect conductors from the sharp edges of the armor. This ring can be cut off when nonmetallic cable is used.

When boxes do not have internal cable clamps, separate fastening devices must be installed in the knockouts to secure the cable. The NEC allows nonmetallic cable to be connected in a nonmetallic box

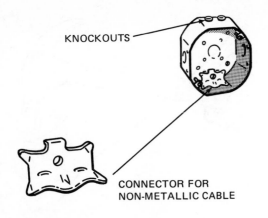

KNOCKOUTS

CONNECTOR FOR
NON-METALLIC CABLE

TOP OR BOTTOM VIEW

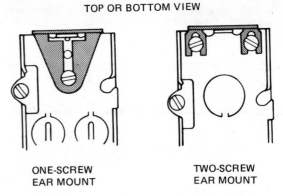

ONE-SCREW
EAR MOUNT

TWO-SCREW
EAR MOUNT

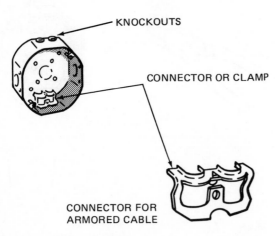

KNOCKOUTS

CONNECTOR OR CLAMP

CONNECTOR FOR
ARMORED CABLE

Figure 8-5. Cable clamps.

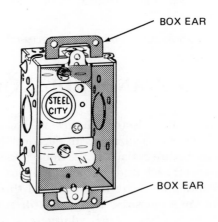

BOX EAR

BOX EAR

Figure 8-6. Box support ears.

without using a cable clamp, provided the cable is secured by a strap or staple within 8 inches of the box and the cable projects at least 1/4 inch into the box. In all other cases, either an internal or external cable clamp must be used to connect cable to a box. The installation of cable and conduit clamps is covered in Chapters 6 and 7.

Mounting Ears

Most wall boxes have screw-mounted brackets at the top and bottom that provide one way of securing the box to a wall. These brackets are known as *ears*. Each ear is mounted by one or two screws that can be tightened on slots on the ears (Fig. 8-6). The slots allow the box to be mounted at the correct depth for the wall material being used. The ears can be reversed so that the boxes can be flush mounted. The use of ears for mounting is covered in the section **Mounting Boxes in Old Work**.

Grounding Screws

Most boxes have a threaded hole tapped for a green-tinted screw to connect the equipment grounding wires (Fig. 8-7). In metal boxes the threaded hole is at the back of the box. Some nonmetallic boxes have a

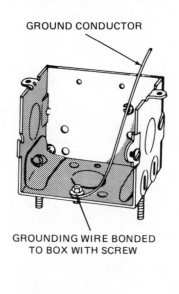

GROUND CONDUCTOR

GROUNDING WIRE BONDED
TO BOX WITH SCREW

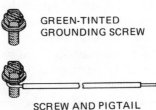

GREEN-TINTED
GROUNDING SCREW

SCREW AND PIGTAIL

Figure 8-7. Grounding screws.

steel plate mounted on the box. The hole tapped for the green-tinted screw is located on the plate. The wiring is the same for either type of box (Fig. 8-8). A jumper is connected to the grounding wire. If the box is used to mount a switch, receptacle, or fixture, another jumper is connected to the green-tinted screw on the device. A wire nut is then used to join these two jumpers and the bare or green-insulated grounding wire in each cable that enters the box. If the box does not contain a green-tinted screw, the jumper that would go to the screw is connected instead to a steel grounding clip that is fastened on the edge of the box (Fig. 8-9).

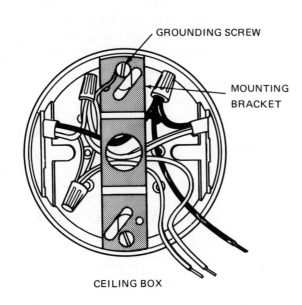

CEILING BOX

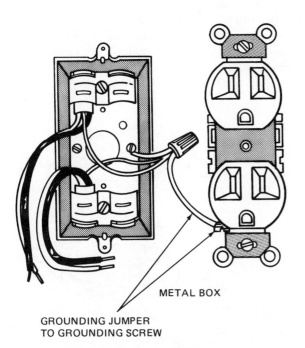

METAL BOX

GROUNDING JUMPER
TO GROUNDING SCREW

Figure 8-8. Grounding screw connections.

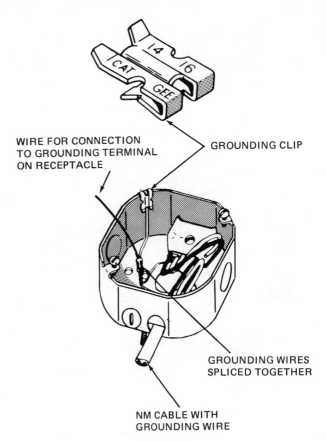

WIRE FOR CONNECTION
TO GROUNDING TERMINAL
ON RECEPTACLE

GROUNDING CLIP

GROUNDING WIRES
SPLICED TOGETHER

NM CABLE WITH
GROUNDING WIRE

Figure 8-9. Grounding clip connections.

Ganging

Some metal wall boxes have removable sides to permit ganging. Ganging is done when two or more outlets or switches must be mounted side by side. Adjoining single box sides are removed and discarded. The screw and brackets that secured the side can then be used to join the boxes (Fig. 8-10). In place of ganging, double, triple, and even larger boxes are available as complete units (Fig. 8-11). These boxes are stronger and therefore preferable to ganging, but ganging is permitted by most codes and may be used when necessary.

Nonmetallic boxes are of one-piece contruction (covers are a separate piece) so they cannot be ganged. Larger sizes must be used to mount several switches or outlets in one location.

Boxes Used with Exposed Wiring

Wall boxes with rounded corners are used in areas where the box remains exposed, such as basements and garages. These are known as utility boxes (Fig. 8-12). The base of the box is of one-piece construction, so ganging is not possible. Otherwise they are similar to the wall boxes used with concealed wiring. Cover plates are made for utility boxes used for switches or receptacles. A blank cover is used if the utility box is a

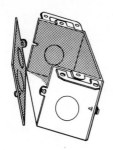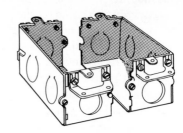

Figure 8-10. Ganging boxes.

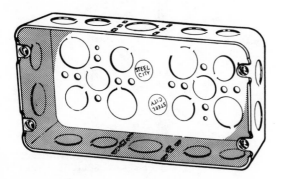

Figure 8-11. Gang box.

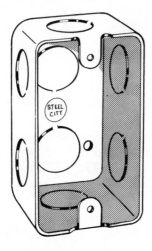

Figure 8-12. Utility box.

junction box. The same types of boxes are used in ceiling locations in both concealed and open wiring.

Box Covers and Extensions

A large assortment of covers is available for wall and ceiling boxes. Covers for all boxes are used mainly with utility boxes to cover switches or receptacles in unfinished areas. Solid covers are used if the utility box is being used only to join cables. Wall boxes used to mount switches and receptacles in finished areas are covered by faceplates. Faceplates may be metallic or nonmetallic and are made in colors and finishes to match interior design. Covers for square, octagonal, and round boxes provide properly positioned mounting holes for switches and receptacles (Fig. 8-13).

One type of cover, known as a tile cover, has a projecting section that matches the thickness of added wall material, such as tile in bathrooms. This cover brings the mounting surface for switches and receptacles flush with the surface of the tile.

Boxes can be enlarged and mounting surfaces can be extended by adding extension rings (Fig. 8-14). Extension rings look like boxes with the back removed. The box mounting screws are used to attach these rings.

• BOX SIZES AND CONDUCTOR CAPACITY •

Standard box sizes range from a rectangular wall box 3 × 2 × 1-1/2 inches to a square ceiling box 4-11/16 × 2-1/8 inches. The smaller box has about 7-1/2 cubic inches of usable interior space. The ceiling box has about five and one-half times as much interior space. Obviously, both boxes cannot handle the same number of conductors. The fact is, the NEC specifies the maximum number of conductors that can be brought into each standard trade-size box. NEC Table 370-6(a) lists 24 trade-size boxes and the maximum number of

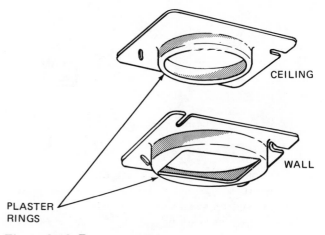

Figure 8-13. Box covers.

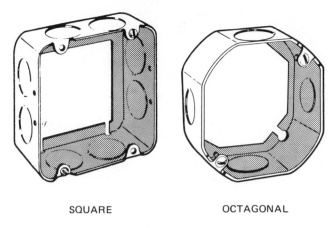

SQUARE   OCTAGONAL

Figure 8-14. Box extension rings.

conductors of each size that may be brought into them. For example, in the two boxes mentioned above, the smaller can be used for a maximum of three no. 14 conductors. The larger box can handle twenty-one no. 14 conductors.

There are a number of rules to be used when counting conductors. These rules are given in footnotes to the table. The rules generally reduce the permitted number of conductors when the box contains some item that takes up part of the usable space. For example, one is deducted from the permitted number if the box is used to mount a switch, receptacle, or other device; one is deducted if the box has internal cable clamps; and one is deducted if the box contains a fixture mounting stud. Some wires need not be included in the count. For example, jumpers that are entirely within the box and wires to a fixture mounted on a box are not counted.

• NEW WORK AND OLD WORK •

The words *new work* and *old work* have special meanings in the electrical trade. New work describes any electrical installation that is done before finished walls are in place. Old work is any installation that must be done in buildings having finished walls, ceilings, and floors.

In new work beams, joists, and studs are accessible (Fig. 8-15). Installing cable or conduit and mounting boxes can be done easily and quickly, using simple brackets, nails, and screws. In old work beams, joists, and studs are covered—by wallboard, plaster and lath, or paneling. Mounting devices must be used that can be installed through wall and ceiling openings. Special hardware has been designed to make the job easier, but installations in old work are more difficult and take more time than they do in new work. When walls and ceilings must be opened to install boxes and wiring in old work, repair of the opening is a specialized task that may be done by other trades. This chapter provides

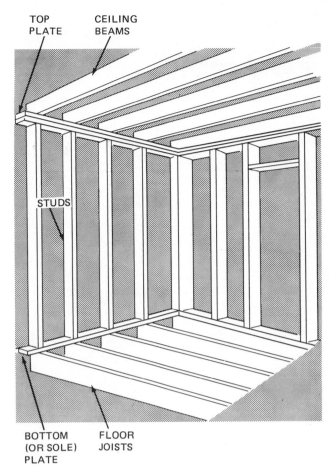

TOP PLATE   CEILING BEAMS

STUDS

BOTTOM (OR SOLE) PLATE   FLOOR JOISTS

Figure 8-15. Wood frame construction.

some basic information an electrician should know about repair of walls and ceilings.

• BOX LOCATIONS •

The locations of receptacles, switches, and fixtures are shown on the building plans. Local codes frequently specify the minimum number and spacing of receptacles, ceiling fixtures, and special lighting. The building architect interprets these requirements in terms that fit the building design. The electrical layout drawing and the electrical portion of the building specification give details on the number and type of electrical devices that must be provided. An electrical box must be installed for each device. In some locations it may be possible to mount several devices in a single ganged box. The electrical drawing shows device locations in a plan view (Fig. 8-16). The height above floor level is covered by notes on the electrical drawing or may be set by local codes or local custom. In the living areas of homes wall receptacles are usually located 12 inches above the floor and switches are located 48 inches above the floor. In kitchens, basements, garages, and workrooms receptacles are often mounted at the 48-inch level.

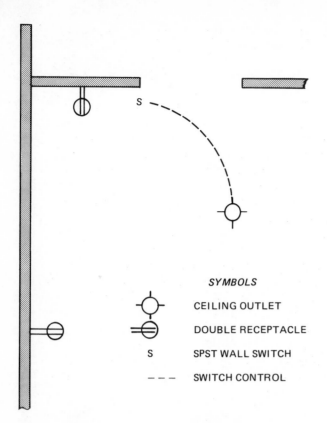

SYMBOLS

⊕ CEILING OUTLET

⊖ DOUBLE RECEPTACLE

S SPST WALL SWITCH

– – – SWITCH CONTROL

Figure 8-16. Portion of an electrical plan.

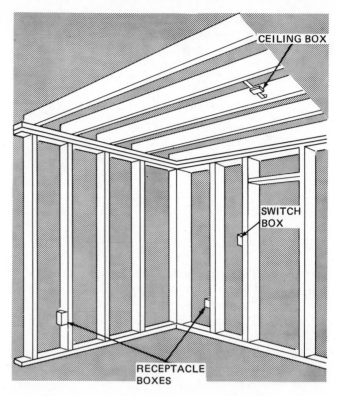

Figure 8-17. Actual location of switches and receptacles shown in Fig. 8-16.

Ceiling fixtures are usually centered over the floor area they illuminate.

Whether these dimensions or others are used, the mounting location of the electrical boxes must be marked on studs, beams, and joists in wood frame buildings after framing is complete. The locations are shown by arrows marked on studs with a lumber crayon or other easy-to-see marking device (Fig. 8-17). The arrow points to the side of the stud on which the box is to be mounted. The point of the arrow is at the vertical midpoint of the box. Ceiling box locations are determined by measuring to find the center of the floor area to be illuminated, and then marking this location on the rough flooring. Drawing symbols and numbers indicating the size and type of box may also be marked on the lumber. Drawing symbols are explained in Chapter 12.

One additional dimension is needed before the boxes can be installed. This is the distance that the front edge of the box will be from the surface of the finished wall. The NEC specifies that if walls or ceilings are concrete, tile, or other material that cannot burn, the box may be set back a maximum of 1/4 inch. If walls or ceilings are wood or some other material that can burn, the boxes must be flush with the finished wall (Fig. 8-18). It is a good practice to flush-mount boxes under all conditions. The electrician must know the thickness of wall and ceiling materials in order to know how much the edge of the box should extend beyond the stud or joist.

When all locations and dimensions are known, the actual mounting of boxes is a simple matter. There are many types of mounting brackets and other hardware that can be used. The most common methods of mounting wall and ceiling boxes in new and old work are covered in the two sections that follow.

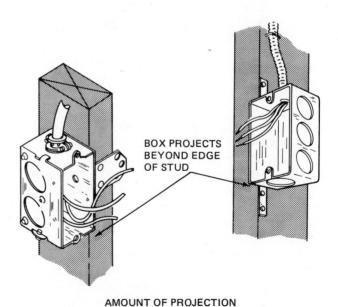

BOX PROJECTS BEYOND EDGE OF STUD

AMOUNT OF PROJECTION DEPENDS ON THICKNESS OF FINISHED WALL

Figure 8-18. Wall box position on stud.

130 Electrical Wiring Fundamentals

• MOUNTING BOXES IN NEW WORK •

In wood frame construction the simplest method of mounting wall boxes is to nail them to studs. However, the NEC specifies that nails that pass through the interior of the box must not be more than 1/4 inch from the back of the box. Boxes that have external mounting brackets can be secured by nailing through the bracket (Fig. 8-19). For safety and secure mounting, nail ends that project through studs should be clinched over. If a wall box must be located between studs, box supports are available that can be mounted on the studs to support the box (Fig. 8-20).

Ceiling boxes are usually mounted on bar hangers that are nailed to beams (Fig. 8-21). To locate the ceiling box accurately, the box location marked on the rough flooring must be transferred to a ceiling point above it. The easiest way to do this is with a plumb bob. Center the plumb bob over the floor mark. Mark the ceiling beams either side of the plumb line. Mount the box hanger between the two marked points. Use the

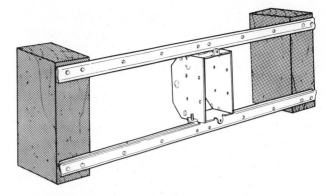

Figure 8-20. Box mounting between studs.

plumb bob again to find the point on the bar hanger that is over the floor mark. The mounting stud on the bar hanger can then be fastened in the correct position.

Straight bar hangers are mounted by nailing them to the sides of the beams. The hangers are adjustable in length to fit the space. The hanger must be far enough above the lower edge of the beam so that the ceiling box will be flush with the finished ceiling.

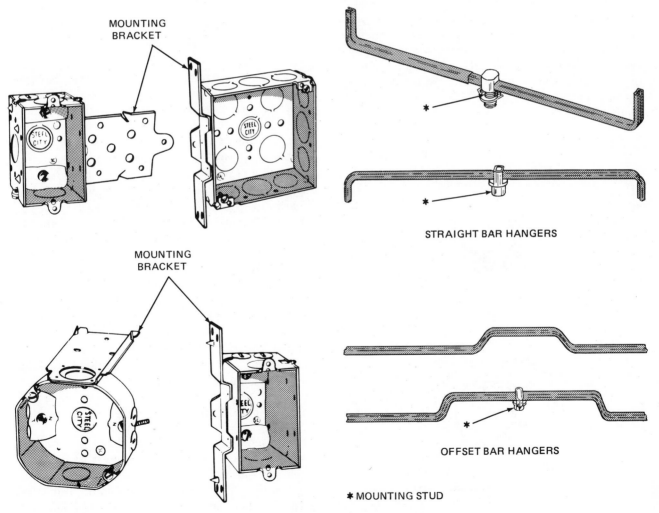

MOUNTING BRACKET

MOUNTING BRACKET

Figure 8-19. Box mounting brackets.

STRAIGHT BAR HANGERS

OFFSET BAR HANGERS

✱ MOUNTING STUD

Figure 8-21. Ceiling box bar hangers.

Offset bar hangers are mounted by nailing them to the lower edge of the beams. To avoid interference with plasterboard ceiling material, the beams should be notched.

Ceiling boxes are also made with mounting brackets. These boxes can be mounted by nailing them directly to the side of a beam. Ceiling boxes are usually intended as a mounting place for a ceiling fixture. For this purpose, the box must contain a threaded fixture stud. If a bar hanger is used, the mounting stud on the hanger not only holds the box but projects into the box enough to serve as a fixture stud or as a mounting place for other hardware. First, the mounting stud knockout is removed from the ceiling box. To use one type of hanger (Fig. 8-22), remove the stud from the hanger and then replace it and secure it with the stud screw to hold the box. In another type of hanger, the stud is permanently attached to the hanger. In this case, the box is placed over the stud and secured with a locknut. When ceiling boxes are mounted by brackets nailed to beams, a different type of fixture stud is used (Fig. 8-23). This stud is bolted inside the box.

In masonry construction electrical boxes must be installed as the masonry work progresses. It is possible to cut into blocks and fish cable through the block spaces. However, it is difficult and time-consuming to do this on a large scale. The more efficient method involves preassembling box and cable conduit units and then dropping them into place as the masonry work reaches the proper level for receptacles or switches. Ceiling boxes must be set into the forms before concrete is poured.

Figure 8-23. Fixture stud.

An alternate method of wiring in masonry construction is to use surface-mounted boxes and conduit. The devices described in Chapter 7 for mounting conduit in masonry can also be used to mount boxes. This method of masonry wiring is particularly well suited to construction in which the masonry walls are finished by covering them with wall paneling or plaster applied to wire mesh on furring strips. In this construction the conduit is concealed by the wall finish. Box sizes can be used that will be flush with the finished wall surface.

• MOUNTING BOXES IN OLD WORK •

Mounting electrical boxes when the building framework is covered by finished walls and ceilings presents special problems. The problems are caused by having limited access to the usual mounting surfaces. In addition, damage to walls and ceilings must be kept to a minimum to reduce the cost of repair. There are many kinds of mounting hardware available to make the job easier. Several of the most frequently used types of hardware are described in this section.

Electrical work in finished buildings can often be done more efficiently and at lower cost with proper planning. For example, electrical work is frequently done along with other remodeling. If this is the case, it may be possible to cut away large sections of plaster or wallboard or break into masonry walls and mount boxes as you would in new work. After mounting the box, new wall or ceiling material can be installed or repaired as part of the renovation. In cases like this, repair of the structure and painting or wallpapering may be done by other trades.

Selecting a Mounting Place

Before mounting a new electrical box in a finished wood-frame building, a clear area must be found between joists or studs. The procedures that follow describe several different methods of finding clear areas. Keep in mind that these procedures may be interchanged. The procedures given for locating a clear area

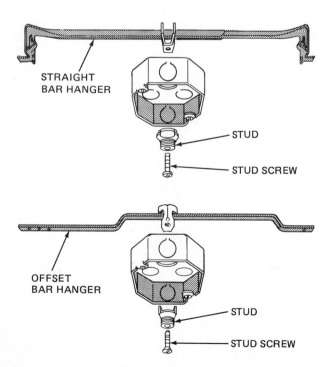

STRAIGHT BAR HANGER

STUD

STUD SCREW

OFFSET BAR HANGER

STUD

STUD SCREW

Figure 8-22. Ceiling box on bar mounting stud.

in a wall may also be used to find a clear area in a ceiling and vice versa. When boxes are being mounted in plaster-and-lath walls special care must be taken in making the opening so that the wall will not be weakened and the box will be securely mounted. After testing to find a wall area clear of studs, some additional tests should be made to locate lath positions. Chip away plaster above and below the test hole until you have exposed one lath. Make a mark in the center of the lath and lay out the wall opening so the mark on the exposed lath will be in the center (Fig. 8-24). This will make the open cut only part way into the lath above and the lath below the one exposed. This assures a solid mounting.

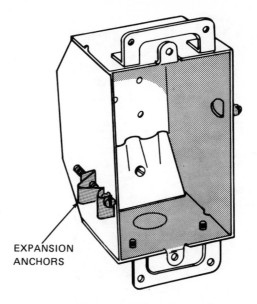

Figure 8-25. Box with expansion anchors.

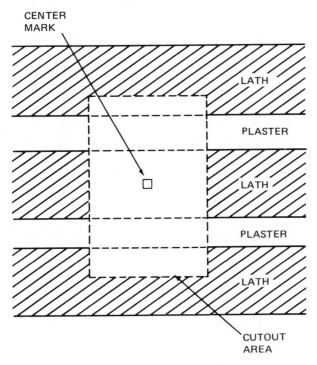

Figure 8-24. Box opening in plaster and lath.

Mounting Wall Boxes

METHOD 1 • One type of wall box designed for old work has expansion anchors on either side (Fig. 8-25). To use this type of box, a clear area of wall must be used. Use a stud finder, or make a series of test holes with a thin nail to locate a stud near where the box is to be mounted. Standard stud spacing is 16 inches on centers. This means there is a bit more than 14 inches of clear space either side of the stud.

Step 1. Lay out the box outline on the wall. Make certain the outline does not include the box ears.

Step 2. Use a drill to bore starter holes for sawing at each corner of the outline.

Step 3. Use a keyhole saw to cut out the wall section. The outline for this box will have two half-round openings on either side to allow the expansion anchors to be tightened.

Step 4. Fish the cable or cables out through the wall opening, and install the cable in the box (Chapter 6).

Step 5. Secure the cable clamps, making sure the conductors projecting into the box are long enough to make connections to the switch or receptacle.

Step 6. Push the box part way into the wall opening, and adjust the box ears to hold the front edge of the box flush with the wall surface.

Step 7. Hold the box in place and tighten the anchor screws. The screws cause the anchors to fold so they extend outward, pulling the box into the wall. The ears press against the outer wall surface to hold the box in place.

METHOD 2 • Another mounting method for old work can be used with standard wall boxes. This method uses two sheet metal hangers to hold the box (Fig. 8-26).

Step 1. Make the box cutout as described in Method 1.

Step 2. Make the cable connections and push the box into the wall opening, adjusting the ears as necessary to bring the edge of the box flush with the wall.

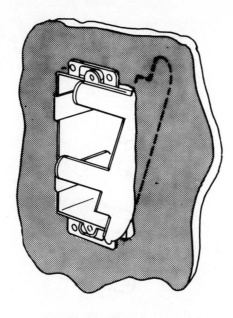

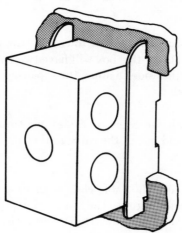

Figure 8-26. Flat metal box supports.

Step 3. Holding the box in place, slip a hanger between the side of the box and the wall opening. Insert the top of the hanger first and then push in the bottom.

Step 4. Slide the hanger down until the arms are centered. Pull the hanger toward you so it presses against the inside of the wall.

Step 5. Bend the arms and fold them into the box.

Step 6. Install another hanger on the other side of the box in the same way. Be sure the hanger arms are flat against the inside of the box.

METHOD 3 • A third method of mounting boxes requires the addition of a spring clamp to the back of a standard box (Fig. 8-27).

Step 1. Make the wall cutout and cable attachment as before.

Figure 8-27. Box support clamp.

Step 2. To mount the box, press the sides of the clamp together against the sides of the box. Push the box into the wall opening until the ears are seated against the wall. The sides of the clamp spring outward as soon as it clears the wall thickness.

Step 3. Tighten the clamp screw further to get a firm mounting.

Mounting Ceiling Boxes

Three different situations commonly come up when ceiling boxes are installed in old work. The simpler situation is the one in which the area above the new box location is an unfinished attic or crawl space. In this case the same hardware used in new work can be installed. Most of the work can be done in the unfinished area. In another instance the attic or crawl space may have rough flooring. When the new box location is between finished floors, the job becomes a bit more difficult. All work must be done from the finished side of the ceiling. Special hardware is available for this type of installation.

UNFINISHED AREA ABOVE BOX LOCATION •

Step 1. Check the unfinished area, noting the direction of joists and the location of any obstruction that may complicate box installation or wiring.

Step 2. On the finished ceiling, mark the center of the desired box location.

Step 3. Check the area first by tapping the surrounding area with your knuckles. The area directly under a joist will have a solid sound; the space between joists will sound hollow. If the area above the desired location sounds hollow, go on to Step 4. If the area does not sound clear, make a second mark in a clear area as close as practical to the original location. Keep in mind the direction of the joists. Relocate the mark between joists.

Step 4. Drill a small (1/8-inch) pilot hole in the center of the box position. The drill should penetrate easily whether the ceiling is plasterboard or plaster and lath. If the drill does not penetrate easily and sawdust is visible when the bit is removed, the box location is beneath a joist or other obstruction. A new location must be chosen and tested. Unused pilot holes can be filled with patching material.

Step 5. Locate the pilot hole in the attic or crawl space. If the pilot hole is at least 4 inches from a joist, the location is satisfactory and a box can be installed. If there is less than 4 inches clearance between the pilot hole and a joist, a box that mounts directly on the joist may be used or a new location may be chosen. Boxes mounted directly on joists are somewhat less accessible and harder to work with than hanger-mounted boxes. If possible select a new location with enough clearance for a hanger mount.

Step 6. When a usable location has been found, the next step is to cut the full box opening. Use the box as a template, centering it over the pilot hole, and trace the box outline on the ceiling.

Step 7. Using a 3/8-inch bit, drill holes at each corner of the box outline (Fig. 8-28). Drill slowly. Do not push the drill through; allow it to cut through. This will prevent cracks or tears from damaging the finished ceiling.

Step 8. Working from the finished side, use a key-hole saw to cut between the drilled holes to make the cutout. If the ceiling is plaster and lath, special care must be taken to avoid cracks spreading outward from the opening. Saw slowly with firm, even strokes to avoid jarring or flexing the ceiling material. Whenever your work is above your head and you must look up at it, wear protective goggles.

Step 9. The box and hanger may now be installed from the attic or crawl space (Fig. 8-29). The installation is made as described previously for new work. Cable connections can be made when the box is installed. Routing of cables in old work is covered in Chapter 15.

PARTIALLY FINISHED AREA ABOVE THE BOX LOCATION •

If the attic or crawl space has rough flooring, some additional work must be done.

Step 1. Refer to the procedures for an unfinished area and perform steps 1 through 4.

Step 2. After drilling the pilot hole (step 4 preceding), check for clearance by inserting a piece of stiff wire bent into an L shape into the hole and rotating it (Fig. 8-30). The leg of the L that rotates should be about 4 inches long. If the wire can be rotated freely, the area is clear. If it cannot, a new pilot hole must be drilled.

Step 3. When a clear area is found, enlarge the pilot hole to 3/4 inch.

Step 4. Add an 8-inch extension to the drill, and drill a hole in the flooring above the ceiling. Drill through the flooring. The hole in the flooring will locate the board in the flooring that will have to be removed to install the box (Fig. 8-31).

Step 5. Remove the floorboard by prying up the entire board or, if that is not practical, by cutting out a section.

Step 6. Complete the box installation as described in steps 7, 8, and 9 of the procedure for an unfinished area.

Figure 8-28. Drilling a box cutout.

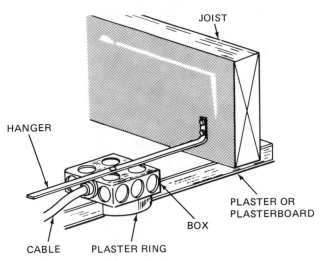

Figure 8-29. Box installed in unfinished area.

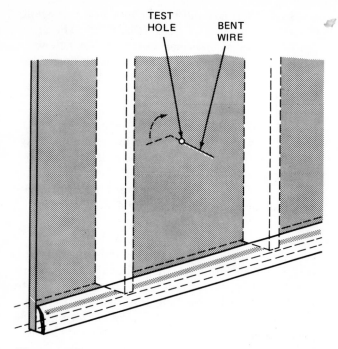

Figure 8-30. Checking clearance using bent wire.

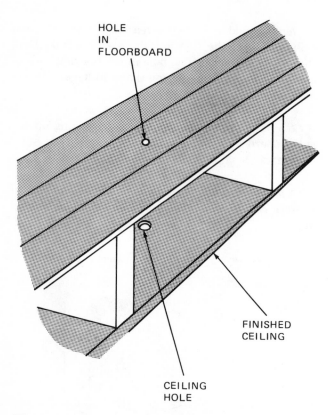

Figure 8-31. Locating floorboard to be removed.

FINISHED AREA ABOVE THE BOX LOCA-TION • The procedure for installing a ceiling box between finished floors depends on the ceiling material and the fixture to be installed.

MOUNTING BOXES FOR LIGHTWEIGHT FIX-TURES • Lightweight fixtures (15 pounds or less may be used as a rule of thumb unless local codes specify other limits) can be supported by either plasterboard or plaster and lath alone. Joist mounting is not necessary. Two styles of mounting hardware can be used for lightweight fixtures. One is similar to an oversize toggle bolt. The other mount is a simple bar with a movable mounting stud (Fig. 8-32).

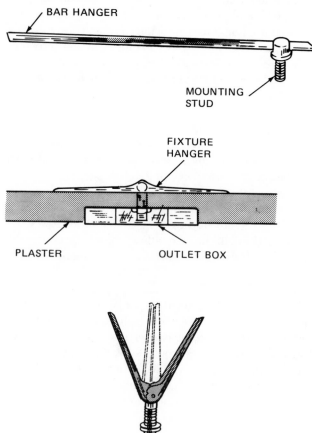

Figure 8-32. Supports for lightweight fixtures.

Step 1. To use either of these mounts, find a clear area by performing steps 1 and 2 in the procedure for partially finished areas preceding.

Step 2. Enlarge the pilot hole to about 2 inches. A paddle bit can be used in plasterboard, but only a keyhole saw should be used in plaster and lath. In either case, drill slowly and carefully to avoid damage to surrounding areas.

Step 3. Bring the fixture cable through the opening. Cable routing in old work is covered in Chapter 15.

Step 4 (A). If a toggle-style hanger is used, insert it through the hole with the arms raised. Let the arms fall to a horizontal position and lock them in place by rotating the mounting stud. If the ceiling is plaster and lath, make sure the support arms are at right angles to the laths.

Step 4 (B). To use a bar hanger, angle the bar through the opening. Slide the stud so that it is positioned about in the center of the ceiling opening and is as near as possible to the center of the bar. In plaster and lath ceilings, the bar, too, should be at right angles to the laths.

The NEC allows boxes as shallow as 1/2 inch to be used if the fixture has a canopy that provides additional space for cable connections and completely covers the box opening. If the base of the fixture mounts flush on the box, the minimum box depth is 15/16 inch.

Step 5. With either type of box, first secure the fixture cable to the box using an external cable clamp. Mount the box on the hanger stud with a locknut. Tighten the locknut firmly to secure the box to the ceiling. The procedure for mounting a fixture on the box is covered in Chapter 9.

MOUNTING BOXES FOR HEAVY FIXTURES •

If a heavy fixture is to be mounted on the ceiling box, a bar hanger fastened to the joists must be used. This requires opening the ceiling. If the fixture weighs more than 50 pounds, it must have an additional support directly from the fixture to a structural part of the building. If the ceiling is plasterboard, a section can be cut out to gain access to the joist.

Step 1. When a clear location has been found, cut an opening large enough for your hand to pass through.

Step 2. Using a small rule, measure the distance from each edge of the opening to the joist on that side. Add 3/4 inch to each of these measurements and mark these points on the ceiling. These marks represent the approximate middle of each adjacent joist.

Step 3. With these marks as a guide, lay out a section approximately 16 × 16 inches square (Fig. 8-33). Score the sides of the rectangle that are centered under each joist.

Step 4. Use a sharp, wide-blade chisel and a hammer to cut through the plasterboard along each scored line. Drill two saw openings in the plasterboard and cut the other two sides with a keyhole saw. Remove the section of plasterboard.

Step 5. Install a fixture hanger and box between the joists as you would in new work. Connect the fix-

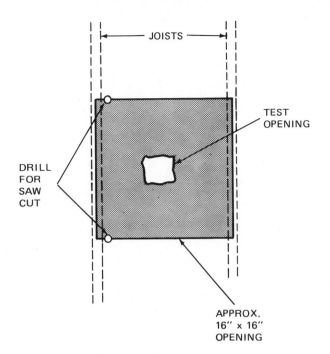

Figure 8-33. Plasterboard ceiling opening.

ture cable. Secure the box on the stud so that the edge of the box is flush with the finished side of the ceiling.

Step 6. To close the ceiling opening, cut a piece of plasterboard exactly the size of the opening. Locate the box position on the piece and make a cutout for the box.

Step 7. Put the new piece of plasterboard into the opening and nail it onto the joists. Use spackling compound to fill the cracks, then cover the cracks with joint tape. Finish with a coat of joint cement.

Step 8. When the cement is dry, sand the edges smooth and paint or paper over the patch.

In a plaster-and-lath ceiling, select a clear location for the box and cut the box opening as you would if the area above were unfinished.

Step 1. Use a cold chisel and hammer to chip away plaster to make a channel from each side of the box opening to the adjacent joist.

Step 2. Cut and remove a section of lath from each channel. Cut the lath at a point just past each joist so that the opening is slightly longer than the space between the joists.

Step 3. Install an offset bar hanger by nailing it to the exposed edges of the joists (Fig. 8-34). Connect the fixture cable to the box and mount the box on the hanger. The hanger can be concealed by patching plaster.

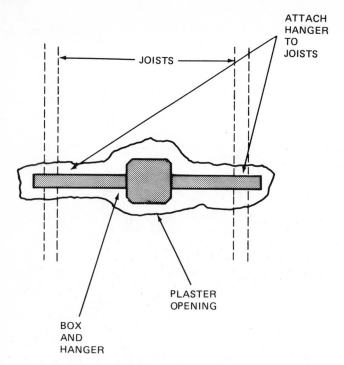

Figure 8-34. Hanger in plaster and lath.

Step 4. Fill the channel and the spaces around the box with the patching mixture. To match the ceiling surface, it may be necessary to keep the patching plaster to a level below the ceiling surface and finish with spackling compound or joint cement.

Step 5. When the patch is dry, sand it smooth and paint it.

• WEATHERTIGHT BOXES •

When switches, receptacles, and fixtures are to be mounted in locations exposed to the weather, weathertight boxes and covers must be used (Fig. 8-35).

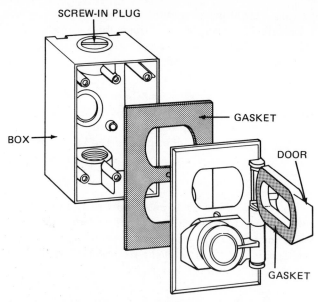

Figure 8-35. Weathertight box.

Weathertight boxes are made of cast metal. Threaded openings are provided for conduit connections. Covers are mounted with countersunk screws. A gasket between box and cover makes a weathertight seal. If a fixture is mounted on the box, the fixture must be suitable for exposed locations and a gasket must be included between the fixture and the box.

Weathertight boxes can be mounted on wood surfaces using corrosion-resistant nails or screws. On foundations or other masonry surfaces, the masonry fasteners described in Chapter 7 can be used. For flush mounting on poured concrete in new work, the boxes can be set in position on the forms before concrete is poured. Make certain all box and conduit openings are sealed to prevent concrete from entering.

For flush mounting in finished masonry, an opening must be made through the wall and sealed after the box is installed. Concrete patching material containing an epoxy bonding ingredient makes a durable seal.

• REVIEW QUESTIONS •

1. Electrical boxes used to mount ceiling fixtures and as junction boxes are usually what shape?

2. Electrical boxes used to mount switches and receptacles are usually what shape?

3. What are box knockouts used for?

4. In most cases cables coming into a box must be secured with either internal or external cable clamps. What exception is permitted by the NEC?

5. Most metallic and nonmetallic boxes contain a green-tinted screw. What is this screw used for?

6. What is ganging? When is it done?

7. The NEC specifies the maximum number of conductors that may be joined in each trade-size box. Why is it often a good idea to use a larger box than the NEC requires?

8. The mounting height for receptacles and switches is usually specified on building plans. What is the most common height for wall receptacles? For switches?

9. Some ceiling box hangers used for old work are supported only by the ceiling material. The use of these hangers is limited. What is the weight limitation?

10. In new work wall boxes are mounted by nailing them to studs. What usually supports boxes mounted in old work?

11. What article and table in the NEC cover the maximum number of conductors in standard boxes?

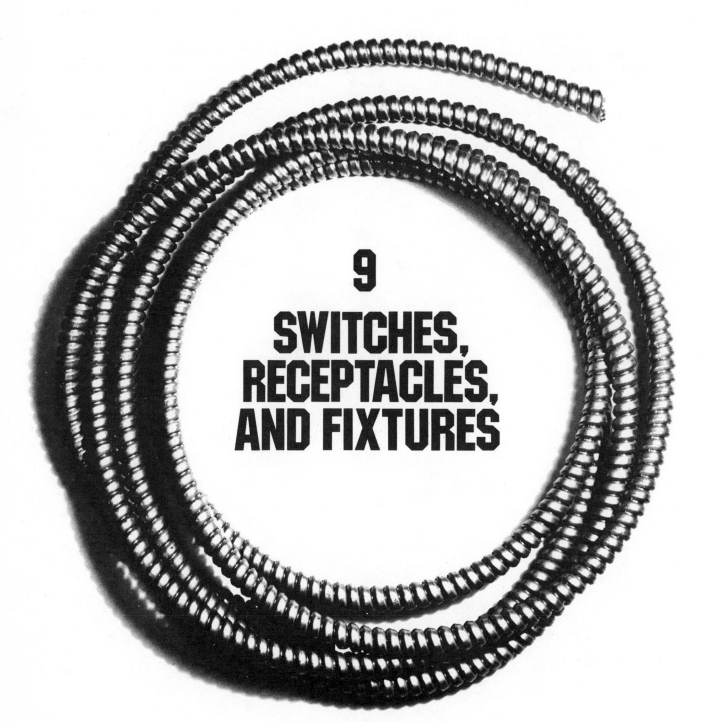

9
SWITCHES, RECEPTACLES, AND FIXTURES

• INTRODUCTION •

Chapters 6 and 7 describe the types of wire, cable, and conduit that are usually used in residential electrical systems. Conductors within the cable or conduit carry electrical power from the point where it enters the building to the points where it will be used. The cable or conduit is connected to electrical boxes at these points. Chapter 8 describes the kinds of boxes used in residential wiring. This chapter describes the final link in the power chain: the switches, receptacles, and fixtures that are mounted on the electrical boxes and connected to the conductors. These are the devices that enable power to be used in the residence.

Switches provide control of electrical power. Receptacles provide a means of connecting lamps and appliances to the power lines. Fixtures provide general area illumination and are often decorative, as well. Dozens of types of switches and receptacles are manufactured for use in homes. Choosing the right device for each location and installing it properly is essential to a good electrical installation. This chapter tells you how to recognize and work with the types that are most frequently used. You will learn what each device does and how it does it. You will learn how to mount switches, receptacles, and fixtures on boxes, and how to connect them to the power conductors.

• SWITCHING •

Switching requires a combination of electrical action and mechanical action in a single device. First, we will consider the electrical action.

Electrical Switching

Switches are described electrically by the number of conductors that are switched, and the number of positions to which the switch can be set. The words that are used to describe these switch characteristics are *pole* and *throw*. These words come from the simplest type of switch, the knife switch. The number of terminals that can be switched is the number of poles; the number of positions to which the switch can be thrown is the number of throws.

SINGLE-POLE SINGLE-THROW (SPST) • The simplest and most widely used switch is a single-pole single-throw switch (Fig. 9-1). This is abbreviated SPST. The standard, wall-mounted toggle switch used in residential wiring is an SPST switch. The switch is clearly marked with ON and OFF positions and has terminals for three wires. Two of these terminals are brass-colored; the third is green-tinted. The brass-colored terminals are used for the switch "hot" wire.

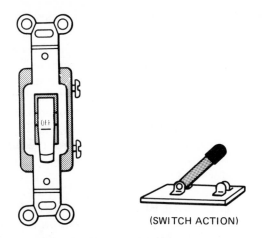

(SWITCH ACTION)

Figure 9-1. Single-pole single-throw switch.

Usually this is the wire that has black or red insulation. The bare or green-insulated grounding wire is connected to the green-tinted terminal. The white- or gray-insulated wire is never connected to a switch terminal. *Note*: A single exception to this rule is made for certain wiring situations. The exception applies, however, only when the white- or gray-insulated wire is being used as part of the hot wire. The general rule holds that the power ground is *never* switched. Switch wiring, including the exception, is covered in Chapter 13.

SINGLE-POLE DOUBLE-THROW (SPDT) • Another type of switch has two positions in which connections are made. This switch is known in the trade as a three-way switch. Electrically, it is a single-pole double-throw (SPDT) switch (Fig. 9-2). This switch is used in pairs for special lighting circuits in which the same light (or lights) is controlled from two different locations. The circuit is so wired that whenever either switch is moved from one position to the other, the condition of the light is changed from off to on or on to off. Consequently, these switches do not have on and off

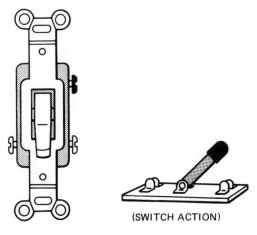

(SWITCH ACTION)

Figure 9-2. Single-pole double-throw switch.

positions marked on them. Connections are made to three brass-colored terminals and a green-tinted screw. One of the brass-colored terminals is darker than the other two. This is the common terminal. The word COMMON or a C may be molded into the switch housing near the terminal. When the switch is operated, the common terminal is alternately connected to each of the other two terminals. As always, the green-tinted terminal is for the bare or green-insulated grounding wire. Chapter 13 describes in detail the wiring for this circuit.

SPECIAL DOUBLE-POLE DOUBLE-THROW (DPDT) • A third type of switch has two positions and also does not have on or off positions marked because both positions are actually on positions. This switch is known as a four-way switch and can be identified by four brass-colored terminals. Electrically, this is a special double-pole double-throw (DPDT) switch (Fig. 9-3). When the switch is moved from one position to the other, the four connections are interchanged or crisscrossed. These switches are used where

a light (or lights) is to be controlled from three or more locations. Of course, four-way switches also have a green-tinted terminal for the grounding-wire connection. Four-way switches are called *special* double-pole double-throw switches to distinguish them from true double-pole double-throw switches. True DPDT switches are equivalent to two SPDT switches and have six brass-colored terminals. Many true DPDT switches also have a center OFF position.

DOUBLE-POLE SINGLE-THROW (DPST) • For some large appliances it is necessary to switch two hot wires (red and black, for example) at the same time. A double-pole, single-throw (DPST) switch is made for this use (Fig. 9-4). This switch has four brass terminals, and OFF and ON positions are marked on it. It is equivalent to two SPST switches.

All switching in residential wiring can be handled by these four switch types, used either singly or in combinations. Switch wiring is covered in Chapter 13.

Mechanical Switching

Switches used in residential wiring do the electrical switching previously described in a number of different ways. Many switches are designed to combine some additional related action with the switching. A review of a few electrical principles will help in understanding how switches work and how they are rated.

When switch contacts are closed to complete a circuit, an arc jumps from one contact to the other just before they meet. Similarly, when switch contacts are opened to turn off power, an arc jumps between the contacts just after they open. This constant arcing causes the switch contacts to become burned and pitted. This, in turn, causes the resistance to current flow to increase at the switch contacts. Resistance to current flow produces heat. This heat shortens the life of the switch and, in extreme cases, can cause fire.

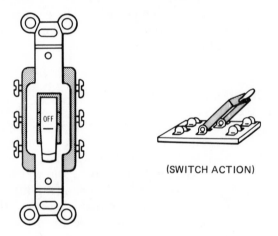

TRUE DOUBLE-POLE DOUBLE-THROW SWITCH

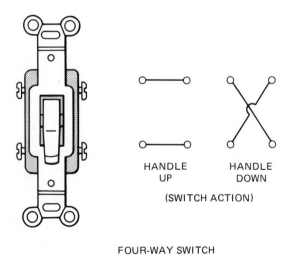

FOUR-WAY SWITCH

Figure 9-3. Special double-pole double-throw switch.

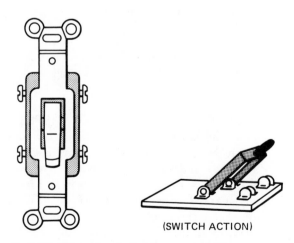

Figure 9-4. Double-pole single-throw switch.

Switches must be designed to keep arcing to a minimum. This can be done in two ways: first, by making the switch action positive and, second, by using for the contacts special metal alloys that are good conductors and are resistant to burning. Even with contacts of special alloys, however, there is a tendency for an insulating layer to form on switch contacts over a period of time. This layer resists current flow and causes the contacts to become hot, which further increases resistance and shortens the useful life of the switch. Formation of this insulating layer can be prevented by coating the switch contacts with a microscopically thin layer of gold. This is often done in some high-quality switches.

• SWITCH TYPES •

Toggle Switches

The most widely used switch type is the toggle switch. The simple up-down handle movement provides good mechanical switching with generally trouble-free service.

At one time, switches were designed to operate on both ac and dc power. To keep arcing to a minimum with dc power, extremely high-speed switch action was required. These switches made contact by moving an L-shaped armature so that it made contact with both terminals in the on position, and moving it to the center position when the switch was off (Fig. 9-5). The wiping action as the armature moved across the terminals helped to keep the contact area clean and, therefore, electrically efficient. To be suitable on dc circuits the make-break action was assisted by a spring. When the handle was near the center position, the spring was released and it snapped the switch on or off with a noise which many people found annoying. This led to the development of the "quiet" switch.

Quiet Switches

Because almost all residential electrical service is now ac, switches designed for ac use only have been introduced. These are known as ac-only switches and are by far the most common type in use. The nature of alternating current makes arcing less of a problem when switches are used only on ac. Switch contacts need not be snapped open or snapped closed as rapidly as on dc circuits. AC-only switches simply push contacts apart or allow them to close (Fig. 9-6). This action is much quieter than the spring action of ac-dc switches.

Another type of quiet switch is turned off and on by pressing a button. No switch positions are marked because the button always returns to the same position. Each time the button is pressed the switch changes from off to on or on to off (Fig. 9-7). When the button is pressed it rotates a ratchet wheel that alternately pushes the contacts apart or allows them to close.

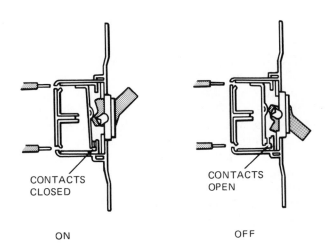

Figure 9-6. AC-only quiet switch.

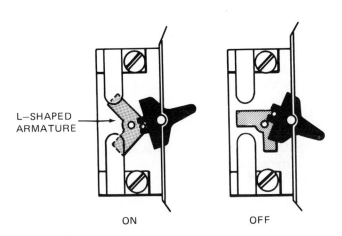

Figure 9-5. AC-DC toggle switch.

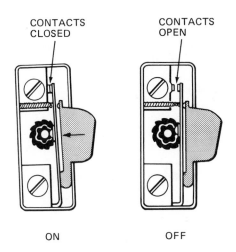

Figure 9-7. Push-button switch.

Dimmer Switches

These switches control the level of illumination in a room, while providing a way to turn lights on and off. Modern dimmers use solid state devices that can control the period of time that current flows during each cycle. For maximum light the control knob is set so that current flows normally, that is, continuously during each cycle. When the knob is turned to reduce the light, the solid state device delays the start of current flow in each half cycle. In this way average current flow is reduced and the light level is correspondingly lower. One type of dimmer switch is made for use with incandescent lamps and another type for fluorescent lamps. Either type may be installed in any dry location. Most often they are used in dining rooms and recreation rooms.

Both incandescent and fluorescent dimmers are available in models to replace either standard on-off switches or three-way switches. A three-way dimmer switch can replace one of the three-way switches on a circuit. Dimming control and on-off control is then available at the dimmer switch. Only on-off control is available at the other switch.

INCANDESCENT • Incandescent dimmer switches consist of a small box with two pigtail leads, rather than screw terminals, for electrical connections (Fig. 9-8). A keyed shaft projects from the front of the switch. The switch mounts in an electrical box just as a standard toggle switch does. Solderless connectors are used to connect the pigtail leads to the circuit. Standard toggle switch faceplates can be used with low-wattage (600-watt) dimmers. Higher-wattage dimmers require a special finned faceplate to carry off internal heat.

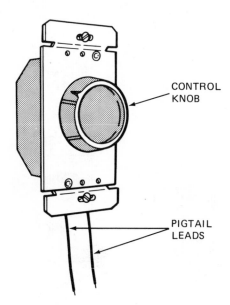

Figure 9-8. Dimmer switch.

The dimmer control knob fits on the keyed shaft. The knob is large enough to conceal the rectangular opening in a standard faceplate. The light under control is turned on and off by push-push action. The level of illumination is controlled by rotating the knob. Clockwise rotation increases illumination. Dimmer switches are rated in terms of the maximum illumination wattage they can control. Standard ratings are 600, 1000, 1500, and 2000 watts. Dimmer switches are suitable only for control of illumination. They cannot be used for motor or heat control. Although dimmers fit in standard switch boxes, they are somewhat larger than standard toggle switches. In new work, to avoid crowding, a larger box than the minimum required by the NEC or local code should be installed where dimmers will be located.

FLUORESCENT • Controlling the illumination level of fluorescent lamps is more complicated than controlling incandescent lamps. Changes must be made to the fluorescent fixtures when dimmers are installed. This subject is covered in the section **Fluorescent Fixtures**.

Time Delay Switches

Switches which have a built-in time delay are often used in locations where, for reasons of safety, illumination is required for a short time after the light switch is turned off. Stairways, basements, and garages are typical locations. The standard delay is 30 to 45 seconds after the switch is turned off. There is no delay when the switch is turned on. Time delay switches are slightly larger than standard toggle switches, but are installed in exactly the same way.

Night Light Switches

Toggle switches with illuminated handles are easy to find in the dark, and are useful in bathrooms, bedrooms, and hallways. A small neon bulb in the handle provides the illumination. The bulb is wired internally in parallel with the switch contacts. When the switch is off, the bulb is illuminated. The bulb has a high resistance that limits the amount of power used. A typical value is 1/25 watt. These switches are installed in exactly the same way as standard toggle switches.

Pilot Light Switches

Another type of illuminated switch is often used when the light controlled by the switch is not visible from the switch location, such as upper-level switches for basement lights. Two types of pilot light switches are generally available. One type has an illuminated handle, like a night light. Another type has a a separate lamp mounted below the switch (Fig. 9-9). To operate

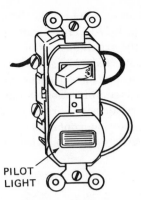

Figure 9-9. Pilot light switch.

as a pilot light, the lamp in the switch must be in parallel with the light controlled by the switch. To wire the pilot lamp in parallel with the load, white- (or gray-) insulated power ground conductor must be available in the box in which the pilot light switch is to be mounted (Fig. 9-10).

Pilot light switches have two brass terminals for the black conductor connections and a silver terminal for the power ground connection. The lamp is wired internally to the load side of the switch, and a jumper is placed between the joined white wire and the silver terminal on the switch. If the pilot lamp stays on all the time, regardless of the switch position, power to the circuit should be turned off and the black connections reversed. The lamp should then operate correctly.

Mercury Switches

Almost all toggle switches currently available for residential wiring are the ac-only, quiet-action type previously described. These switches are much quieter than ac-dc snap switches, but they do make some noise. If complete quiet is required, mercury switches must be used. These switches have a sealed vial containing a small quantity of mercury mounted within them in such a way that the movement of the switch handle rotates the vial (Fig. 9-11). Mercury is a liquid metal— sometimes called quicksilver—that stays in a liquid form at usual room temperatures. The switch contacts consist of two pieces of metal that project into the vial. When the vial is rotated, the contacts are immersed in the mercury and current can flow between them. Of course, this action is completely silent. Because gravity keeps the mercury at the bottom of the vial, these switches must be vertically mounted, with the proper side up, to work correctly.

• RECEPTACLES •

The terms *outlet* and *receptacle* are often used interchangeably in the electrical trade. To the NEC and most manufacturers, however, the terms have different meanings. An electrical outlet is a point in a circuit where other devices can be connected, that is, any place

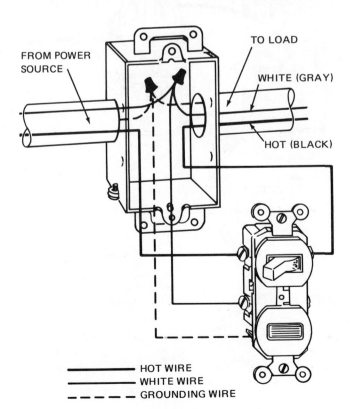

Figure 9-10. Pilot light wiring.

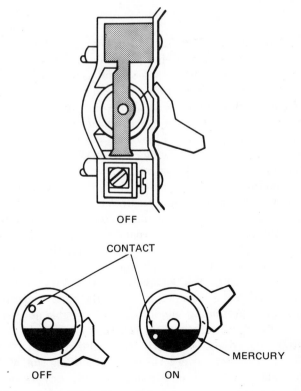

Figure 9-11. Mercury switch.

Switches, Receptacles, and Fixtures **145**

where cable or conduit is connected to a box. A receptacle is the device that is installed in a box at an outlet point to provide for the connection of cord-and-plug power lines (Fig. 9-12).

Electrical receptacles are passive devices, that is, they do not consume power. They provide a convenient place to connect active devices, such as lamps and appliances. Receptacles should make good electrical contact with the mating plugs and should be designed to prevent accidental contact with live surfaces. Most receptacles also have provision for an equipment-grounding connection. To prevent accidental connection of a low-voltage appliance to a higher receptacle voltage, receptacles and plugs are keyed so that only rated combinations are possible.

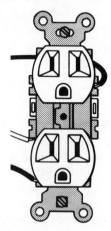

Figure 9-12. Receptacle, 15- to 20-ampere, three-prong.

Good electrical contact with the mating plug is provided by making receptacle contact surfaces as large as possible, and by forming the metal parts so that they press against the plug prong. The metal contacts are made of alloys that retain shape and springiness for long periods of time. *Accidental* contact with current-carrying parts of the receptacle is prevented by having the current-carrying parts recessed in nonconducting material. Of course, if any object made of conducting material is pushed into the slots of a receptacle, shock or fire can result. Various kinds of slot covers and closures are available to prevent this from happening (Fig. 9-13). A connection for equipment grounding is provided by a U-shaped slot (Fig. 9-14) that connects an inserted plug prong to a green-tinted screw located on the bottom of the receptacle. When the receptacle is installed in a box, the bare or green-insulated grounding conductor in the cable is connected by jumper to the green-tinted screw. If the grounding circuit is properly installed throughout the electrical system, every U-shaped grounding slot will provide a solid ground connection for any device plugged into it. As an added safety feature, the

grounding prong on the plug is slightly longer than the power prong (Fig. 9-15). This means that when the plug is inserted, the grounding connection is made before power is connected. Also, when the plug is removed, the grounding connection is maintained until power is disconnected. The most common receptacle formats used in residential wiring are shown in Fig. 9-16. The receptacle rating not only covers voltage and amperage, but also lists poles and wires. The number of

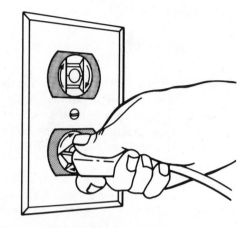

Figure 9-13. Covered receptacle.

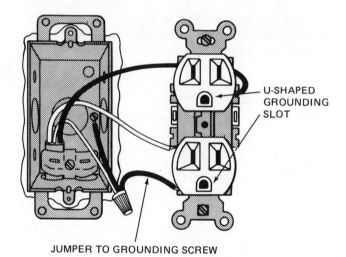

U-SHAPED GROUNDING SLOT

JUMPER TO GROUNDING SCREW

Figure 9-14. Standard duplex receptacle wiring.

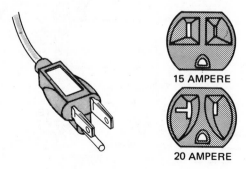

15 AMPERE

20 AMPERE

Figure 9-15. Three-prong plug.

RATING/APPLICATION	WIRING	RECEPTACLE
15A 125V **2-POLE 2-WIRE** FED. SPEC. STYLE A FOR REPLACEMENT.	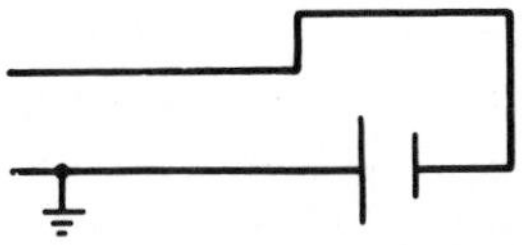	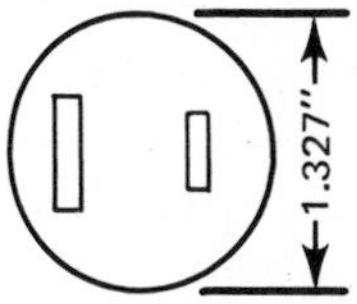 1.327″
15A 125V GROUNDING **2-POLE 3-WIRE** FED. SPEC. STYLE D STANDARD FOR RESIDENTIAL, COMMERCIAL, INDUSTRIAL.	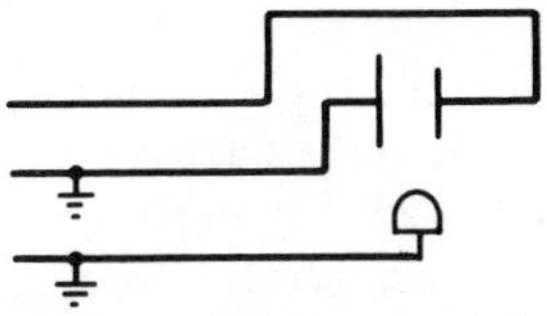	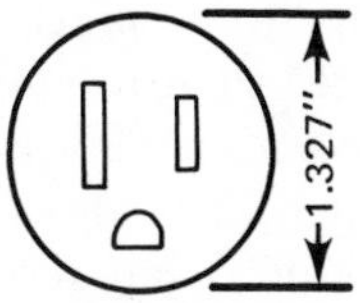 1.327″
20A 125V GROUNDING **2-POLE 3-WIRE** FED. SPEC. STYLE X ROOM AIR-CONDITIONERS, KITCHENS, HEAVY DUTY PORTABLE TOOLS AND APPLIANCES — RESIDENTIAL, COMMERCIAL, INDUSTRIAL.	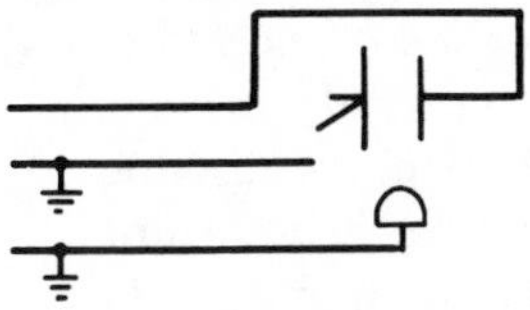	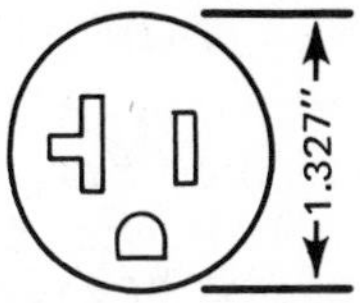 1.327″
15A 250 V GROUNDING **2-POLE 3-WIRE** FED. SPEC. STYLE H ROOM AIR-CONDITIONERS, HEAVY DUTY PORTABLE TOOLS, COMMERCIAL APPLIANCES.	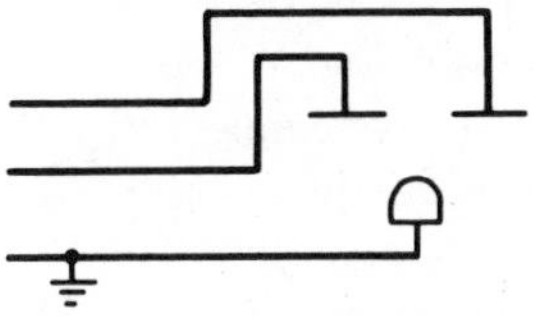	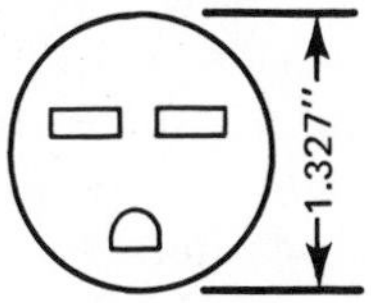 1.327″
20A 250V GROUNDING **2-POLE 3-WIRE** ROOM AIR CONDITIONERS, HEAVY DUTY PORTABLE TOOLS, COMMERCIAL APPLIANCES.	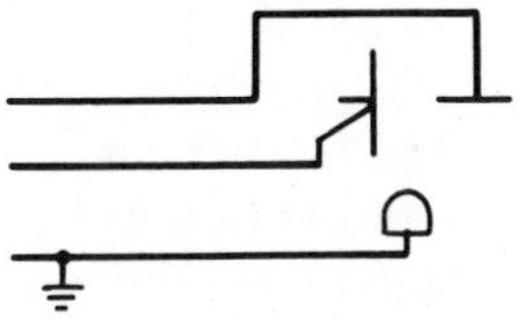	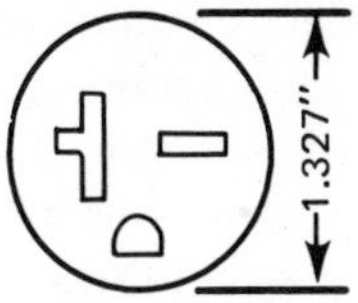 1.327″
30A 125/250V **3-POLE 3-WIRE** FED. SPEC. STYLE S CLOTHES DRYERS IN RESIDENCES, HEAVY DUTY EQUIPMENT IN COMMERCIAL AND INDUSTRIAL BUILDINGS. NOT FOR EQUIPMENT GROUNDING.	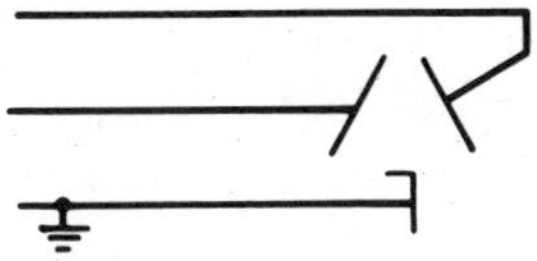	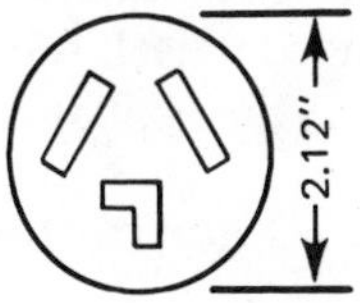 2.12″
50A 125/250V **3-POLE 3-WIRE** FED. SPEC. STYLE T RANGES IN RESIDENCES, HEAVY DUTY EQUIPMENT IN COMMERCIAL AND INDUSTRIAL BUILDINGS. NOT FOR EQUIPMENT GROUNDING.	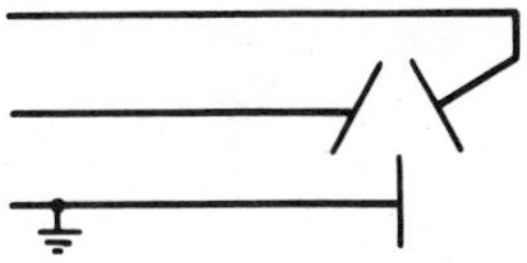	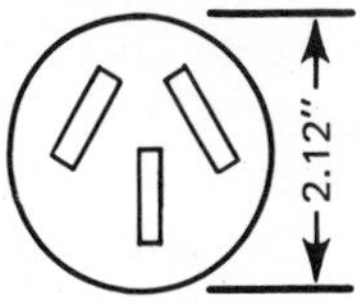 2.12″
30A 125/250V GROUNDING **3-POLE 4-WIRE** PROVIDES GROUNDING PROTECTION FOR CLOTHES DRYERS AND HEAVY DUTY EQUIPMENT.	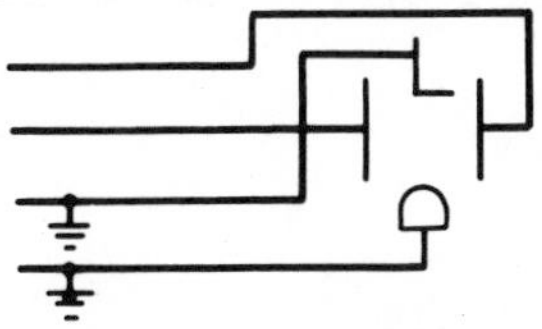	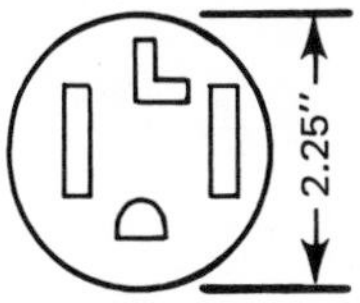 2.25″
50A 125/250V GROUNDING **3-POLE 4-WIRE** PROVIDES GROUNDING PROTECTION FOR RANGES AND HEAVY DUTY EQUIPMENT.	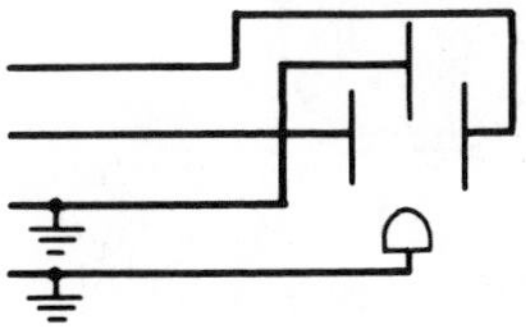	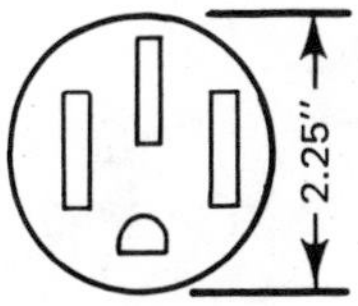 2.25″

Figure 9-16. Receptacle types and ratings.

poles is the number of conductors normally carrying current that are connected to the receptacle. The number of wires is the sum of pole wires plus a grounding wire. If the number of poles and wires is the same, no grounding slot is provided.

Polarized Receptacles

Note (Fig. 9-16) that the power slots of the 125-volt 15- and 20-ampere receptacles are not the same size. One slot is longer than the other. The plugs on some appliances have both a wide power blade and a narrow one. Because the wide blade can be inserted only in the larger slot, the plug and receptacle must always be connected the same way. This is called polarization. Receptacles are made with the longer slot connected to the silver-colored terminals. These terminals, in turn, are connected to the white (or gray) power ground conductor. Polarization is necessary because some electronic appliances (TV sets, radios, high-fidelity components, etc.) have exposed metal parts that are connected to one side of the input power line. When a polarized plug is used, these parts are always connected to the wide plug blade and, therefore, are always connected to power ground (Fig. 9-17). At one time, all 120-volt appliances had plugs with two identical blades. These plugs could be inserted in a receptacle in either of two ways. In one plug position the exposed metal parts of the appliance were connected to the hot side of the power line. The appliance would work normally but the user could receive a severe, perhaps fatal, shock by touching the exposed metal while in contact with any grounded object, such as a water faucet. This cannot happen if the appliance has a polarized plug.

Specialized Receptacles

Specialized receptacles (Fig. 9-18) may be installed in homes to provide connection points for electric ranges, clothes dryers, and air conditioners. These receptacles are the three-pole, three-wire type rated at 30 to 50 amperes, 125/250 volts. If copper wire is used, circuits for these devices must be wired with no. 6 or no. 8 conductors. Conductors this large require larger

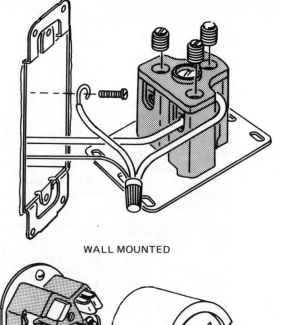

WALL MOUNTED

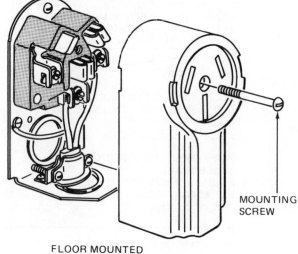

FLOOR MOUNTED

MOUNTING SCREW

Figure 9-18. Specialized receptacles.

terminals and more working space. In residences specialized receptacles are always single-connection type; they may be mounted in wall boxes or surface-mounted on the floor. (Some local codes prohibit floor surface mounting.)

Wall-mounted 30- and 50-ampere receptacles have compression-type connections. To make connections, loosen the setscrews, insert the wires in the opening beneath, and then tighten the setscrews. The white- (or gray-) colored conductor is inserted in the opening marked WHITE. The red and black conductors are connected to the other terminals.

Surface-mounted receptacles are used wherever wall mounting is not practical, particularly in old work. The receptacle and box are a single unit with knockouts provided for cable entrance. Conductors are connected to marked screw terminals and the box is secured to the floor. An insulating cover is fastened to the box with one or more mounting screws. Note that these receptacles have the same number of poles and wires, so no slot is provided for the grounding wire.

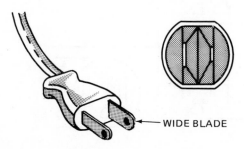

← WIDE BLADE

Figure 9-17. Polarized receptacle and plug.

If the receptacle is wall-mounted, connect the bare or green-insulated grounding wire to the grounding screw terminal at the back of the box. Surface-mounted receptacles have a similar grounding screw terminal on the metal frame. The white- (or gray-) insulated power ground wire provides the grounding connection as well for exposed metal parts of appliances such as ranges, ovens, and clothes dryers. This is possible because large appliances are on individual circuits and the power ground conductor runs without a break from the power source to the appliance.

• MARKINGS ON SWITCHES AND RECEPTACLES •

Switches and receptacles are rated by manufacturers for the maximum current and voltage at which they should be used. Testing organizations such as Underwriters' Laboratories test switches and receptacles in accordance with the manufacturer's rating and list the item for use if it performs satisfactorily. In addition to mechanical and electrical performance, switches and receptacles are rated for use with copper, copper-clad aluminum, or aluminum conductors. All important information as to use is marked on the metal mounting yoke (Fig. 9-19) and is defined below.

1. Testers' Symbol. Some form of the Underwriters' Laboratories symbol will be marked on the device if it has been tested and is listed for use. Only listed switches and receptacles should be used. Many devices also have a CSA monogram. This is the symbol of the Canadian Standards Association, another testing organization.
2. Type of Current. Switches for ac only must be marked either AC alone or AC-ONLY. AD-DC switches have no current marking. Likewise, receptacles do not have ac or dc markings.
3. Voltage and Amperage Rating. The maximum safe voltage and current that can be handled by a switch or receptacle is marked on it. The marking sometimes gives alternate combinations. Typical markings are "10A 120V - 5A 250V," "15A 120V," or "15A 120-277V." The first marking means the device may be used at either combination. The second marking means the device is for use on 120-volt circuits only. The third marking indicates a range of voltage, but the maximum current remains constant.
4. Conductor Material. Switches and receptacles are marked to show which conductor materials may be safely connected to them. The markings vary with the current rating, as shown in Table 9-1.

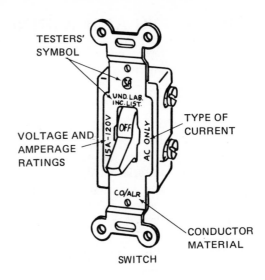

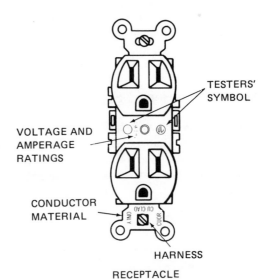

Figure 9-19. Markings on switches and receptacles.

TABLE 9-1. CONDUCTOR MATERIAL CODES.

Rated Current (amperes)	Marking	Permitted Conductor Materials
15-20	No marking	Copper
15-20	CU-AL*	Copper
15-20	AL-CU*	Copper
15-20	CU or CU CLAD ONLY	Copper or copper-clad aluminum
15-20	CO-ALR	Copper, copper-clad aluminum, or all aluminum
30 and higher	No marking	Copper or copper-clad aluminum
30 and higher	CU-AL	Copper, copper-clad aluminum, or all aluminum

*These markings originally meant that copper-clad aluminum and all-aluminum conductors could be used. These markings are now obsolete. It is recommended that devices with this marking be wired with copper wire only.

• MOUNTING AND WIRING AIDS •

Mounting Yoke Ears

You will recall that electrical boxes have attachments known as ears that are used to hold the box flush with plaster surfaces when boxes are mounted in plaster-and-lath or plasterboard walls. Box ears also provide threaded holes for switch and receptacle mounting screws. Projections on the mounting yoke of switches and receptacles are also known as ears and have a related function (Fig. 9-20). When boxes are flush-mounted, switches and receptacles can be mounted flush also. When boxes are recessed, the switch or receptacle must still be flush with the wall surface to fit properly in the face plate. The yoke ears project out far enough to hold the device on the surface. Yoke ears can be broken off, if necessary, when flush-mounted boxes are used.

Mounting Yoke Screw Slots

Mounting screws for switches and receptacles fit through slots rather than holes. The slots allow the mounting angle of the device to be adjusted to vertical position, even if the box or the stud it is mounted on is not vertical.

Terminal Color Code

All listed receptacles have terminals coded to include the color of the conductor that should be attached. The hot side of the power line—usually black- or red-insulated—is connected only to brass- or copper-colored terminals. The ground side of the power line—always white- or gray-insulated—is connected only to the silver-colored terminal. Switches always have brass-colored terminals and are always connected to the hot wire (red or black) of the power line. There is a single exception to this rule. The exception is covered in Chapter 15. Both switches and receptacles have green-tinted terminals for the grounding-wire connection. The bare or green-insulated grounding wire is connected to the green-tinted terminal.

Push-in Terminals

Some switches and receptacles have openings in the back for conductor connections. The bare conductor is simply pushed into the opening to make the connection (Fig. 9-21). A spring-loaded blade grips and holds the conductor. The conductor can be released by inserting a small screwdriver in a slot next to the conductor opening. The standard switch and receptacle color code must be observed, and the NEC permits only copper or copper-clad aluminum to be connected to devices having push-in terminals. Push-in terminals may not be used with bare aluminum wire.

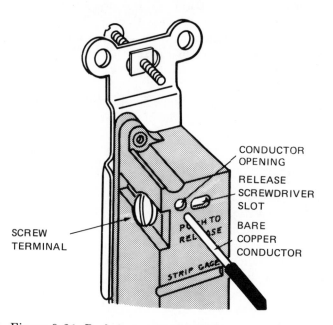

Figure 9-21. Push-in terminals.

Two-Circuit Wiring

Duplex receptacles have four power terminals, as well as a green-tinted grounding screw. Two terminals are brass-colored and two are silver-colored. In most cases in residential wiring both receptacles will be on the same circuit. The two brass-colored terminals are connected together by a strip of metal. A similar strip connects the silver-colored terminals. It is necessary only to connect red or black wire to one of the brass

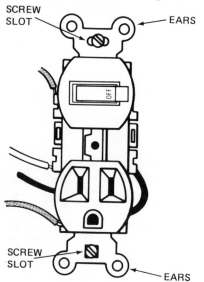

Figure 9-20. Mounting yoke.

terminals and a white or gray wire to one of the silver terminals to provide power to both halves of the receptacle. In some cases it may be desirable to wire each half of the receptacle on a different circuit. To do this just break off the connecting metal between each pair of terminals (Fig.9-22). Now all four terminals are independent and may be connected as separate circuits.

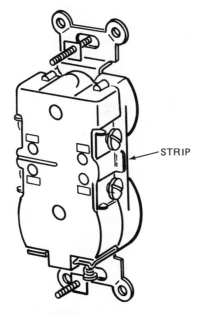

Figure 9-22. Break-off strip.

• TESTING SWITCHES AND RECEPTACLES •

Switches

Switches can be tested using a continuity tester or (on the low scale) an ohmmeter. To test a standard SPST toggle switch, connect the tester or the ohmmeter to the switch terminals and operate the switch (Fig. 9-23). The continuity light should always light when the switch is in the on position and always go out

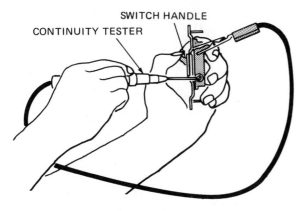

Figure 9-23. Testing switches.

when the switch is in the off position. If an ohmmeter is used, the ohmmeter should always indicate zero when the switch is on and always indicate infinity (INF or ∞) when the switch is off. Any other test result indicates a defective switch. Note particularly any flickering or dimming of the continuity tester light or any erratic movement of the ohmmeter needle. These conditions indicate faulty switch action, and the switch should be replaced. Testing three-way (SPDT) and four-way (DPDT) switches is a bit more complicated, but the good and bad indications are basically the same.

To test a three-way switch, connect one lead of the tester or ohmmeter to the darker-colored common terminal and the other lead to one of the brass-colored terminals. Operate the switch. The tester or the ohmmeter should indicate an on-off condition as it did in the SPST test. Move the second test lead to the other brass-colored terminal and repeat the test. The same on-off indication should be obtained, but in reverse. Connect the test leads to the two brass-colored terminals and operate the switch. The tester should not light, the ohmmeter needle should remain on infinity. Any other condition indicates faulty switch action, and the switch should be replaced.

To test four-way switches, connect one test lead to any terminal. Connect the other to either terminal on the opposite side of the switch. Operate the switch. An on-off indication should be obtained. Move one lead only to the other terminal on the same side. Operate the switch. Again, an on-off indication should be obtained. Move the other lead to the second terminal on the same side. Repeat the on-off test. As a final check, connect the test leads to terminals on the same side and operate the switch. No indication of continuity should be obtained.

Receptacles

Receptacles do not often cause trouble. When trouble does occur, it is most often caused by poor contact or intermittent contact between a plug blade and the contact within the receptacle. You can check for this condition by plugging a test light or extension light into the receptacle and noting how the light works. Move the cord around so that the plug is stressed *lightly* in each direction. If the test light flickers or goes out, the receptacle is not maintaining good contact with the plug blade. Replace the receptacle.

If necessary, continuity tests can be made on receptacles. Insert one test lead in a test slot and the other on the appropriate terminal. In standard 120-volt receptacles the wide slot is connected to the silver-colored terminals. The narrow slot is connected to the brass terminals (Fig. 9-24). If a short is suspected, check for continuity between the brass and silver terminals. If the receptacle is shorted, some continuity indication will be obtained.

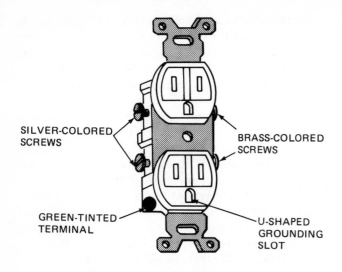

SILVER-COLORED
SCREWS

BRASS-COLORED
SCREWS

GREEN-TINTED
TERMINAL

U-SHAPED
GROUNDING
SLOT

CONTINUITY:

WIDE SLOT TO SILVER-COLORED TERMINAL
NARROW SLOT TO BRASS-COLORED TERMINAL
GROUNDING SLOT TO GREEN-TINTED TERMINAL

Figure 9-24. Polarized receptacle connections.

Receptacles made for voltages higher than 120 volts can be checked by referring to a manufacturer's diagram and making appropriate continuity checks. An example is shown in Fig. 9-25.

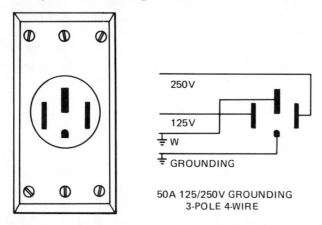

250V

125V

W

GROUNDING

50A 125/250V GROUNDING
3-POLE 4-WIRE

Figure 9-25. High-voltage receptacle diagram.

• INCANDESCENT FIXTURES •

Electrical Connections

Some small single-lamp fixtures have screw terminals for electrical connection, but but these are the exception. The rule is to make electrical connections to fixture wires. You will recall from Chapter 6 that a special type of wire is made for fixtures. This wire is insulated to withstand the heat likely to be generated by the lamps. The wire is flexible and easy to work with. The NEC requires that the fixture wires be

color-coded. The wire connected to the screw shell in the lamp sockets must have white or gray insulation. The other wire must be a contrasting color. Most fixture manufactuers use black insulated wire. When fixtures have more than one lamp, all screw shells must be connected to the white or gray wire. When fixtures are connected to the power source cable, the color code must be maintained. White (or gray) is connected to white (or gray). The other conductors—usually black—are connected together. If the fixture is externally switched, that is, controlled by a wall switch, the black wire is the switch wire. Fixtures that are internally switched also have the switch in the black wire.

Fixture Mounting

SINGLE-LAMP UTILITY FIXTURE • The simplest lighting fixture is the single-lamp type. In garages, work areas, utility rooms, laundry rooms, etc., the standard ceramic fixture (Fig. 9-26) is often used. This fixture may be controlled by a wall switch or an internal pullchain switch. The fixture can be mounted directly on a rectangular or octagonal box on wall or ceiling. Two methods of wiring are used for these fixtures. Some have fixture wire attached; some have screw terminals. Fixture wire can be joined to the power wires in the box by means of solderless connectors. Power lines can be connected directly to the screw terminals. Observe conductor and screw terminal color coding; white (or gray) to white; or white (or gray) to silver.

These fixtures can usually be mounted by two screws. Holes in the ceramic base line up with the threaded box tabs. Some single-lamp ceramic fixtures have wider bases that do not line up with the box tabs. These can be mounted by first installing a mounting strap with wider spaced mounting holes on the box and then mounting the fixture to the strap.

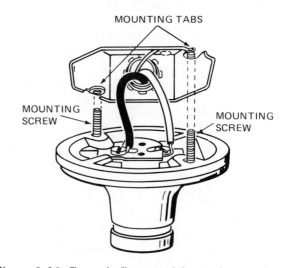

MOUNTING TABS

MOUNTING
SCREW

MOUNTING
SCREW

Figure 9-26. Ceramic fixture wiring and mounting.

SINGLE-LAMP DECORATIVE FIXTURE • Decorative single-lamp fixtures are often mounted on walls and ceilings in foyers, hallways, and small rooms. These fixtures have a metal base—called a canopy—in which a metal or ceramic socket shell is mounted. After making electrical connections, the metal canopy can be attached directly to the ceiling box by screws in the same way as the utility fixture. Single-lamp decorative fixtures for wall mounting are designed to mount on a fixture stud. The fixture canopy fits flat against the wall and has an opening in the center. The lamp socket is mounted on a bracket so that the lamp is parallel to the wall. To mount this fixture, the wall box must have a fixture stud either mounted in the box or added by installing a strap on the box (Fig.9-27) and adding a fixture stud to it. The stud is adjusted to project through the fixture canopy just enough to allow a capnut to be threaded securely on the stud. Typically, 1/4 to 3/8 inch is sufficient.

MULTIPLE-LAMP FIXTURES • Multiple-lamp ceiling fixtures are mounted by means of a threaded nipple. If a stud is in the box, as it usually is when the box has been mounted on a hanger, a reducing nut and a nipple can be threaded onto the stud (Fig. 9-28). If the box has no stud, a strap and nipple can be attached to the box to hold the fixture. To mount lightweight fixtures that fit against the ceiling, a nipple length should be used that is just long enough to project through the fixture about 1/4 to 3/8 inch when the fixture is in place. A capnut is threaded onto the nipple to hold the fixture. The fixture wires are joined to the power wires with solderless connectors before the fixture is secured by the capnut.

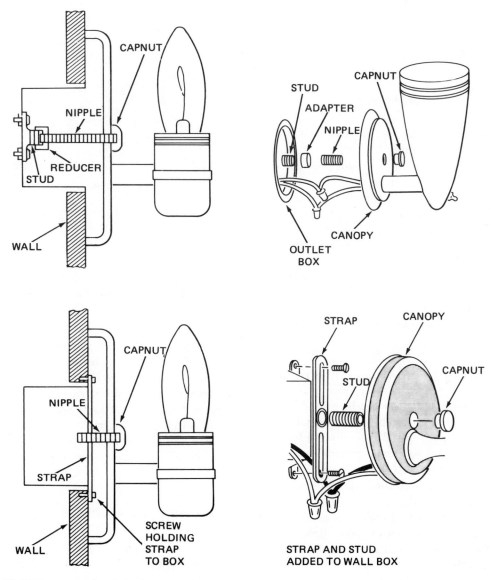

Figure 9-27. Wall fixture on box stud.

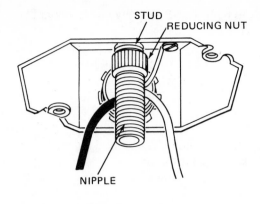

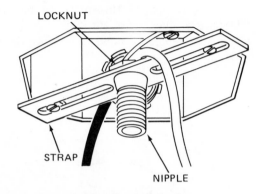

Figure 9-28. Installing fixture nipples.

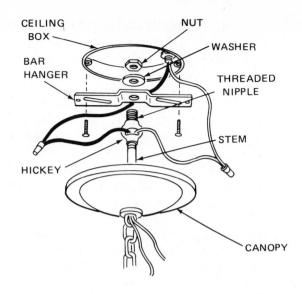

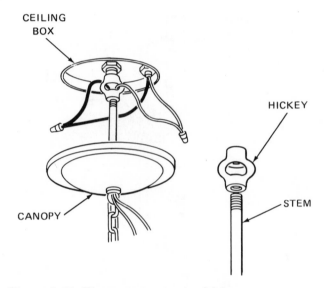

Figure 9-29. Fixture mount using hickey.

Larger hanging fixtures have a canopy that fits over the ceiling box. A threaded stem fits through the canopy. The stem supports the fixture and provides an opening to bring the fixture wires into the box (Fig. 9-29). For these fixtures a hickey is used, rather than a reducing nut. The hickey is threaded at each end, but has openings in between so the fixture wires can be brought out for connection to power wires. (Do not confuse this hickey with the hickey used to bend conduit discussed in Chapter 7.) As noted in Chapter 8, ceiling fixtures weighing more than 50 pounds must be supported by a building structural member (such as a joist). A box and hanger should not be the only support for fixtures this large.

Troubleshooting Incandescent Fixtures

Faults in incandescent fixtures can be caused by either short circuits or open circuits. Shorts within a fixture will cause a circuit breaker to trip or a fuse to blow when power is applied to the fixture circuit or when the fixture is turned on. If the fixture has an open circuit, it will be impossible to turn on one or more of the lamps. In either case, the first step is to turn off power to the fixture and then disconnect the black and white fixture wires from the power conductors.

As a general rule tests for shorts are made with no lamps installed in the fixture; tests for opens are made with lamps installed. Both troubles can be located by making a close visual inspection of the fixture wiring and, if this does not show the cause of trouble, by testing with a continuity tester or an ohmmeter. When making the visual inspection, look for scorched or discolored wire insulation or exposed conductors as the source of short circuits. Look for loose or broken wires as the source of open circuits. Some basic procedures that can be used to locate open circuits or short circuits, using a continuity tester or an ohmmeter, are given below.

OPEN CIRCUITS, SINGLE-LAMP FIXTURES •

Step 1. Check unswitched, single-lamp fixtures for open circuits by checking for continuity with a lamp installed in the fixture.

Step 2. If the tester indicates no continuity, remove the lamp and check each side of the power connection (Fig. 9-30). There should be continuity between the screw shell of the socket and the white lead. There should be continuity between the center terminal at the base of the socket and the black fixture wire.

Step 3. If either leg shows no continuity, check for breaks in the fixture wire or poor connections to the socket.

OPEN CIRCUITS, MULTIPLE-LAMP FIXTURES •

Step 1. Check unswitched multiple lamp fixtures by installing a lamp in one socket at a time and performing the single-socket test.

Step 2. If no continuity is found in any test, disassemble the fixture, if necessary, and check internal wiring and connections to sockets.

NOTE: The internal wiring of many multiple-lamp fixtures is visible and can be checked when the fixture is removed from the box. Larger hanging fixtures, such as are often used in dining rooms, must be taken apart to check internal wiring. Such a large variety of types and styles of hanging fixtures are made that detailed disassembly instructions are not practical. However, one method of manufacture is widely used and you should be familiar with it: Many fixtures are assembled around a threaded rod. The rod runs through the center of the fixture. Decorative pieces and the socket mounting pieces are held in place on the rod by capnuts attached to the top and bottom. When the bottom capnut is removed, the lower part of the fixture can be removed to make internal wiring accessible.

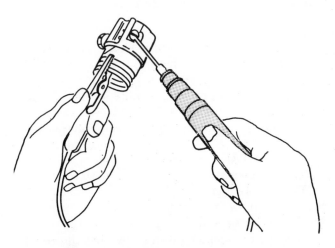

Figure 9-30. Testing single lamp fixture.

Switched fixtures can be tested the same way as unswitched, except that the switch must be turned off and on during the continuity test to check switch action. The switch must always be wired in the black lead.

Some fixtures have switches that allow various levels of illumination to be selected by turning on lamps in sequence or in groups. Fixtures such as this often have multiple-position rotary switches. If some lamps light and others do not, and inspection shows no wiring breakage, a faulty switch is a likely source of trouble. The switch action can be checked using a continuity tester, if the switching sequence is known. In a new fixture, the switching sequence may be shown on the carton or on installation data sheets packed with it. Sometimes markings on the switch itself indicate switching combinations. The sequence for testing a simple combination switch used in many fixtures is shown in Fig. 9-31.

Switch position	Touch probe to wire listed	Tester light
1	RED	OFF
	GREEN OR BLUE	OFF
2	RED	ON
	GREEN OR BLUE	OFF
3	RED	OFF
	GREEN OR BLUE	ON
4	RED	ON
	GREEN OR BLUE	ON

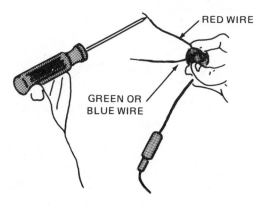

TO TEST, CONNECT ONE CONTINUITY TEST LEAD TO THE BLACK LEAD. MOVE THE SWITCH THROUGH ALL FOUR POSITIONS. AT EACH POSITION, TOUCH THE PROBE TO EACH OF THE OTHER TWO LEADS. IF THE SWITCH IS OK, THE TESTER WILL LIGHT OR NOT LIGHT AS INDICATED IN THE CHART ABOVE. ALTHOUGH THE COLOR OF THE THREE LEADS MAY VARY, THE ON-OFF PATTERN FOR ANY THREE-BULB LAMP SWITCH WILL BE THE SAME. IF THE TESTER FLICKERS WHEN IT SHOULD BE ON, DOES NOT LIGHT WHEN IT SHOULD, OR LIGHTS WHEN IT SHOULD NOT, THE SWITCH IS DEFECTIVE. REPEAT THE TEST TO BE CERTAIN THE RESULTS ARE CORRECT.

Figure 9-31. Test sequence for four-position fixture switch.

SHORT CIRCUITS, SINGLE-LAMP FIXTURES •

Step 1. Check unswitched single-lamp fixtures for short circuits by checking for continuity between the black and white wires with no lamp in the fixture.

Step 2. If the tester indicates continuity, check each side of the power connection. There should be no continuity between the screw shell of the socket and the black lead; nor between the center terminal at the base of the socket and the white fixture wire.

Step 3. If either leg shows continuity, check for breaks in the fixture wire insulation.

SHORT CIRCUITS, MULTIPLE-LAMP FIXTURES •
Check unswitched multiple-lamp fixtures for shorts by performing the single-socket test on each individual socket after disconnecting either the black or white lead to the socket. If the leads to the sockets cannot be readily disconnected, the fixture should be replaced.

Switched fixtures can be tested the same way as unswitched, except that the switch must be turned off and on during the continuity test to check switch action. If the fixture has a metal frame, check for continuity between the frame and the switch leads. If a continuity indication is obtained, the switch is shorted internally. This same test can be used to locate shorts in multiple-position rotary switches. Move the switch through all positions while making the test.

• FLUORESCENT FIXTURES •

Fluorescent lighting provides a bright, even illumination that is desirable in many areas of the home. It is more complex and, initially, more costly than incandescent lighting. Fluorescent lamps, however, produce more light per watt of power used and they last four or five times longer than incandescent lamps. The long-range cost of fluorescent lights, then, is significantly lower than the cost of incandescents. More light per watt of power used is not only more economical, but is important in the conservation of energy.

The life of fluorescent lamps is determined primarily by how often they are turned on and off; the less this occurs, the longer they last. Fluorescents, then, are best suited to areas where lights stay on for relatively long periods of time. This would include recreational areas, workshops, and kitchens.

How They Work

Fluorescent lights and incandescent lights operate on entirely different principles. All fluorescent lamps contain a small filament (similar to the filament of an incandescent lamp) at each end. The glass tube is filled with a gas (mercury vapor) and the inner surface of the tube is coated with a phosphorescent substance.

When the current is turned on, power is applied to the filaments, causing them to heat up. The hot filaments vaporize the gas in the tube, making it a good conductor of electricity. The filaments are then turned off and a high-voltage surge of power is momentarily applied to the tube. The surge starts current flowing through the tube. Once the current flow is established, it continues with only normal line voltage applied. In fact, current flows so easily in the vaporized gas that it must be limited by a device called a ballast.

The flow of current through the gas produces ultraviolet light. Although it is barely visible to human eyes, the ultraviolet light causes the phosphorescent coating on the tube to emit strong and visible light.

The ballast is an inductive device similar to a transformer. It produces the high-voltage surge necessary to start current flowing in the fluorescent tube. Once the current flow has been established, the ballast limits the current through the tube to the rated value. You will recall from Chapter 3 that when current flows through an inductive device a voltage is induced that opposes changes in current flow. As the alternating current is constantly changing, the ballast continually opposes this change, thus limiting current flow. The limiting action is required because, when current is flowing in a fluorescent lamp, the lamp's internal resistance drops to a low value. If not limited, the current would destroy the lamp in a short time.

The starter found in some older fluorescent fixtures is a small metal canister that fits in a socket on the fixture. The starter does the switching needed to turn the filaments on and off, apply the high-voltage surge to start the lamp, and switch in the ballast to limit the current.

Three Most Common Types

RAPID-START • (Fig. 9-32). This is currently the most widely used type of fixture. It lights almost immediately when switched on. Rapid-start fixtures also have the advantage of being readily adapted for use with fluorescent dimmer switches. Lamps for rapid-start fixtures have two pin-type connectors at each end.

INSTANT-START • (Fig. 9-33). This type of fluorescent fixture lights a second or two after it is switched on. It requires a higher initial voltage surge than other types. Lamps for instant-start fixtures have one pin-type connector at each end. On this type of fixture, the lamp holder contains a built-in switch that allows high voltage to be applied only when the fixture contains a lamp.

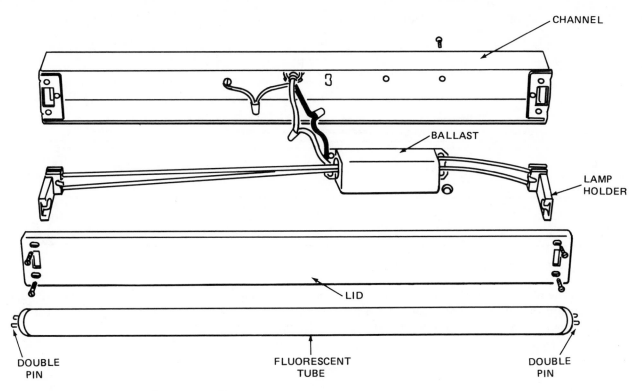

Figure 9-32. Rapid-start fluorescent fixture.

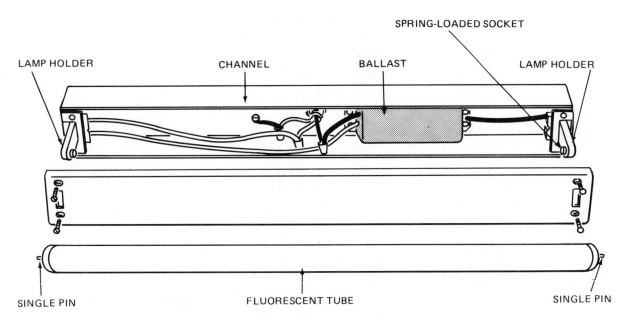

Figure 9-33. Instant-start fluorescent fixture.

STARTER-TYPE • (Fig. 9-34). This type of fixture has a separate starter. It uses lamps that have two pin-type connectors at each end. The starter, like the lamps, has a specified life, but is replaceable. It is located in a socket near one of the lamp holders. For replacement, power to the fixture is turned off, and the lamp is removed. Then the starter is twisted and pulled out of its socket. Replacement starters must match the wattage of the lamp.

Fixture Installation

Fluorescent fixtures consist of a metal channel with one or more sockets mounted at each end. The channel provides a mounting place for the ballast. The channel can be readily disassembled by removing a sheet metal cover. The back of the channel has knockouts at several locations to provide different types of mounting. To mount the fixture on a wall or ceiling box containing a

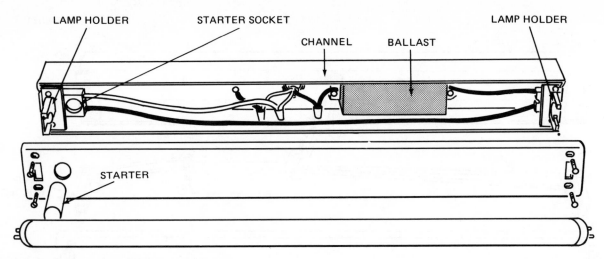

Figure 9-34. Starter-type fluorescent fixture.

nipple, thread the fixture wires through a washer and locknut, through the channel knockout, and then through the nipple (Fig. 9-35). Make electrical con- nections. Place the fixture over the nipple, put the washer on the nipple, and secure the fixture with the locknut.

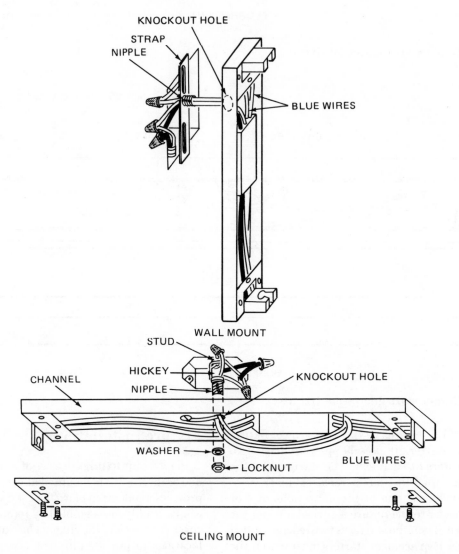

Figure 9-35. Small fluorescent fixture mounting.

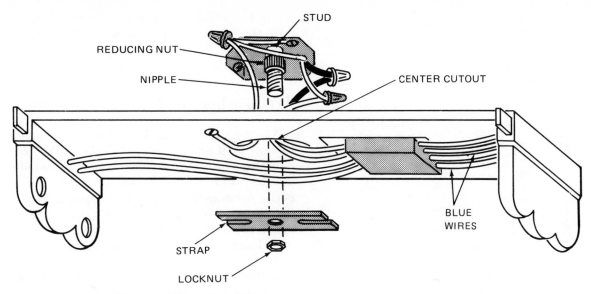

Figure 9-36. Large fluorescent fixture mounting.

Larger fluorescent fixtures have a mounting cutout. The cutout is large enough to allow electrical connections to be made without running the fixture wires through the nipple. When connections are made, use a strap to span the cutout and secure the strap with a locknut (Fig. 9-36).

Circular fluorescent fixtures can be mounted on ceiling boxes in which a mounting stud, hickey, and nipple have been installed. When electrical connections have been made, the fluorescent base is placed over the nipple and secured with a capnut or collar (Fig. 9-37).

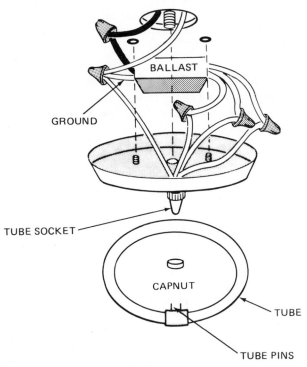

Figure 9-37. Circular fluorescent fixture mounting.

Some small circular fixtures are designed to mount directly to the threaded tabs on the ceiling box with two machine screws.

Dimmer Installation

Adding control of illumination level to fluorescent lighting is more complex than it is for incandescent lighting. All parts of the fluorescent lighting system must be matched to make the dimming control work properly. The ballasts in the fluorescent fixtures must be replaced with special dimming ballasts, rapid-start lamps and fixtures must be used, and the dimmer must be suitable for the load. One model dimmer can handle two through ten 40-watt rapid-start lamps. A different group of models is made to handle twelve through forty 40-watt lamps. For still larger numbers of lamps, commercial and industrial dimmers are made. These require special wiring and are not suitable for residential use.

Installation of fluorescent dimmer switches is mechanically the same as installation of incandescent dimmers. Fluorescent dimmers fit into standard wall boxes and have two black pigtail leads for connection in the hot side of the power line to the fixture (Fig. 9-38).

Manufacturers of fluorescent dimmers specify the ballasts that can be used with their controls. All ballasts in the fixtures to be controlled must be removed and new ballasts installed. It is generally recommended that all dimming ballasts used be the same make and model. Dimming ballasts have the same number of leads as regular ballasts, but there may be a difference in color coding. Follow the manufacturer's instructions to install dimming ballasts in the fixtures.

For even illumination, all lamps on a dimming circuit should be of the same type, color, and age. New

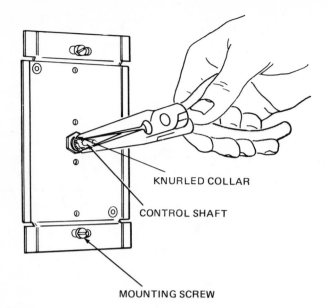

KNURLED COLLAR

CONTROL SHAFT

MOUNTING SCREW

Figure 9-38. Fluorescent dimmer adjustment.

lamps should be operated at full brightness for 100 hours before the illumination level is reduced.

When all ballasts have been replaced, proper lamps installed, and the dimmer wired and mounted, the dimmer must be adjusted. The adjustment is made to keep some illumination at the low end of the illumination range. Adjustment may be made by setting a screwdriver control, or rotating a knurled collar on the control shaft. The adjustment procedure is the same for both.

Step 1. Rotate the control shaft to the highest (brightest) setting.

Step 2. Rotate the adjusting screw or collar fully clockwise.

Step 3. Rotate the control shaft to the lowest setting that still keeps the lamps on.

Step 4. Adjust the screw or collar counterclockwise until the lamps begin to flicker. Then turn back clockwise until the flickering stops. This is the correct adjustment.

Troubleshooting Fluorescent Fixtures

A wider variety of faults can occur in fluorescent fixtures than in incandescent fixtures. Troubles can range from incorrect installation of lamps to failure of a ballast transformer. Of course, as in any electrical device, opens and shorts can also happen. Inspect faulty fixtures carefully for evidence of wiring faults before replacing major parts such as the ballast. Some common fluorescent troubles are described below.

INCORRECT LAMP INSTALLATION • Fluorescent lamps have marks on the end caps to indicate proper positioning in the fixture (Fig. 9-39). A common fault is not rotating the lamp enough to engage both pins properly. If lamps do not light, check lamp installation.

EXCESSIVE NOISE • All fluorescent ballasts generate a hum which can range from barely audible to annoyingly loud. The sound rating should be marked on fixture cartons. It can range from a rating of A (low) to F (high). Fluorescents used in home living areas should have A or B sound ratings. Workshops or utility rooms may have C or D ratings. Excessive ballast hum can occur if the ballast is loose in the fixture or is overheating. Check the ballast mounting and check the marking on the ballast to make certain it is matched to the type of fixture and wattage of lamps. Ballast overheating can happen if airflow around the fixture is restricted.

LAMP DISCOLORATION • Blackening at the ends of the lamp is normal as the tube ages and will be most noticeable near the end of lamp life. Spots due to

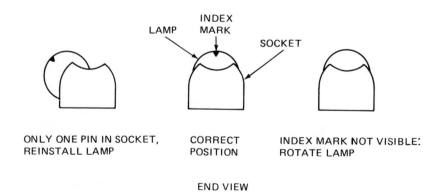

Figure 9-39. Fluorescent installation.

mercury condensation may show up on new or old lamps. Usually the dark spot will be near the center of the lamp, but it can occur anywhere. These spots have no effect on lamp performance (Fig. 9-40).

LAMP BLINKS ON AND OFF • This often happens in new lamps and will stop after a short period of operation. Low line voltage can also cause flickering. Measure the line voltage. If it is low, check with the utility company for the cause. Fluorescent lamps are sensitive to low temperatures (below 50°F, 10°C). If the temperature condition is permanent, a special low-temperature ballast can be installed.

SHORT LAMP LIFE • Frequent on-off operation shortens lamp life. In some two-lamp fixtures, failure of one lamp will shorten the life of the other lamp. In rapid-start and instant-start fixtures, burned-out lamps should be replaced promptly.

ACRID ODOR OR SMOKE FROM FIXTURE • These symptoms clearly indicate ballast failure. Check the wiring. If the ballast is correctly wired, and is overheating, it has reached the end of its useful life and should be replaced.

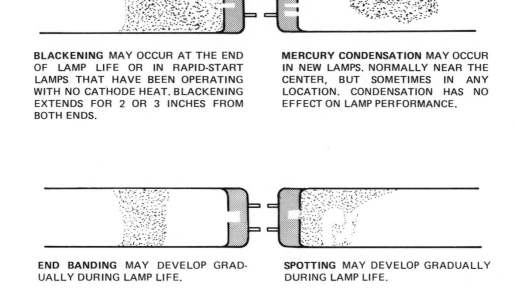

BLACKENING MAY OCCUR AT THE END OF LAMP LIFE OR IN RAPID-START LAMPS THAT HAVE BEEN OPERATING WITH NO CATHODE HEAT. BLACKENING EXTENDS FOR 2 OR 3 INCHES FROM BOTH ENDS.

MERCURY CONDENSATION MAY OCCUR IN NEW LAMPS. NORMALLY NEAR THE CENTER, BUT SOMETIMES IN ANY LOCATION. CONDENSATION HAS NO EFFECT ON LAMP PERFORMANCE.

END BANDING MAY DEVELOP GRADUALLY DURING LAMP LIFE.

SPOTTING MAY DEVELOP GRADUALLY DURING LAMP LIFE.

Figure 9-40. Fluorescent lamp discoloration.

• REVIEW QUESTIONS •

1. Identify the switches diagramed below in terms of poles and throws.

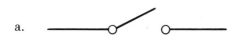
a.

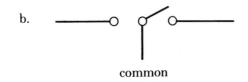
b.

common

2. Why do some toggle switches not have on and off positions marked on them?

3. What is the disadvantage of ac-dc toggle switches?

4. What areas of a house are suitable for time delay switches?

5. What two types of switches have illuminated handles? When is the handle illuminated in each type?

6. Why must mercury switches be mounted vertically?

7. A three-prong plug mates with 120-volt 15- and 20-ampere grounding receptacles. The grounding prong on the plug is longer than the others. Why?

8. What does the marking 10A 125V - 5A 250V on a switch mean?

9. What does the marking CO-ALR mean on a 15-20 ampere switch or receptacle?

10. What two functions are performed by the ballast in fluorescent fixtures?

11. What article in the NEC covers general use ac-only snap switches?

10
OVERCURRENT
PROTECTION

• INTRODUCTION •

An essential part of every circuit in an electrical installation is a device to provide what the NEC describes as overcurrent protection. All parts of an electrical system are designed to operate safely within specified limits of current and voltage. Usually maximum limits are specified. Any condition that causes these limits to be exceeded means trouble. The troubles can range from the minor inconvenience of flickering lights or an unstable TV picture to serious fire and shock hazards.

The devices which protect against overcurrent conditions are fuses and circuit breakers. Shock hazard can also exist without an overcurrent condition. Devices known as ground fault circuit interrupters protect against the severe shock that can result from this form of abnormal current flow.

Every circuit in an electrical system must have overcurrent protection. It is convenient and logical, and an NEC requirement, to install this protection at the point where electrical power is brought into a building. A metal enclosure, called a service panel, provides a place to divide incoming power to supply the individual circuits in the building and to group the overcurrent protection devices.

The service panel is part of a section of the electrical installation known as the service entrance. The service entrance is discussed in Chapter 11. This chapter deals with the types and sizes of overcurrent devices that are used in service panels. The general rules for overcurrent protection as specified by the NEC and local codes are summarized.

The types and uses of ground fault circuit interrupters are also covered. These devices—usually abbreviated GFCI—are relatively new in electrical wiring. They provide protection against a source of shock to which people are increasingly exposed. They are presently required in several parts of an electrical installation. Their use seems certain to be expanded in future revisions of the NEC.

• CAUSES OF OVERCURRENT CONDITION •

Protection against overcurrent conditions in a circuit consists, simply, of sensing excess current and turning off power to the affected circuit when more than the rated current flows. Excessive current flow can occur in a circuit under three general conditions. One of these excess current conditions is normal; the other two are not.

The normal condition of excess current is the surge that occurs when some appliance, especially a motor-driven appliance, is turned on. For the first few seconds after turnon, motors such as those used in refrigerators, freezers, dishwashers, and washing machines can draw from six to ten times the current they will draw when they reach normal running speed (Fig. 10-1). Large lighted areas—such as luminous ceilings—also allow a high initial current to flow. A normal surge can also happen if several high-current appliances are turned on simultaneously. These conditions are normal and present no danger, if the duration of the surge is brief and if the circuit was not operating close to an overcurrent condition before the surge occurred. Overcurrent protection, then, should be designed to tolerate short-term, moderate current surges without turning off the circuit.

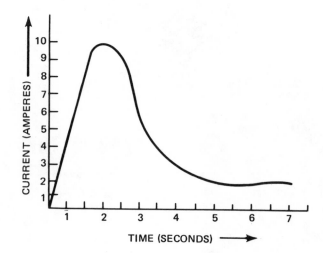

Figure 10-1. Motor starting current curve.

One condition of abnormal overcurrent results from connecting too large a current load to a circuit. All devices connected to a circuit are in parallel with all other devices on the same circuit. We know from our study of parallel circuits that the more loads (resistors) that are wired in parallel, the lower the effective resistance across the line becomes. With voltage constant, lowered resistance means increased current flow (Fig. 10-2). If this type of overload condition occurs only rarely, perhaps when some device such as an electric heater is used, the condition is not serious and no major corrective action is necessary. If this overcurrent condition happens frequently, however, a need for additional circuits is indicated.

The third cause of overcurrent is potentially the most dangerous. This is the sudden occurrence of a low-resistance path between the hot wire and ground. This, of course, is what is known as a short circuit (Fig. 10-3). Current flow in a short circuit can reach values of 10,000 amperes or more. The effect of lowered resistance on current flow can be seen from Ohm's law.

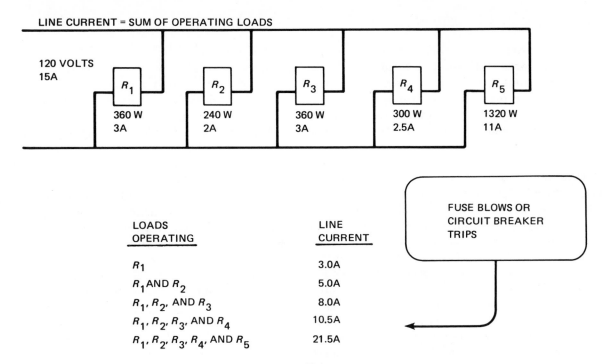

LINE CURRENT = SUM OF OPERATING LOADS

120 VOLTS
15A

R_1 — 360 W, 3A
R_2 — 240 W, 2A
R_3 — 360 W, 3A
R_4 — 300 W, 2.5A
R_5 — 1320 W, 11A

LOADS OPERATING	LINE CURRENT
R_1	3.0A
R_1 AND R_2	5.0A
R_1, R_2, AND R_3	8.0A
R_1, R_2, R_3, AND R_4	10.5A
R_1, R_2, R_3, R_4, AND R_5	21.5A

FUSE BLOWS OR CIRCUIT BREAKER TRIPS

Figure 10-2. Too many loads can cause an overcurrent condition.

The current flow I in a circuit is equal to V/R. Note that as R becomes smaller, the value of I becomes greater. (For example, $120/10 = 12$ amperes, $120/1 = 120$ amperes, $120/.01 = 12,000$ amperes, etc.) This huge current flow generates extremely high temperatures, melts metal so rapidly it almost explodes, and vaporizes many plastic materials. Effective protection against short circuits, then, requires rapid turnoff, before current flow can reach the levels where serious damage and fire can result. Note that a short circuit can occur when a low-resistance path exists between a hot (red- or black-insulated) wire and either the power-ground (the white- or gray-insulated) wire or any point connected to the grounding conductor. This includes the bare or green-insulated wire, a cold water pipe, or any conductive material connected to these points. When the short occurs between a hot wire and some grounding point, the surge of current flows from the hot wire to ground (Fig. 10-4). There is no abnormal current flow in the white- or gray-insulated wire. Overcurrent protection, then, must be located in the hot wire. This location agrees with the rule that the white- or gray-insulated line must never contain any means of interrupting the flow of current.

EQUAL SAFE CURRENT FLOW IN BOTH LINES

NORMAL LOAD (HIGH RESISTANCE)

EQUAL UNSAFE CURRENT FLOW IN BOTH LINES

SHORT CIRCUIT (LOW RESISTANCE)

Figure 10-3. Power line short circuit.

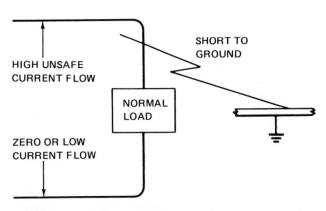

HIGH UNSAFE CURRENT FLOW

NORMAL LOAD

SHORT TO GROUND

ZERO OR LOW CURRENT FLOW

Figure 10-4. Short circuit to ground.

Overcurrent and Overload

The words *overcurrent* and *overload* are often used interchangeably, but they really have different meanings. Overcurrent describes a condition in which a circuit is carrying more than its rated current flow. As we have seen, the seriousness of an overcurrent condition depends on the amount of excess current flowing and the period of time the overcurrent continues.

Overload applies most often to the operation of electrical motors. Motors require heavy current flow when starting and much less when they reach normal running speed. Fuses and circuit breakers must be designed to carry this starting current without turning off the circuit. If a motor has some mechanical fault that increases the turning friction or if the load on the motor is too heavy, the high-current condition will continue. This is an overload. An overload will cause a current flow below the cutoff level of the fuse or circuit breaker but above the safe level for the motor. Under overload conditions, motors can get dangerously hot and motor windings can burn out. The NEC, therefore, requires special overload protection for most large (1 hp or more) motors (Fig. 10-5). Portable motors of 1 horsepower or less that can be plugged into 15- or 20-ampere receptacles are an exception and do not need separate overload protection. Other, smaller motors that are used in major appliances generally have built-in overload protection consisting of a heat-sensing element that shuts off the motor if the temperature becomes too high. These motors are marked THERMALLY PROTECTED on the nameplate. The heat-sensing element in these motors can be reset after the motor has cooled. The motor can then be started again. This should be done only after the cause of the overload has been determined and the problem corrected. In the great majority of cases the overload will be found to result from worn bearings, either in the motor or in the device being driven. Worn bearings cause binding on drive shafts and gearing. The resulting extra load causes the motor to overheat.

Large motors require special starting devices called controllers and have special wiring requirements. This subject is beyond the scope of a text on residential wiring; however, it is covered in NEC Article 430.

• FUSES •

Fuses provide overcurrent protection by adding a strip of metal in series with the hot wire of a circuit. The strip of metal has a low melting point. The size of the metal strip determines how much current it can carry before heating to the melting point. This is the ampere rating of the fuse. The rated current can flow through the metal strip indefinitely. When a greater amount of current flows, the strip becomes hot and melts. This opens the circuit.

Fuse Characteristics

The current that can flow through a fuse without causing the strip to melt is the ampere rating of the fuse. This is the most important rating in selecting fuses for residential circuits. However, fuses do have other ratings that an electrician should be familiar with. One of these is the fuse voltage rating. The voltage rating of a fuse must be equal to or greater than the voltage of the circuit protected by the fuse. The voltage rating is the ability of a fuse to extinguish the arc that occurs as the fuse melts, and to maintain an open circuit after the element melts by preventing arcing across the open space in the element. Standard fuse ratings are 600 volts, 300 volts, 250 volts, and 125 volts. Fuses having higher voltage ratings than the circuit voltage can be used, but lower ratings should not be used.

Another measure of fuse performance is called the interrupting rating. As previously noted, when a short circuit occurs, current flow can be hundreds or even thousands of times greater than normal current. The fuse must be able to react to this surge of current and operate properly to open the circuit. The maximum short circuit current that can flow at the point where power lines enter a building is determined by the maximum current that can be drawn from the utility lines. This, in turn, is determined by the characteristics of the pole transformer. The pole transformer short circuit current rating can be obtained from the utility company. Typical short circuit current from utiltiy transformers ranges from 25,000 to 75,000 amperes. Most fuses listed by Underwriters' Laboratories have interrupting ratings equal to or greater than these values. The rating is abbreviated AIC, for amperes, interrupting current. The maximum short circuit current at other points in a residential electrical installation is about 10,000 amperes.

The NEC requires that all overcurrent protective devices be capable of opening a circuit before extensive damage can be done to circuit components. In alternating current circuits the greatest surge of current occurs on the next half cycle after the short occurs. To protect other circuit components, fuses should open the circuit in less than one-half cycle. Many manufacturers

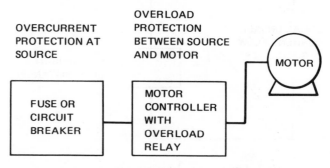

Figure 10-5. Overload protection.

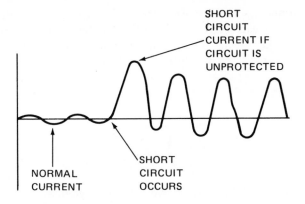

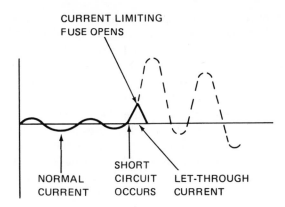

Figure 10-6. Fuse let-through current curve.

specify the peak let-through current in amperes or the time to shutoff in fractions of a cycle (Fig. 10-6). Less than one-quarter cycle or approximately 4/1000 of a second (4 milliseconds) will limit short circuit current to acceptable amounts.

Fuse Types

The strip of metal that melts—the fuse element—is enclosed in several different types of casing for insertion into the circuit.

PLUG FUSES • These fuses have a screw shell base (known as an "Edison base"), like a standard incandescent lamp (Fig. 10-7). They are inserted in a circuit by screwing them into a socket. The top of the plug has a transparent cover, making the fuse element visible. A fuse that is open (a "blown" fuse) can be located in a fuse panel by discoloration or clouding of this window (Fig. 10-8). Plug fuses used in residences are rated at 15, 20, 25, and 30 amperes. They can be used on any existing installation in which the voltage to ground does not exceed 150 volts. (New panel installations must use type S fuses, described below.) These fuses can, therefore, be used on 120/240-volt service because the voltage *to ground* is 120 volts.

Time Delay Fuses. Two variations of the standard plug fuse are frequently used. One is a time delay fuse, designed to carry a temporary overload without blowing (Fig. 10-9). This fuse has a standard fuse element that will open quickly on short circuit current surges. To provide a time delay on temporary overloads, the end of the element is embedded in a small

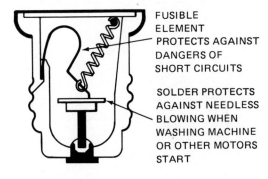

Figure 10-8. Reading blown plug fuses.

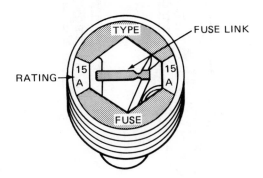

Figure 10-7. Plug fuse.

FUSIBLE ELEMENT PROTECTS AGAINST DANGERS OF SHORT CIRCUITS

SOLDER PROTECTS AGAINST NEEDLESS BLOWING WHEN WASHING MACHINE OR OTHER MOTORS START

Figure 10-9. Time delay fuse cross section.

block of solder. A coil spring is linked to the fuse element at the point where it is joined to the solder block. The spring is under tension and is secured to the fuse housing. The melting point of the solder determines the length of time the fuse will carry an overload, and the amount of the overload. When the solder becomes hot enough to soften, the spring pulls the fuse element free, opening the circuit. Time delay fuses are especially useful on circuits serving large, motor-driven appliances. The time delay fuse will not blow during the period of heavy starting current drawn by the motor.

Type S (Nontamperable) Fuses. These fuses have the time delay feature just described, but have, in addition, a mechanical design that prevents large-size fuses from being inserted in circuits designed for lower ratings (Fig. 10-10). Type S fuses can be substituted for standard plug fuses in any fuse panel. Each type S fuse has a special adapter that will accept only a fuse of equal rating. The adapter is inserted into the fuse holder and is so designed that it cannot be removed, once it is inserted. With the adapter in place, only one size fuse can be inserted in that fuse holder. This prevents accidental or deliberate insertion of a higher rated fuse on any circuit. Type S fuses are made in 15-, 20-, and 30-ampere sizes. Both 25- and 30-ampere fuses can be used with the 30-ampere adapter.

CARTRIDGE FUSES ● Cartridge fuses are manufactured with the same ratings and features as plug fuses, but in addition, are made in sizes designed to handle much higher current. Cartridge fuses are the only type available for circuits rated over 30 amperes. Cartridge fuses for 30- to 60-ampere circuits have ferrule contacts (Fig. 10-11). Above 60 amperes the fuses have knife-blade contacts (Fig. 10-12). Cartridge fuses with a time delay feature are available at all

amperage ratings (Fig. 10-13). The length and diameter of cartridge fuses increases in steps with the amperage rating. This limits, but does not completely eliminate, the possibility of replacing a fuse with one of the wrong size. Make certain blown fuses are replaced with new fuses of the proper value.

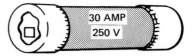

Figure 10-11. Ferrule contact cartridge fuse.

Figure 10-12. Knife-blade contact cartridge fuse.

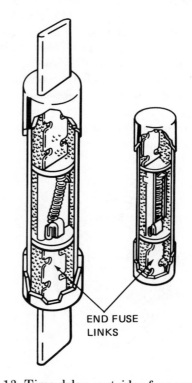

Figure 10-13. Time delay cartridge fuse cross section.

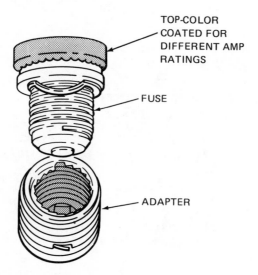

TOP-COLOR COATED FOR DIFFERENT AMP RATINGS

FUSE

ADAPTER

Figure 10-10. Type S fuse.

The cartridge fuses used in residential wiring provide no visible evidence of a melted fuse element, as plug fuses do. Cartridge fuses must be checked with a continuity tester or ohmmeter (Fig. 10-14). To make the check you must remove the suspected fuse from the fuse panel. Always use a fuse puller to remove fuses from the panel (Fig. 10-15). If the fuse has recently been carrying heavy current, or if it has just blown, it may be hot enough to cause a painful burn.

Some cartridge fuses can be reused by installing a new fuse element in the fuse cylinder. Disassemble the fuse by unscrewing an end cap. Remove the remains of the blown link and insert a renewable link. It is important to tighten the mounting screws firmly when the new link is in place (Fig. 10-16). Looseness in the link mounting can cause overheating at the terminal.

To summarize, fuses are a simple, highly reliable, and inexpensive way of providing overcurrent protection. Fuses have no mechanical parts to fail. They do not age or wear out. The only important shortcoming fuses have is the amount of time and effort necessary to replace them when they blow.

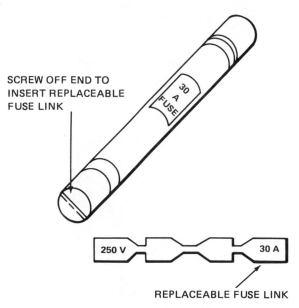

Figure 10-16. Replaceable link cartridge fuses.

• CIRCUIT BREAKERS •

Circuit breakers combine the functions of a switch and a fuse in a single device. They provide overcurrent protection as a fuse does and, in addition, provide a means of turning power to the circuit on and off. When installed in a circuit breaker service panel, they look much like ordinary toggle switches or pushbutton switches (Fig. 10-17).

Circuit breakers are available in ratings of 15 to 200 amperes for residential use. Larger sizes are made for commercial and industrial applications. Circuit breakers, like fuses, are also rated for voltage and interrupting current.

The internal mechanism of circuit breakers consists of a bimetallic strip and spring-loaded contacts (Fig. 10-18). The bimetallic strip is made of two different types of metal—such as steel and bronze—fused together. The strip acts as a latch to hold the contacts together. When more than the rated current flows through the circuit breaker, the heat makes the two metals expand, by different amounts and at different rates, causing the fused strip to bend (Fig. 10-19). The spring-loaded contacts are released and current flow is interrupted.

The spring-loaded contacts can also be opened by moving the toggle handle to the off position. The bimetallic shutoff requires some time to heat up and trip the circuit breaker to off. This provides a time delay feature. Most circuit breakers will carry one and one-half times their rated current for about 1 minute and as much as three times their rated current for 5 seconds. This provides enough delay to allow a motor-driven appliance to be used without tripping the circuit breaker.

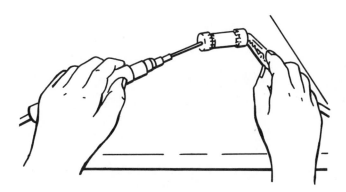

Figure 10-14. Checking a cartridge fuse.

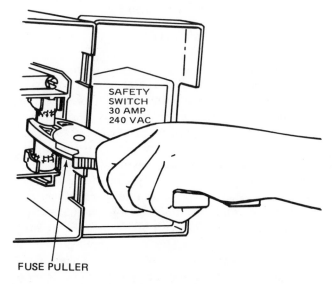

Figure 10-15. Using a fuse puller.

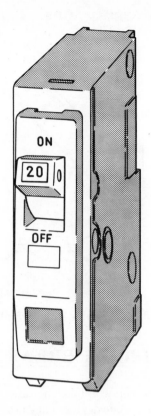

Figure 10-17. Types of circuit breakers.

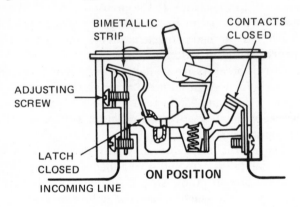

Figure 10-19. Bimetallic strip.

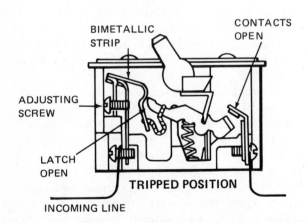

Figure 10-18. Circuit breaker cross section.

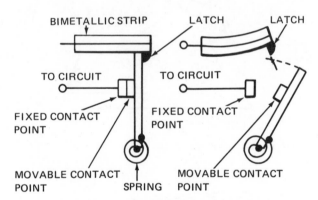

The handle on some toggle-type circuit breakers has four positions (Fig. 10-20). Under normal conditions the handle is up. When an overcurrent condition occurs, the handle moves to a midpoint position. To reset, the handle must be moved down as far as possible and then back up to the on position. When manually switched off, the handle must be moved past the midpoint to the off position. Other types of circuit breakers have only two positions and can be operated just as a toggle switch is for both resetting and manual switching. The NEC requires circuit breakers to show clearly whether they are on or off. Toggle types do this by marking the positions. Push-button types have an on-off indicator visible through an opening on the front of the circuit breaker.

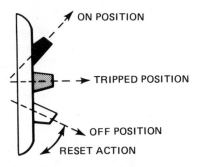

Figure 10-20. Circuit breaker handle positions.

• GROUND FAULT CIRCUIT PROTECTION •

Strictly speaking, ground fault protection is not *overcurrent* protection. Ground faults can cause leakage paths for quite small amounts of current and yet can be extremely dangerous.

A short circuit is defined as a low- (or zero-) resistance path for current flow. The path can be between the hot wire and the power-ground line or between the hot wire and any grounded point. A short circuit, however, need not be a low-resistance path. High-resistance shorts can also occur. These are generally referred to as leakage paths because the amount of current is small. Recall from the chapter on safety that heavy current flow is not required for a shock to be fatal. As little as 1/10 ampere (100 milliamperes) can cause a fatal shock. This can happen, however, only if the victim is in contact with ground or a grounded conductor.

For example, if an electric hair dryer has a break in insulation near the plug on the power cord, current can flow from the break to any ground point. In bathrooms, many exposed points such as water faucets, metal wash basins, and, often, decorative metal trim are all possible ground points. A person making contact between any of these points and the wire exposed at the break may receive a shock (Fig. 10-21). The severity of a shock depends as much on the length of time the shock current flows as on the amount of current. A shock of 20 to 30 milliamperes, while below the fatal level, can cause muscles to freeze, so the victim cannot let go of the live conductor. This small increase in current flow is not sufficient for a circuit breaker to trip or a fuse to blow. Under these conditions the flow of current will continue and perhaps cause serious injury.

Devices known as ground fault circuit interrupters (abbreviated GFCI) have been developed to protect against this type of shock hazard. In new construction, the NEC requires that GFCI protection be installed on all 120-volt 15- and 20-ampere receptacle circuits outdoors and in bathrooms and garages. There are also special GFCI requirements in the Code for swimming pools, fountains, marinas, boatyards, and recreational vehicles.

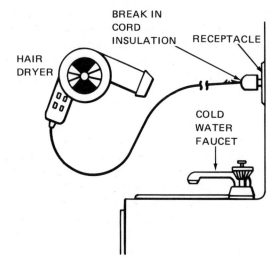

Figure 10-21. Appliance ground fault.

Characteristics of Ground Faults

To understand how GFCI's protect, let us consider what happens in a circuit when a high-resistance short to ground (known as a ground fault) occurs. Under normal conditions the current flow in any two-wire circuit is exactly the same in the hot wire and the white- or gray-insulated power ground. When a ground fault happens, current flow can follow two paths, power ground and the fault path. As in any parallel resistance circuit, the current flow will divide, with the heavier current flowing through the lower resistance and vice versa. In the example previously described, when the user provided a path for current flow from the insulation break to a ground point, current could flow through the device (the hair dryer) to power ground and also through the fault path (the user) to the grounded point (Fig. 10-22). This second path required

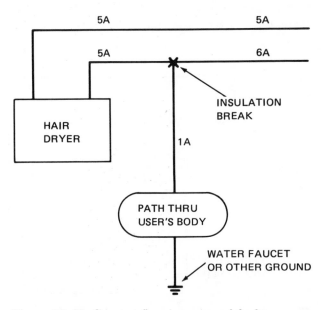

Figure 10-22. Current flow in a ground fault.

an increase in the hot wire current flow. The current flow in the hot wire and the power ground wire were then no longer equal. This unbalance in current flow that results from a ground fault is sensed by the GFCI, and causes the circuit to be turned off before serious injury can occur.

How GFCI's Work

GFCI's contain a differential transformer, a sensing and test module, and a magnetic switch (Fig. 10-23). The differential transformer consists of a circular iron-core secondary winding. The circuit power conductors act as the primary of the differential transformer. These conductors pass through the center of the circular core. When current flows in a conductor, a magnetic field is created around the conductor, the strength of the field being proportional to the current flow. When the current in both conductors is equal, the field around each conductor is equal and opposite. The fields cancel and no current flows in the secondary. When the current in the hot wire becomes greater than the current in the ground line, the field of the hot wire becomes greater and current is induced in the differential secondary. This secondary output is sensed and amplified by the circuit-sensing and test module. The module output activates a magnetic switch that cuts off power to the load. An unbalance of as little as 4 to 6 milliamperes will cause cutoff. GFCI's also contain a test circuit to check operation of the module and the switch.

GFCI Types

There are three basic types of GFCI's, all having some common features. A test button is provided on GFCI's to simulate a leakage condition and to check that proper turnoff occurs. A reset control restores current flow after a test shutoff or an actual ground

fault shutoff. Some visible indication of shutoff is provided by a fault indicator light or by the position of the reset control. Manufacturers generally recommend that GFCI's be tested after installation, and thereafter once a month. GFCI's are manufactured with various current and voltage ratings. The rating of the GFCI must match the circuit on which it is installed.

The simplest form of ground fault protection requires no special installation. It consists of a small rectangular unit with standard plug prongs on the back (Fig. 10-24). It is necessary only to plug the unit into a receptacle. The front of the unit contains the test and reset buttons and either one or two 3-prong receptacles. Any appliance plugged into the receptacle has ground fault protection. This unit has no effect on any other receptacle or device on the circuit. These GFCI's are available for both two-wire 120-volt and three-wire 120/240-volt circuits in current ratings up to 30 amperes. They have the advantage of simplicity and portability, but if many receptacles require ground fault protection, use of these units at every receptacle would be quite costly.

To provide ground fault protection for several receptacles on the same circuit more economically, another type of GFCI unit is installed in an electrical box in place of a standard receptacle (Fig. 10-25). It provides ground fault protection, not only for the devices

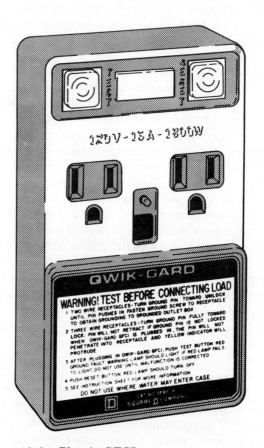

Figure 10-24. Plug-in GFCI.

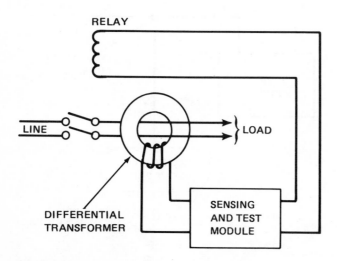

Figure 10-23. GFCI circuit.

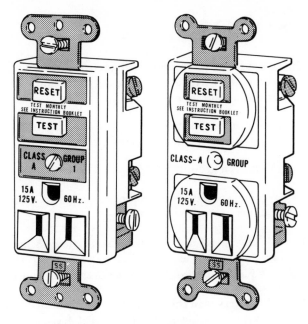

Figure 10-25. Receptacle-type GFCI.

plugged into it, but for all devices plugged into receptacles between it and the end of the branch circuit. These receptacles are known as "feed-through" units. Details on choosing a location and wiring these units are provided in Chapter 13. These GFCI's fit into any electrical box 1 1/2 inches or more deep. They are available for two-wire 120-volt circuits of 15 or 20 amperes.

For complete ground fault protection on any circuit, combination units are made which include both circuit breaker overcurrent protection and ground fault protection in a single unit. These units fit into circuit breaker service panels (Fig. 10-26). (Service panels are described in the following chapter.) These GFCI's have a white pigtail lead that must be connected to a neutral terminal in the service panel. The circuit power-ground conductor must also be connected to the unit. Combination GFCI's are manufactured for two-wire 120-volt and three-wire 120/240-volt circuits in 15- to 30-ampere ratings.

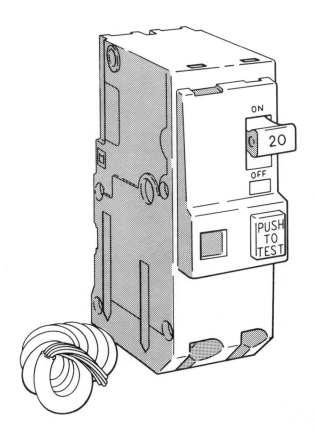

Figure 10-26. Service panel combination unit.

• REVIEW QUESTIONS •

1. Why is it desirable for overcurrent protection devices to allow short term, moderate overloads without cutting off power to a circuit?

2. Why is it necessary for overcurrent protection devices to shut off power to a circuit rapidly when extremely large current surges occur?

3. What is the main difference between an overcurrent condition and an overload condition?

4. In addition to a current rating, fuses have a voltage rating. Why?

5. What special advantage do plug fuses have over cartridge fuses?

6. What is the name of the fuse type that provides the characteristics referred to in question 1?

7. What is the advantage of using type S fuses?

8. Circuit breakers provide overcurrent protection combined with another necessary function. What is the second function?

9. What indicates whether or not a circuit breaker has tripped, turning off power to the circuit?

10. What is a ground fault?

11. Why is special protection needed against ground faults?

12. GFCI's sense unbalanced current flow in a hot line and a power ground. What two basic electrical principles are used to sense this unbalance?

13. What article in the NEC specifies the locations in homes (dwelling units) that must have ground fault circuit protection?

11
SERVICE ENTRANCE

The place where the utility company lines are tapped for connection to a building's electrical system is called the service entrance. The service entrance has four principal parts. Part one is the *service head*. This is the point where the power lines from the pole are attached to the building. When overhead lines are used, this point must be high enough above ground to provide adequate clearance to pedestrians and vehicles, and to satisfy the requirements of local codes. This connection point must also be able to withstand wind and weather. In some areas the lines from the pole to the building are buried underground. This reduces the problems of wind and weather, but requires special protection against moisture. The lines that run from a pole to a building are called the service drop when run overhead, and the service lateral when run underground.

The generation and distribution of electrical power is a business. Customers must pay for the power they use. For billing purposes, then, the power consumed must be measured. The second item in a service entrance is a *meter* to record power used over a period of time.

For safety and servicing convenience, a way must be provided to turn off all electrical power in a building. Part three of the service entrance is a *main disconnect switch*, installed for this purpose.

The cable, switches, receptacles, fixtures, and appliances that make up an electrical system are wired in groups called circuits. The incoming power must be divided to provide appropriate amounts of energy to each circuit. This splitting of power is done by the fourth element in the service entrance, the *service panel*. This is the point where all the individual circuits originate. The service panel also is the place where overcurrent protective devices are installed to guard against shock and fire. In many respects, the service entrance is the most important part of an electrical installation; it is the source of all power in the building and the point where automatic safety shutoff occurs.

The installation of the service drop from pole to building is usually the responsibility of the local utility company. All the remaining service entrance installation is done by the electricians wiring the building. The types and ratings of service entrance equipment are determined by local codes, utility company regulations, and the expected level of power that will be used by the building residents.

This chapter tells you what things must be considered when each part of the service entrance is installed, what function the part has, and what calculations are made when service entrances are planned.

In Chapter 2 we learned that transformers can be used to change the levels of voltage and current in ac circuits. Step-up transformers raise voltage and reduce current. Step-down transformers reverse the process; voltage is lower and current is increased. The power (voltage times current) remains about the same in both the primary and secondary windings of step-up and step-down transformers. It is the relationship of voltage and current that changes. Note that the power induced in transformer secondary windings is approximately equal to the power in the primary winding, but not exactly equal to it. Transformers, like other electrical devices, are not 100 percent efficient; some power loss occurs. Present-day transformers are, however, among the most efficient electrical devices, so the small losses are more than offset by the transmission advantages of stepping up and stepping down voltage levels.

Electric utility power generating stations produce power for large areas. The power generated must be transmitted over long lines to reach the end user. We know that all conductors offer some resistance to current flow. This resistance causes a drop in voltage, known as the *IR* drop, across transmission lines. A simple example will show how this line loss can be reduced by transformer action. A 100-watt generator has an output of 100 volts at 1 ampere. If this power is transmitted a distance of 2 miles over two conductors having a resistance of 5 ohms per mile, the total line resistance is 20 ohms. This results in an *IR* drop of 20 volts. The power available at the end of the line is 80 volts at 1 ampere, or 80 watts. Twenty watts have been wasted in transmission. If we apply the generator output to a 10:1 step-up transformer, the power in the transformer secondary is 1000 volts at 0.1 ampere. When carried over the same lines, power in this form results in an *IR* drop of 2 volts (20 ohms × 0.1 ampere). The power available at the end of the line is 998 volts at 0.1 ampere, or 99.8 watts. Only 0.2 watt of power has been lost.

To hold transmission power losses to the lowest practical level, utilities generate power in ac form, step it up by transformer action to very high levels (sometimes as much as 750,000 volts), and then send it over the high power lines you often see in rural areas. As the power nears towns and cities, transformer action is again used to reduce the voltage. The reduction is done in steps rather than all at once, so that the highest practical voltage level can be maintained for as long as possible. The final voltage reduction takes place on the line transformers mounted on utility poles that carry power to the end user (Fig. 11-1). In many communities power lines are underground. Of course, neither the

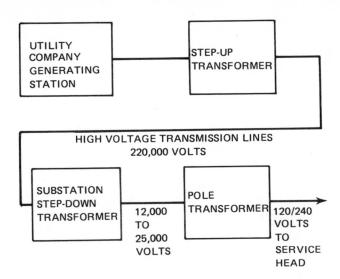

Figure 11-1. Power distribution from generator to pole.

lines nor the transformers are visible in these areas, but the scheme of distribution is the same.

The last step-down pole transformer reduces the voltage to the level used in the cables, switches, receptacles, and fixtures installed in residences. In most areas this power level is 120/240 volts at a frequency of 60 Hz. This power is carried from the pole transformer to the building on three wires: two hot wires and one neutral wire (Fig. 11-2).

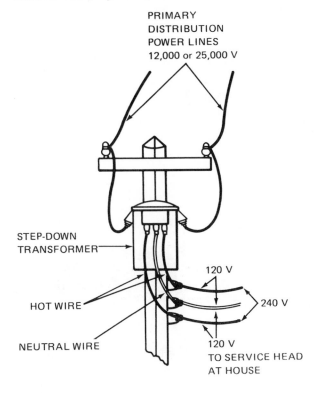

Figure 11-2. Power from pole transformer to end user.

Regulations

The location, the construction, and the wiring at the service entrance must be in accordance with the regulations established by the local utility company, as well as local building and electrical codes (Fig. 11-3). Most utility companies specify that the service drop or lateral must be as short as possible. The lines from the pole transformer should run to the closest point on the building to be served. One service drop is installed for each residence even though the building may house several families, and a separate meter will be installed for each family. Commercial buildings that require more than one form of power may have several service drops.

Many utility companies install overhead service drops at no cost to the user, but require that the user pay all or part of the cost if underground service connections are made. Of course, exceptions can be made to these and other rules when good and sufficient reason exists for some other arrangement. It is important to discuss any special service entrance requirements with utility company representatives as far in advance of actual installation as possible. Changes in service entrance location, meter location, and length and location of service drops may require permission of local building and zoning authorities, as well as the utility company.

Wiring

The cables most often used for service entrance wiring are types SEC or SER. SEC (service entrance concentric) consists of a stranded neutral conductor wrapped around red- and black-insulated, hot-wire conductors (Fig. 11-4). A tough, weather-resistant sheath covers the neutral conductor.

SER (service entrance round) also has a tough, weather-resistant sheath (Fig. 11-5). SER contains two black-insulated hot-wire conductors, one white- or gray-insulated neutral conductor, and a stranded, bare conductor.

When conduit is used for the service entrance wiring, all fittings must be watertight (Fig. 11-6). Type TW or equivalent wire can be used for electrical connections.

Installation Procedure

The general procedure for service entrance installation requires that all wiring and hardware up to the service head (sometimes called the entrance cap or weatherhead) be done by the building electrical contractor. Utility company inspectors, or some other authorized inspectors, then check the installation. If it

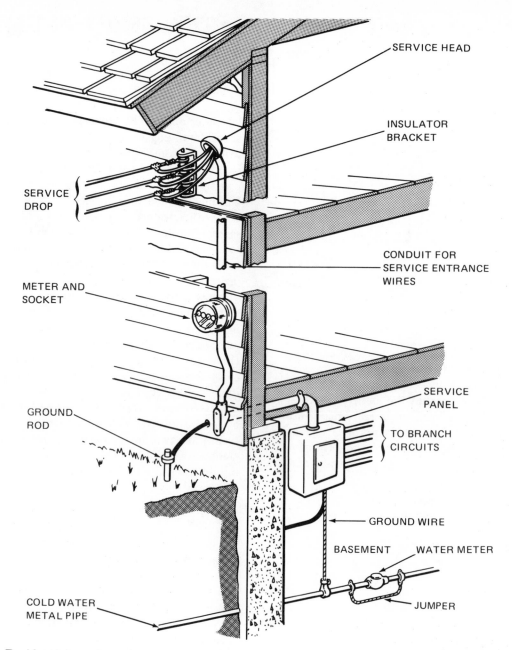

Figure 11-3. Residential service entrance.

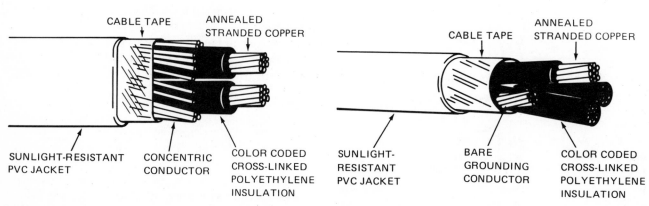

Figure 11-4. Service entrance concentric cable.
(General Cable Corporation)

Figure 11-5. Service entrance round cable.
(General Cable Corporation)

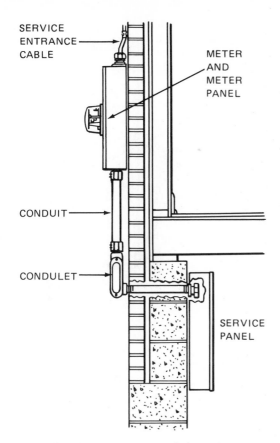

Figure 11-6. Service entrance conduit.

is satisfactory, utility company personnel install the service drop and make the connection to the service entrance cable conductors. The building's electrical contractor installs a meter socket as part of the service entrance. When incoming lines are in place and all inspections have been made and satisfactorily passed, the utility company personnel install a meter in the socket and turn on power to the building.

The materials used in the service entrance are determined by the electrical specification for the building, local code requirements, and utility company regulations. In some instances other trades may be responsible for part of the job, particularly if a mounting pole or timber is needed to bring in the service drop at the required height.

Installation of other parts of the service entrance follows general NEC rules and good trade practice. Service entrance cable must be supported within 12 inches of the service head and all other devices and enclosures used in the service entrance. Cable must be supported at least every 4 1/2 feet in all other cases. Intermediate and EMT conduit must be supported within 3 feet of the service head and other devices, and at least every 10 feet otherwise. The support of rigid conduit depends on conduit diameter. Refer to NEC Table 347-8.

Meter sockets must be securely fastened to the building. In wood frame construction, the socket should be stud-mounted. In masonry buildings, any of the masonry fasteners described in Chapter 7 may be used. Service panels are usually mounted to basement walls. Again masonry fasteners can be used. Service panels installed above ground level can be stud-mounted.

• SERVICE DROP •

The service drop can be attached to the building in a number of ways. The NEC specifies that conductors for 120/240-volt service must be 10 feet above grade or ground level at the point where they are attached to the building. If a building is two or more stories high, the lines can generally be attached directly to the structure (Fig. 11-7). For single-story buildings it may be necessary to install a mast of galvanized steel pipe or a wood timber on the building to obtain the required ground clearance (Fig. 11-8). The NEC and most local codes require greater overhead clearance when service lines cross driveways and roads. Figure 11-9, which is reproduced from a local code, illustrates other clearance requirements. The clearances shown are in accordance with the NEC (Section 230-24).

The service lines can be attached to the building or to a wood timber by individual or ganged, screw-mounted insulators (Fig. 11-10). If attached directly to a wood frame building, the insulator must be secured to

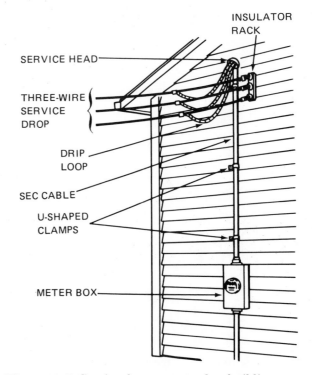

Figure 11-7. Service drop connected to building.

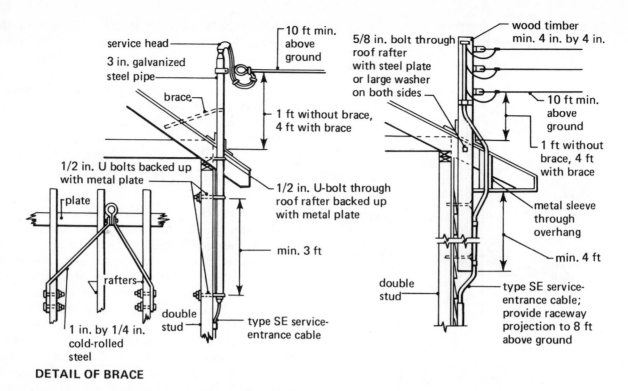

Figure 11-8. Service drop mast or timber installation. (Illustration is from the *Code Manual for the New York State Building Construction Code* and is used with permission.)

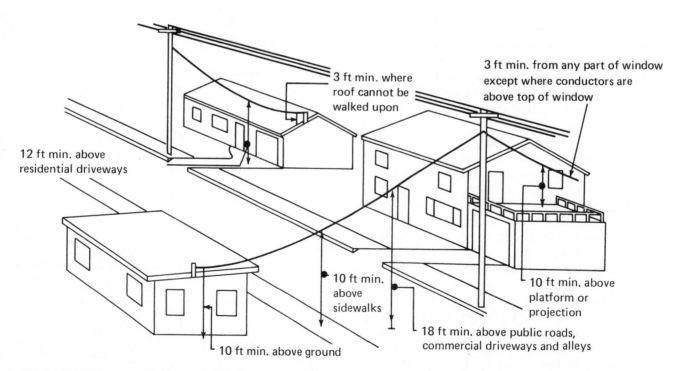

Figure 11-9. Service drop overhead clearances. (Illustration is from the *Code Manual for the New York State Building Construction Code* and is used with permission.)

Figure 11-10. Service drop insulators.

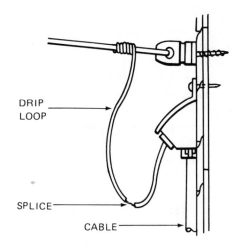

Figure 11-12. Drip loop at service head.

a stud (Fig. 11-11). If a steel pipe is used to obtain required ground clearance, a clamp can be attached to the pipe to secure the power lines. Some codes may require that the pipe be braced if it extends more than 1 foot above the roof. A service head is mounted on the building, pipe, or timber above the level of the service drop connection. The service entrance cable conductors extend out of the service head.

When the service entrance installation is complete, and has passed inspection by both building and utility company inspectors, the service entrance cable conductors will be spliced to the service drop. This connection is usually made by utility company personnel. The service entrance conductors must be long enough to provide drip loops at the service head (Fig. 11-12). The drip loops prevent water from flowing along the wire into the service head.

• METER SOCKET •

Meters that measure the electrical energy consumed over a period of time are supplied by the local utility company. The electrical contractor must supply a socket in which the meter can be mounted (Fig. 11-13). Plug-in meters are used for quick removal and replacement, if necessary. For residential service up to 200 amperes the meters are series-connected, that is, all power consumed in the building must pass through

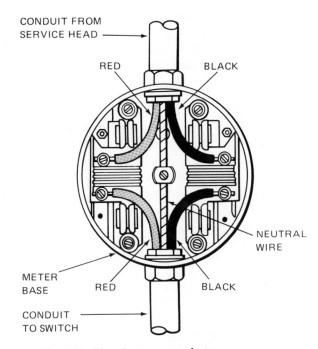

Figure 11-13. Plug-in meter socket.

the meter. The meter socket can be installed indoors or out, as specified by the utility company. In most locations, residential meters are located outdoors. The meters are, of course, weather-resistant and the outdoor location makes meter reading easier and faster. The general requirements for socket installation are the same, whether indoors or out. The socket must be at eye level. This is usually interpreted to mean between 4 and 6 feet above ground or floor level. The meter socket jaws must be at a true vertical. They should be checked with a plumb line to be sure of proper position. There must be at least 3 feet of clear space in front of the meter. Meter sockets are rated at 125 amperes and 200 amperes. The socket used must match or exceed the service to be supplied. Meter sockets for the customary three-wire 120/240-volt

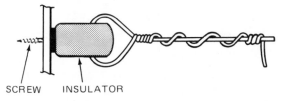

Figure 11-11. Insulator connected directly to wood frame building.

service have four jaws for meter plug-in connection. Each jaw has a setscrew connector. The conductors from the service head are normally brought in at the top of the socket. The two hot wires are connected to the two top jaws. The two hot wires to the service panel or service switch are connected to the two bottom jaws. The neutral conductors are joined at a center connector. Meter sockets for underground service installations are basically the same as overhead sockets except that provision is made for cable entrance and exit at the bottom of the socket.

• MAIN DISCONNECT SWITCH •

The NEC and most codes require that means be provided to disconnect all conductors in the building from the source of supply. The NEC allows a maximum of six switches to disconnect all power. In residential wiring this requirement is met in two ways. In some localities a main cutoff switch, sometimes referred to as EXO (for externally operated) is installed between the meter and the service panel (Fig. 11-14). The switch consists of a metal box with a handle on one side and a hinged front door. The handle operates a double-pole single-throw switch that can open or close both incoming hot wires. The box is made so that when the handle is in the on position, the door is locked in the closed position. The door can be opened only when the handle is in the off position.

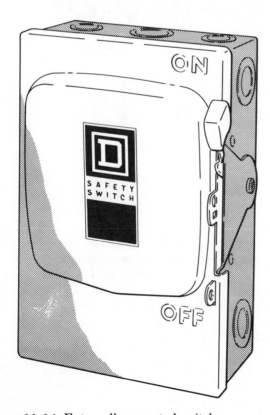

Figure 11-14. Externally operated switch.

A more common way of providing the main disconnect is to incorporate it in the service panel. This feature is described in the section on service panels that follows.

• SERVICE PANEL •

The prime requirement for any service panel is that it be adequate to handle the building's electrical system. The building specifications or local codes will determine the type of circuit protection to be provided. Either fuses or circuit breakers may be used. (These and other protective devices are covered in Chapter 10.) The panel must be large enough to hold wiring for all the branch circuits in the building. It is generally desirable to install a panel that has some space available for additional circuits that may be added at a later date. Some service panels have power takeoff terminals that make it possible to install a service subpanel to handle circuits that may be added.

Service panels have some of the features of the small electrical boxes used on branch circuits. They have knockouts that are removed to bring in cables, they can be wall-mounted, and they must be covered when wiring is complete (Fig. 11-15). Service panels also provide a means of securing incoming power lines, a source of power for branch circuits, mounting for fuses or circuit breakers, and a grounding connection to the neutral power conductor.

Circuit Breaker Panel

A typical circuit breaker panel for residential service is shown in Fig. 11-16. The service entrance cable is brought in through the top of the panel. The cable is secured by a clamp at the enclosure knockout. The

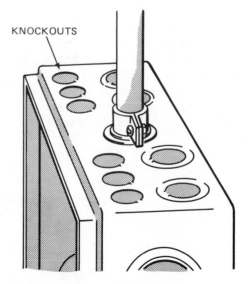

KNOCKOUTS

Figure 11-15. Service panel knockouts.

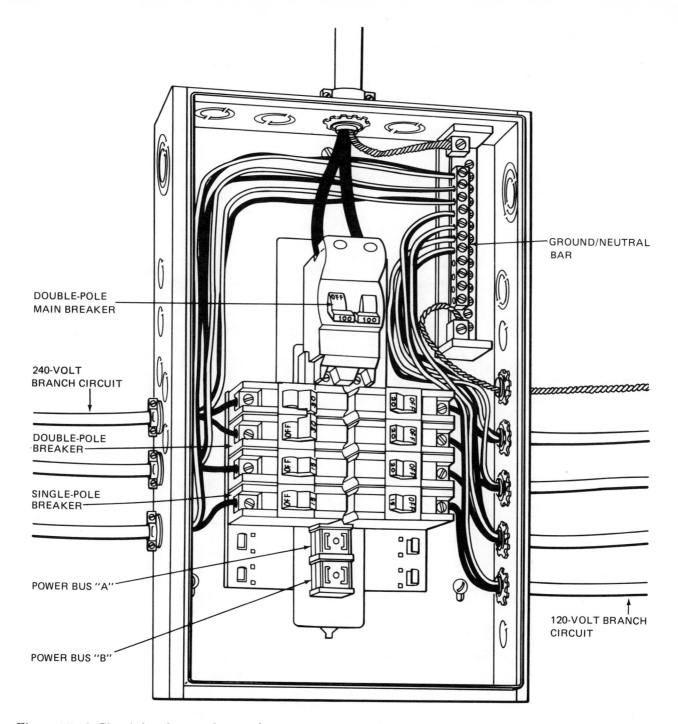

GROUND/NEUTRAL BAR

DOUBLE-POLE MAIN BREAKER

240-VOLT BRANCH CIRCUIT

DOUBLE-POLE BREAKER

SINGLE-POLE BREAKER

POWER BUS "A"

POWER BUS "B"

120-VOLT BRANCH CIRCUIT

Figure 11-16. Circuit breaker service panel.

neutral conductor is connected to a terminal strip by means of a setscrew connector. The two hot wires are attached, also by setscrews, to the two inputs of a double-pole main circuit breaker. These circuit breakers can serve as the main disconnect. Each of the two hot wires is connected through the circuit breakers to two bus bars that are mounted vertically on insulated standoffs. Bus bars are solid pieces of metal that can conduct current with almost zero resistance. These solid bars are drilled and tapped for screw mounting the circuit breakers. The screw mount into the bus bar

provides one hot-wire connection to the circuit breaker; the double-pole main circuit breaker provides the disconnect feature required by the NEC. It is customary to arrange the two power busses so that the circuit breakers are connected to alternate wires from top to bottom. If we call one bus the black-wire bus and the other the red-wire bus, the top circuit breaker might be connected to the red bus and the next circuit breaker on the same side to the black bus. This practice distributes the branch circuit load between the two hot wires; adjacent circuit breakers can be used for the red and the black wires when connecting 240-volt circuits. The neutral wire is connected to a neutral terminal strip where all white (or gray) power-ground lines can be connected. The neutral terminal strip also provides a place to connect the bare or green-insulated service-grounding conductors.

Fuse Panel

Fuse panels for 100- to 200-ampere service have features similar to circuit breaker panels of the same rating. The metal enclosures are about the same size, have knockouts for cable entrances, and have covers (Fig. 11-17). The incoming conductors are secured to setscrew connectors that are linked to power busses. The main power disconnect, an insulated block with a wire handle, is called a main pullout. Conductors or cartridge fuses in this insulated block carry the power from the two hot wires to the power busses. When the block is pulled out, power is cut off.

If the block itself has no fuses for overcurrent protection, the fuse panels must have provision for power

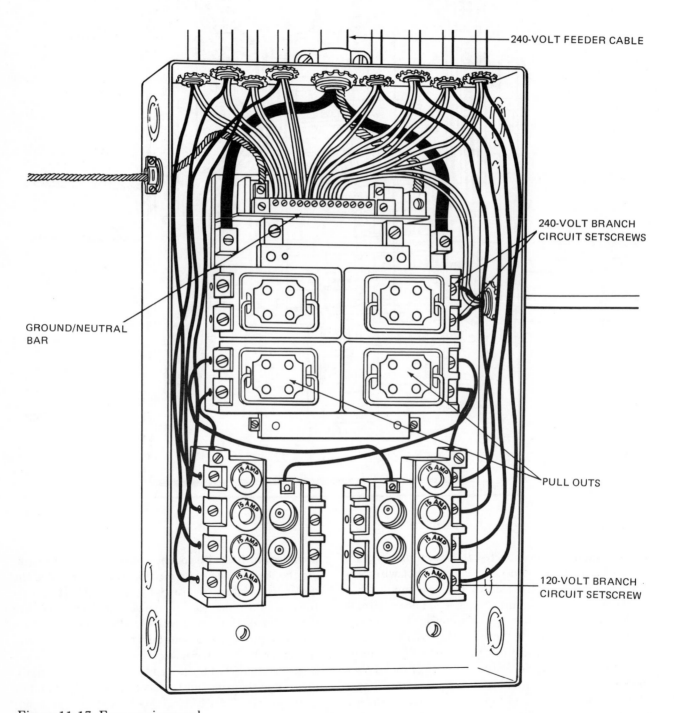

Figure 11-17. Fuse service panel.

line fuses between the hot wire connections and the main pullout. One or more additional pullouts may be included in the panel to cut off power to large appliances, such as electric ranges and air conditioners. Screw-base fuses connect the branch circuits to the power busses. The neutral power line conductor is secured to a setscrew connector and to a neutral terminal strip. As in the circuit breaker panel, all white and bare (or green-insulated) wires from the branch circuits are joined at the terminal strip.

• GROUNDING ELECTRODE SYSTEMS •

What the Code calls a grounding electrode system must be installed at the service panel. The phrase "grounding electrode system" refers to the connection of the neutral power conductor, the power-ground (white) wires in the electrical system, the bare or green-insulated grounding wires, and two solid, dependable connections to earth ground (Fig. 11-18).

The neutral terminal strips in service panels are not automatically grounded to the enclosure because the enclosure may sometimes be used for other purposes. A jumper or grounding screw is provided to make this connection (Fig. 11-19). When the enclosure is used as a service panel, the grounding screw or jumper must be attached to the neutral bar to ground it to the enclosure.

When electrical systems have a separate grounding conductor, the necessary electrical continuity in the service panel is achieved by joining these conductors and the incoming neutral power line at a neutral terminal bar grounded to the enclosure. This method of grounding applies to all nonmetallic cable and flexible conduit having a bare or green-insulated grounding conductor.

Electrical systems using rigid, intermediate EMT conduit do not have a separate grounding conductor. Equipment grounding is provided by the conduit. It is, then, important that the conduit be securely attached to the service panel enclosure. The NEC permits several methods of bonding (Section 250-72). However, the use of grounding bushings and bonding-type locknuts are the methods most often used (Fig. 11-20).

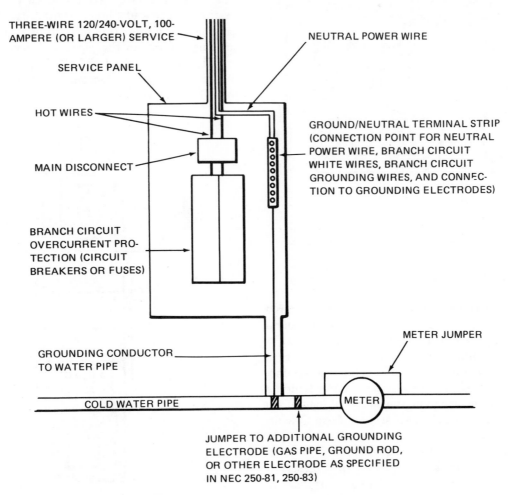

THREE-WIRE 120/240-VOLT, 100-AMPERE (OR LARGER) SERVICE

NEUTRAL POWER WIRE

SERVICE PANEL

HOT WIRES

MAIN DISCONNECT

GROUND/NEUTRAL TERMINAL STRIP (CONNECTION POINT FOR NEUTRAL POWER WIRE, BRANCH CIRCUIT WHITE WIRES, BRANCH CIRCUIT GROUNDING WIRES, AND CONNECTION TO GROUNDING ELECTRODES)

BRANCH CIRCUIT OVERCURRENT PROTECTION (CIRCUIT BREAKERS OR FUSES)

METER JUMPER

GROUNDING CONDUCTOR TO WATER PIPE

COLD WATER PIPE

METER

JUMPER TO ADDITIONAL GROUNDING ELECTRODE (GAS PIPE, GROUND ROD, OR OTHER ELECTRODE AS SPECIFIED IN NEC 250-81, 250-83)

Figure 11-18. Grounding electrode system.

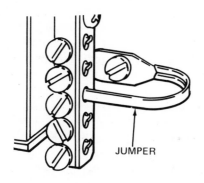

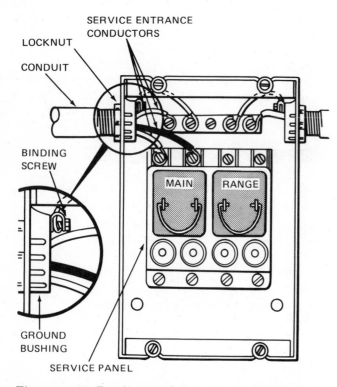

6, or no. 8 copper wire. Use of no. 8 wire, however, requires that the wire be protected by conduit or armor. This is not necessary for larger sizes if they are not located where severe physical damage may occur. It will usually be easier and less expensive to install the larger size wire (no. 4 or no. 6) without conduit or armor.

NOTE: Table 250-94 specifies the conductor size to be used for the grounding electrode. Do not confuse this with the size requirement for the neutral current-carrying conductor. The method of determining the minimum safe size for the neutral conductor is shown in the next section.

This wire—the grounding electrode conductor—must be connected between the neutral bar in the service panel and a grounding electrode system. The system must consist of at least two grounded conductors bonded together. The conductors most generally available in residences are water and gas pipes, and, where necessary, separately installed ground rods.

Two items must be considered in determining the effectiveness of water and gas pipes as good grounds. The water line must have no plastic sections or connections that would break continuity. The gas line must not have insulating sections or joints, nor an insulating coating on the pipe. It is also important to install jumpers around the water and gas meters to maintain the ground continuity if the meters are removed for any reason (Fig. 11-21). If these conditions

Figure 11-19. Neutral terminal strip ground strap.

Figure 11-20. Bonding conduit to service panel.

With ground continuity in the service panel assured, the next step is to make the connection to earth ground. The wire size used to connect the service panel to ground is determined by the wire size of the service entrance cable, the size of which is, in turn, determined by the expected power consumption of the building. The method of determining the electrical load of the dwelling is covered in the next section, Load Calculation.

NEC Table 250-94 lists the grounding electrode conductor sizes required for various sizes of service entrance conductor. For residential service, the grounding electrode conductor will usually be no. 4, no.

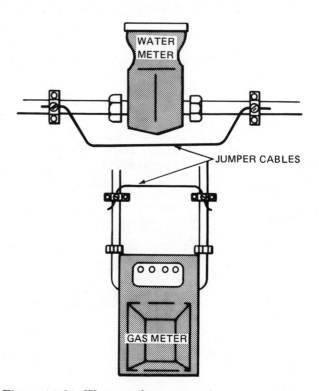

Figure 11-21. Water and gas meter jumpers.

can be satisfied, the grounding conductor can be installed between the service panel neutral bar and the water pipe. A second bonding jumper can be installed between the water and gas pipes. Grounding clamps are generally used to make these connections (Fig. 11-22).

If there is no gas service in the house, or if the gas pipe cannot be used for any reason, an outside ground rod can be installed. The rod must be driven into the ground to a depth of 8 feet (Fig. 11-23). If the rod cannot be driven into the ground because of a rock layer or other obstruction, it can be buried in a trench at least 2 feet deep. It is usually permissible to combine the two methods if *more* than 4 feet of rod have been driven into the ground when the obstruction is encountered. In this case, a trench 2 feet deep must be dug and the rod bent at a right angle so that it lies in the trench.

Galvanized pipe can be used as a ground rod if it is 3/4-inch diameter or larger. A more widely used material is a steel rod with copper coating. This is specially made for use as a ground rod and is UL-listed. When the rod is in place, a clamp and jumper are attached to the rod, and the jumper is run to the water pipe (Fig. 11-24). As ground rods are installed 1 to 2 feet from the foundation, it will be necessary to drill through the masonry to make an entrance place for the cable.

• LOAD CALCULATION •

To determine the size of the service entrance conductors and to establish the service required from the utility company, the expected electrical load for the residence must be calculated. The NEC and most local codes specify three-wire 100-ampere service as the minimum service to be installed in one-family dwellings. For many homes 150- or 200-ampere service is needed. A systematic load calculation will show the service required for any dwelling.

Present versus Future Needs

The load calculation establishes the minimum service that is adequate for a specific size of house and group of appliances. For many years it has been a standard practice to recommend installing service entrance equipment capable of handling a much greater load than was indicated by the load calculation. The reason for the reserve capacity was that usage of electrical energy was continually increasing over the years. This reasoning is no longer valid in every case.

It has become apparent that the ever-increasing generation of electrical energy must be limited, and ways must be found to use electrical energy more efficiently. The use of alternate sources of energy and the design and manufacture of more efficient appliances should cause the electrical energy demand curve to rise

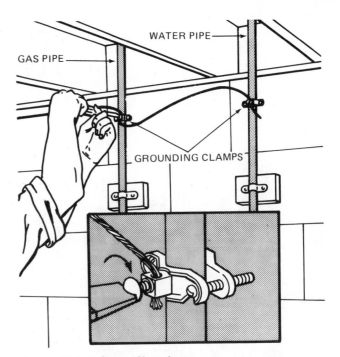

Figure 11-22. Grounding clamps.

Figure 11-23. Installing a ground rod.

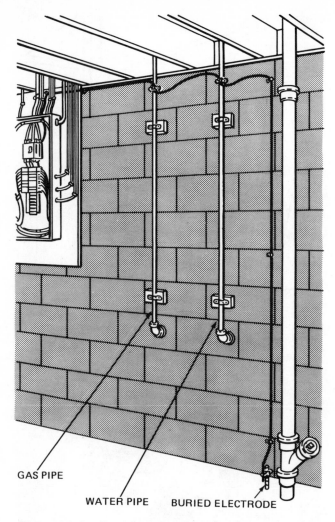

GAS PIPE

WATER PIPE   BURIED ELECTRODE

Figure 11-24. Jumper from ground rod to water pipe.

more slowly and, perhaps, even to take a downward turn. The allowance for increased use of electrical energy must be viewed differently than in past years. In homes where efficient use of energy has been emphasized in design, future needs for electrical energy should show little change. Traditionally designed homes—initially equipped with few electric appliances—will probably need some increase in electrical service in future years.

Sample Load Calculation

An electrical load calculation from a local code is presented at the end of this section to illustrate the general method of calculation. Different rules may be used by other codes. The general requirements are given in NEC Article 220. Keep in mind that the example given is from a local code. It does not conflict with the NEC, but it is more limited in coverage. Refer to the NEC for complete description of load calculations.

Two phrases are used in describing load calculations. They are *demand factor* and *demand load*. Both terms refer to the same basic idea. It is recognized that not all electrical appliances and lights are on at any one time. A demand factor is a percentage that is applied to the maximum load to reduce the figure to a more practical level. A demand load is a similar value for appliance loads, but is stated in table form because it is not a fixed percentage. The example given includes a demand load for household electric ranges. NEC Article 220 provides other demand load tables for clothes dryers, wall ovens, other household cooking appliances, and kitchen equipment.

Note that all calculations made in the example use values of 115/230 volts for the line voltage. Some localities require the use of 115/230 volts even though the actual line voltage is 120/240 volts. This is done to provide an extra measure of safety in the calculation and, also, because the line voltage may drop slightly during peak-use periods.

The last calculation in the example covers determination of the size of the neutral service conductor. The minimum safe size for this conductor in a three-wire 120/240-volt service can be determined by rule of thumb or by a calculation. The rule of thumb is simply to use a conductor one size smaller than the hot wires. The calculation takes a bit of time, but it may show that a much smaller neutral can be safely used.

It may seem contradictory that a neutral wire that serves two hot wires can be smaller than either. To understand why this is so, we must review some points about transformers. The three wires of a 120/240-volt service are the output lines from a center-tapped secondary winding on the utility pole transformer. The transformer is designed so that the voltage across the full secondary is 240 volts (Fig. 11-25). When the center tap on the secondary is grounded, the voltage between the center tap and each outside winding is 120 volts. The voltage across the full secondary is still 240 volts. For these two statements to be true the two 120-volt outputs must have opposite polarity at any instant (Fig. 11-26). When the voltage on the top winding is a positive 120-volt half cycle, the voltage in

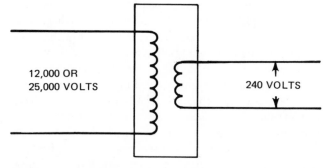

12,000 OR
25,000 VOLTS

240 VOLTS

Figure 11-25. Utility pole transformer.

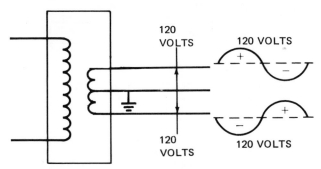

Figure 11-26. Transformer secondary output voltages.

the bottom winding must be a negative 120-volt half cycle in order to have a 240-volt difference across the full secondary.

If each 120-volt power source is connected to an equal load, the same current will flow in each circuit (Fig. 11-27). Because of the opposite polarity of the circuits, the two currents flowing in the neutral line are equal and opposite, or zero. In this case, the neutral line can be removed (Fig. 11-28). Because of the polarity difference, return current flows in each hot wire on alternate half cycles. Of course, in most situations, the current flow in the two hot wires is not exactly equal. However, the neutral line need carry only the net difference. Calculating the maximum net difference indicates the minimum neutral conductor size.

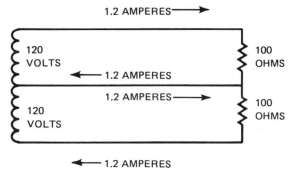

Figure 11-27. Instantaneous current flow in three-wire service.

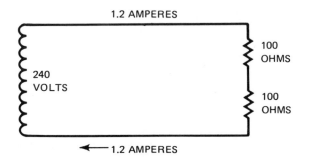

Figure 11-28. Instantaneous current flow in service without neutral conductor.

A sample load calculation from a local code follows. The NEC specifies a similar load calculation. An example from a local code is used because local code requirements almost always take precedence over the NEC. Some notes have been added to explain parts of the calculation. While reading the local code description of load calculations, look at Fig. 11-29 to see how the entry was made. Check the calculations as you go along to be sure you understand the procedure. (Material is from the *Code Manual for the New York State Building Construction Code* and is quoted with permission.)

Electrical Loads

In dwelling occupancies other than hotels the load for general lighting shall be determined on a basis of 3 watts per square foot of floor area. The floor area shall be computed from the outside dimensions of the building, apartment or area involved, and the number of floors, but not including open porches, garages or unfinished and unused space unless adaptable for future use.

To the calculated general lighting load there shall be added a load of 4500 watts for each dwelling unit in order to provide for portable appliances used in kitchen, dining room, and laundry.

A demand factor of 100 per cent shall be applied to the first 3000 watts or less of general lighting load, 35 per cent to the next 117,000 watts, and 25 per cent to any amount over 120,000 watts.

The load for an electric range shall be determined from the table entitled, "Demand Loads for Household Electric Ranges." . . .

The load for fixed appliances other than the electric range shall be the sum of the nameplate ratings of such appliances. Where more than three fixed appliances in addition to an electric range are to be supplied a demand factor of 75 per cent may be applied to the fixed appliance load, but not including the electric range, air conditioning equipment, space heating equipment load or clothes dryer.

The load for a single motor shall be based on 125 per cent of the full-load current rating of the motor. Where two or more motors are supplied, the load shall be based on 125 per cent of the full-load current rating of the highest rated motor in the group, plus the sum of the full-load current ratings of the remainder of the motors in the group. Full-load current ratings shall be as listed in the **National Electrical Code** or as given on the motor nameplate.

NOTE: Motor load is increased to 125 percent of rated value to allow for the heavy starting current.

DEMAND LOADS FOR HOUSEHOLD ELECTRIC RANGES

Requirements		Examples	
Nameplate Rating in Watts	Demand Load in Watts	Nameplate Rating in Watts	Demand Load in Watts
Up to 1750	Nameplate rating	1500	1500
1751 to 8750	80 per cent of nameplate rating	6500	6500 x 0.80 = 5200
8751 to 12,000	8000	11,600	8000
12,001 to 27,000	8000 plus 400 for each multiple of 1000 or major fraction thereof by which the nameplate rating exceeds 12,000	15,750	$8000 + 400 \left(\frac{15,750 - 12,000}{1000}\right)$ $= 8000 + 400 \ (3.75; \text{use } 4)$ $= 9600$

Note: Electric range ratings are usually given in kW. One kW equals 1000 watts.

NOTE: Table is from the *Code Manual for the New York State Building Construction Code* and is reproduced with permission.

In . . . [Fig. 11-29] the basic method for determining the size of service conductors is shown. The application of demand factors is indicated and the resultant total demand load used to calculate the required current-carrying capacity of the conductors. For a three-wire service the neutral service conductor may be of a size smaller than that required for the ungrounded service conductors. As indicated, the minimum size of the neutral service conductor may be determined from the demand load of the maximum connected load between the neutral and any one ungrounded conductor, except for the household electric ranges the maximum unbalanced load shall be considered as 70 per cent of the demand load.

NOTE: The water heater and the clothes dryer require 240-volt input power and are connected only to the two incoming hot wires. Therefore, they add no current load to the neutral conductor and can be excluded from the neutral calculations. Both the water heater and the clothes dryer would have non-current-carrying metal parts connected to a bare or green-insulated equipment-grounding conductor.

• DETERMINING THE NUMBER OF BRANCH CIRCUITS •

The electrical load calculation determines what service the utility company must supply. For the example shown in Fig. 11-30, 150-ampere 120/240-volt service would be required. The next determination to be made is how many circuits must be installed to distribute this power where needed. Each individual circuit is known as a branch circuit. The number and types of circuits needed are based on the NEC and local code requirements.

Electrically, a branch circuit consists of the conductors that run from an individual fuse or circuit breaker on the service panel to one or more outlets for receptacles, switches, or fixtures where power is used. The current rating of the circuit is the rating of the overcurrent protection. Four types of branch circuits are used in residential wiring.

General-purpose Circuit. This is a branch circuit that supplies power to lighting fixtures and receptacles that will be used for lighting or small appliances.

Appliance Circuit. This is a circuit intended for appliances. In residences it usually supplies receptacles installed near kitchen work surfaces. These circuits are not intended for lighting or large appliances.

Load calculations to determine the minimum size of service conductors required for the building shown in the illustration entitled, "Electrical Layout for One-Family Dwelling," Fig. 11-30.

Floor area, 1,500 square feet

General lighting:
1,500 square feet x 3 watts per square foot .... 4,500 watts
Small appliance load .... 3,000 watts
Laundry .... 1,500 watts

Computed load .... 9,000 watts
3,000 watts x 100 per cent .... 3,000 watts
9,000 minus 3,000 = 6,000; 6,000 watts x 35 per cent .... 2,100 watts
Demand load.. 5,100 watts

Electric range:
Nameplate rating.. 15,300 watts
For first 12,000.. 8,000 watts
For excess over 12,000 watts

$$400 \left(\frac{15,300 - 12,000}{1,000} \right) = 400 \,(3.3; \text{use } 3) \dots \dots 1,200 \text{ watts}$$

Demand load.. 9,200 watts

Clothes dryer: 5,000 watts, 230 volts .... 5,000 watts
Furnace blower: ¼HP, 115 volts; 115 x 5.8 .... 667 watts
Largest motor: 115 x 5.8 x 25 per cent .... 167 watts

Fixed appliances:
Dishwasher: 1,200 watts, 115 volts........................... 1,200 watts
Waste disposal unit: 6 amperes, 115 volts .... 690 watts
Water heater: 2,500 watts, 230 volts........................ 2,500 watts

Computed load: .... 4,390 watts
Demand load: 4,390 watts x 100 percent .... 4,390 watts

Total demand load .... 24,524 watts

For 115/230 volt, three-wire service: Current, 24,524 watts ÷ 230 volts = 107 amperes
Minimum size of ungrounded service conductors required, based on type TW wire, is No. 1.

Load calculations to determine minimum size of neutral service conductor.
In this example, water heater and clothes dryer have no neutral connections.

General lighting demand load .... 5,100 watts
Electric range demand load: 9,200 watts
9,200 watts x 70 per cent .... 6,440 watts

Furnace blower... 667 watts
Largest motor .... 167 watts
Fixed appliance load: 4,390 minus 2,500 .... 1,890 watts

Demand load (for neutral conductor determination) .... 14,264 watts
Current, 14,264 watts ÷ 230 volts = 62 amperes
Minimum size of neutral service conductor, based on type TW wire, is No. 4.

Figure 11-29. Sample electrical load calculation. (Illustration is based on an example from the *Code Manual for the New York State Building Construction Code* and is used with permission.)

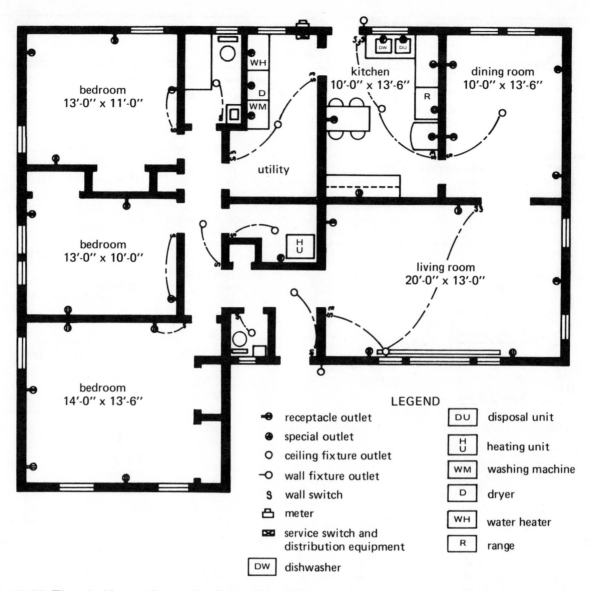

bedroom
13'-0" x 11'-0"

WH

D

WM

utility

kitchen
10'-0" x 13'-6"

dining room
10'-0" x 13'-6"

DW DU

R

bedroom
13'-0" x 10'-0"

H U

living room
20'-0" x 13'-0"

bedroom
14'-0" x 13'-6"

LEGEND

- receptacle outlet
- special outlet
- ceiling fixture outlet
- wall fixture outlet
- S wall switch
- meter
- service switch and distribution equipment
- DW dishwasher

- DU disposal unit
- H U heating unit
- WM washing machine
- D dryer
- WH water heater
- R range

Figure 11-30. Electrical layout for one-family dwelling. (Illustration is from the *Code Manual for the New York State Building Construction Code* and is used with permission.)

Individual Circuit. This is a two-wire 120-volt circuit that supplies one appliance. In residences, individual 120-volt circuits are usually installed in kitchens and utility rooms for major appliances and in basements for heating systems.

Large Appliance Circuit. This may be either a two-wire or a three-wire 240-volt circuit for single large appliances, such as electric ranges or air conditioners.

These branch circuit names are widely used in the electrical trade and are defined in the NEC (Article 100). However, branch circuit electrical characteristics are described (Article 210) only in terms of ampere rating, and whether one outlet or two or more is pro-

vided. Branch circuits supplying two or more outlets can be installed and wired for loads of 15, 20, 25, 30, 40, or 50 amperes. Branch circuits supplying a single load can be rated higher than 50 amperes. Other branch circuit requirements, such as ground fault protection, are also covered by NEC Article 210. Of the four circuit types described previously, the first two must meet Code requirements for branch circuits having two or more outlets. The second two types must meet single-outlet circuit requirements.

The number of each type of circuit that must be provided and the method of calculating the requirement is covered by the local code example below. As in the overall load calculation, a line voltage value of 115/230 volts is used throughout the local code example. Local code requirements and calculations for general-purpose circuits, appliance circuits, individual circuits, and large appliance circuits are summarized following the local code example below. (Material is from the *Code Manual for the New York State Building Construction Code* and is quoted with permission.)

Branch Circuits—Code Example

Branch circuits shall be classified in accordance with the maximum permitted rating or setting of the overcurrent device protecting the circuit. The rating or setting of the overcurrent device for a branch circuit supplying a single nonmotor-operated appliance shall not exceed 150 per cent of the current rating of the appliance served, except that such rating or setting need not be less than 15 amperes. Branch circuits supplying two or more outlets in dwelling occupancies shall be in accordance with the table entitled "Branch Circuit Classification," . . .

The number of branch circuits for the general lighting load shall not be less than that determined from the capacity of the branch circuits to be used and the computed load before demand factors are taken. In addition, at least one 20-ampere branch circuit shall be provided to supply the laundry receptacles and two or more 20-ampere small appliance branch circuits to supply all receptacle outlets in the kitchen, dining room, family room, pantry and breakfast room. Such branch circuits shall have no other outlets.

Example. Calculations to determine the number of branch circuits required to supply the general lighting load for the building shown in illustration entitled, "Electrical Layout for One-Family Dwelling," . . .

Computed load for general lighting 4500 watts
4500 watts ÷ 115 volts = 39 amperes
Branch circuits required:
Three 15-ampere or two 20-ampere branch circuits plus
three 20-ampere branch circuits for portable appliances
in kitchen, dining room and laundry.

Example. The branch circuit for an electric range shall be determined from the demand load calculated in accordance with the table entitled "Demand Loads for Household Electric Ranges," . . .
Branch circuit calculations for electric range circuit:
Electric range rated at 15,300 watts
Demand load from table:

$$8000 + 400 \left(\frac{15{,}300 - 12{,}000}{1000}\right)$$

$$8000 + 400 \ (3.3; \text{use } 3) = 9200 \text{ watts}$$

$$9200 \text{ watts} \div 230 \text{ volts} = 40 \text{ amperes}$$

Branch circuit required: one individual branch circuit with a capacity of 40 amperes.

Branch circuits for fixed appliances shall be determined from the computed load of the appliances supplied. Demand factors do not apply to branch circuit calculations for fixed appliances other than electric ranges. The appliances indicated in the illustration entitled, "Electrical Layout for One-Family Dwelling," . . . [Fig. 11-30] "may be supplied by individual branch circuits or may be grouped where possible on branch circuits indicated

BRANCH CIRCUIT CLASSIFICATION

Rating of Branch Circuit, Amperes	Rating or Setting of Over-current Protective Device, Amperes	Minimum Size, AWG, of Circuit Conductors in Raceway or Cable, Types RH, RUW, RUH, RHW, RHH, T, TW, THW, THHN, THWN, and XHHW	Outlet Devices Supplied	
			Rating of Receptacle, Amperes	Lamp Holders
15	15	14	15	Any type
20	20	12	15 or 20	Any type
30	30	10	30	Heavy duty
40	40	8	40 or 50	Heavy duty
50	50	6	50	Heavy duty

in table entitled, "Branch Circuit Classification." It is recommended that fixed appliances be supplied by individual branch circuits.

The minimum size of conductor for branch circuit wiring shall be no. 14 with the following exceptions: for branch circuit wiring supplying receptacles for small appliances in kitchen, dining room, and laundry, no. 12; for branch circuit wiring supplying electric range rated 8 3/4 kW or more, no. 8 for ungrounded conductors, and no. 10 for the neutral conductor.

GENERAL-PURPOSE CIRCUITS • Three are required. Three 15-ampere circuits, using the local code voltage of 115 volts, provide 3 × 1725 watts or 5175 watts. This allows 675 watts over the code minimum. These circuits can be wired with no. 14 copper wire conductors. The alternative of two 20-ampere circuits would provide 2 × 2300 or 4600 watts. This would satisfy the code, but the excess capacity of only 100 watts is not adequate for additional lighting that may be required from time to time. Three 15-ampere circuits allow for better load distribution. GFCI protection will be included on the circuit supplying bathroom outlets.

APPLIANCE CIRCUITS • Two are required. These are 20-ampere circuits which supply kitchen outlets at work surfaces. One of these circuits is used for the refrigerator. Number 12 copper conductors are used on these circuits. The local code specifies one 20-ampere circuit to supply laundry receptacles. However, the code also recommends individual circuits for fixed appliances. In this example, individual circuits are used for the laundry appliances.

INDIVIDUAL CIRCUITS • Four are required. Many local codes require individual circuits for heating systems. Whether required or not it is desirable to have separate overcurrent protection and turnoff for heating systems. The relatively light load of the furnace blower in the sample calculation (667 watts + 167 watts starting current) can be provided by a 15-ampere circuit. The waste disposal unit, rated 690 watts, and the washing machine (most are 600 to 800 watts) can each be supplied by individual 15-ampere circuits.

LARGE APPLIANCE CIRCUITS • Three are required. The electric range circuit would be a three-wire 40-ampere circuit. As shown in the example, it is usually permissible to use the demand factor when calculating circuit capacity for electric ranges. Demand factor allowances are not usually permitted in calculating any other branch circuit capacity. A three-wire 30-ampere circuit would be adequate for the clothes dryer. Electric water heaters usually require no neutral conductor. The load is a heating element and

current flow in both hot wires is balanced. The heater in the example could be supplied by a two-wire 230-volt, 20-ampere circuit. The range circuit would require no. 8 copper conductors, the dryer circuit no. 10 copper conductors, and the water heater no. 12 copper conductors.

Load Balance

It is desirable to divide the branch circuit load as evenly as possible between the two incoming hot wires. Insofar as possible, the frequency of use and the time of use of the appliances should also be considered in balancing a load. One method of balancing the sample load of Fig. 11-29 is shown in Fig. 11-31.

• THE ELECTRIC METER •

What the Meter Measures

The meter the utility company installs in the meter socket is a kilowatthour meter. The dials on the front of the meter indicate the number of kilowatthours (kWh) used by the residence. A kilowatthour is a unit of measure representing 1000 watts in use for 1 hour. An appliance rated at 500 watts in use for 2 hours consumes 1 kilowatthour of power. An appliance rated at 6000 watts—such as a clothes dryer—in use for 1/2 hour consumes 3 kilowatthours of power.

The meter that measures power consumed is similar to the wattmeter described in Chapter 5. You will recall that the wattmeter measured real watts of power by combining the functions of an ammeter and a voltmeter in one unit. The wattmeter has both fixed and movable coils. The fixed coils (there are two) create a magnetic field proportional to the current flow in a circuit. The current flow in the movable coil is proportional to the voltage. The interaction of the magnetic fields causes a needle to deflect, indicating the product of current and voltage, or watts.

The induction type utility meter operates on a similar principle. However, the voltage and current windings are on a common core that acts as the stator of a motor (Fig. 11-32). The rotor is a disk that rotates in an air gap in the stator. The magnetic fields of the stator windings produce a torque proportional to power that causes the rotor to move. Retarding magnets act as a governor to make the rotor speed proportional to power. A register geared to the rotor records the watthours (Fig. 11-33). Utility meters have been developed to a high degree of accuracy and reliability under a wide range of operating temperatures and changes of load and voltage. Little or no maintenance is required on this meter, and the calibration stays accurate through many years of use.

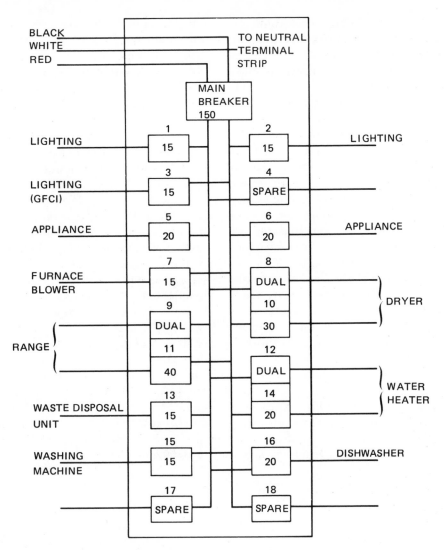

Figure 11-31. Load balancing at the service panel.

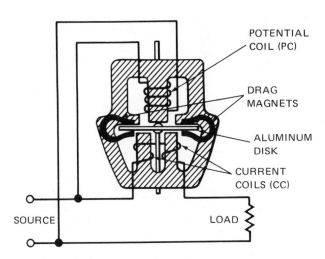

Figure 11-32. Utility meter stator.

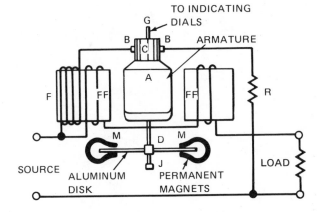

Figure 11-33. Utility meter functional diagram.

The register driven by the rotor consists of a series of geared dials. Each dial represents one digit of a decimal number. The pointer on the right-hand dial moves one

digit when the armature has rotated enough times to equal 1 kilowatt of power used for 1 hour. When the pointer on the right-hand dial makes one complete revolution, the pointer on the second dial moves the space of one digit. The dial on the right shows units, the next dial shows tens, the next hundreds, and so forth.

The kilowatthour unit of measure is being replaced in some areas by a unit representing energy consumption, rather than electrical power consumption. The unit is the megajoule, equal to 1,000,000 joules (pronounced "jools"). A joule is the amount of energy expended in 1 second by an electric current of 1 ampere in a resistance of 1 ohm. From Ohm's law we know 1 volt is required to cause 1 ampere to flow in a resistance of 1 ohm. One volt times 1 ampere equals 1 watt. A joule can be considered as 1 watt of power consumed in 1 second. A megajoule, then, is any combination of watts and seconds equal to 1 million. One watt consumed for 1 hour equals 3600 joules. One kilowatt consumed for 1 hour (this is 1 kilowatthour) equals 3,600,000 joules, or 3.6 megajoules. The megajoule, then, is a smaller unit of measure than the kilowatthour. Conversion of utility company meters to this new unit represents a huge expenditure of money. For this reason complete conversion to the new system will take many years.

How To Read A Meter

The meter reading is determined by the positions of pointers on five dials. Each dial has ten markings, from zero to nine. But the markings on different dials represent different amounts: units, tens, hundreds, and so on. The pointer always moves from 0 to 1, to 2, to 3, etc., and on to zero. However, the dials are marked in different directions. That is, the numbering of the 10,000-unit dial increases in a clockwise direction, the 1000-unit dial counterclockwise, etc.

Determine the direction of a dial rotation by the way the dials are numbered. Not all meters are the same. As long as any electricity is being used in your home, all the pointers will be moving. The amount of movement will be ten times slower on each dial from right to left. Unless power consumption is unusually heavy, movement will be noticeable only on the right-hand dial. To read the meter, note the number the pointer has just passed—keeping in mind the direction of rotation—on each dial, starting from the left.

For example: Fig. 11-34a shows a reading of 03758 or 3758 kWh. Meter dials are not precisely marked. A pointer on any dial may appear to be exactly on a number, but actually be above or below it. To decide which it is, note the pointer on the dial to the right of it. If the pointer to the right has not reached zero, the preceding dial has not reached the nearest number. If the pointer to the right has passed zero, use the next higher number for the preceding dial.

In Fig. 11-34b, the middle dial appears to be on 6, but because the next dial to the right is below zero, 5 is the correct reading for the middle dial. The full reading is 04592, or 4592 kWh.

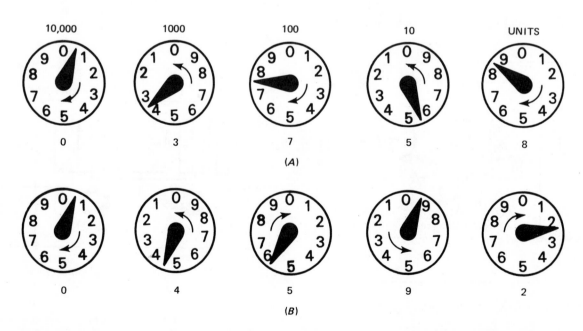

IF READINGS A AND B ABOVE WERE TAKEN ABOUT A MONTH APART, THE POWER CONSUMPTION FOR THAT MONTH WOULD BE THE DIFFERENCE BETWEEN THE TWO READINGS—OR 834 KILOWATTS.

Figure 11-34. Utility meter readings.

1. Utility companies use step-up transformers to raise voltage levels before transmitting power over long lines. Why?

2. What is a service drop? Who is usually responsible for the service drop?

3. Name the four main parts of a service entrance.

4. Aboveground service entrances can be wired two ways. What are they?

5. What does the NEC specify as the minimum height above ground for a service drop connection?

6. When conductors at the service head are connected to the service drop, enough slack is left to form a loop. Why is this done and what is it called?

7. Why are utility meters plug-in devices?

8. Internal hot-wire power is distributed in service panels by means of bus bars. What are bus bars? Why are they used?

9. Every service entrance must have provision for complete power turnoff. This feature is often included in fuse and circuit breaker service panels. How?

10. What must be joined to form a grounding electrode system?

11. Why must ground rods sometimes be installed at service entrances?

12. What is the purpose of making adjustments in load calculation because of demand factor or demand load?

13. Refer to Fig. 11-29 and calculate the general lighting demand load, using the formula given in the sample, for a dwelling having a floor area of 2200 square feet.

14. If the two meter readings shown below were taken one month apart, how many kilowatthours were consumed during the month?

15. What article and table in the NEC specify the general lighting load per square foot for homes (dwelling units) and other types of occupancy?

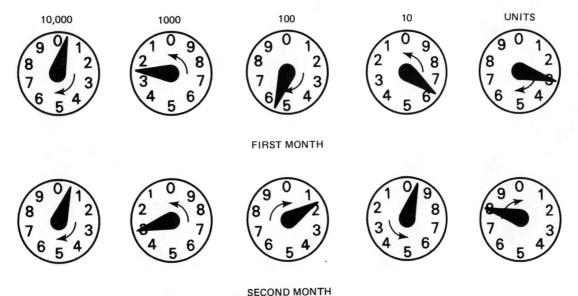

FIRST MONTH

SECOND MONTH

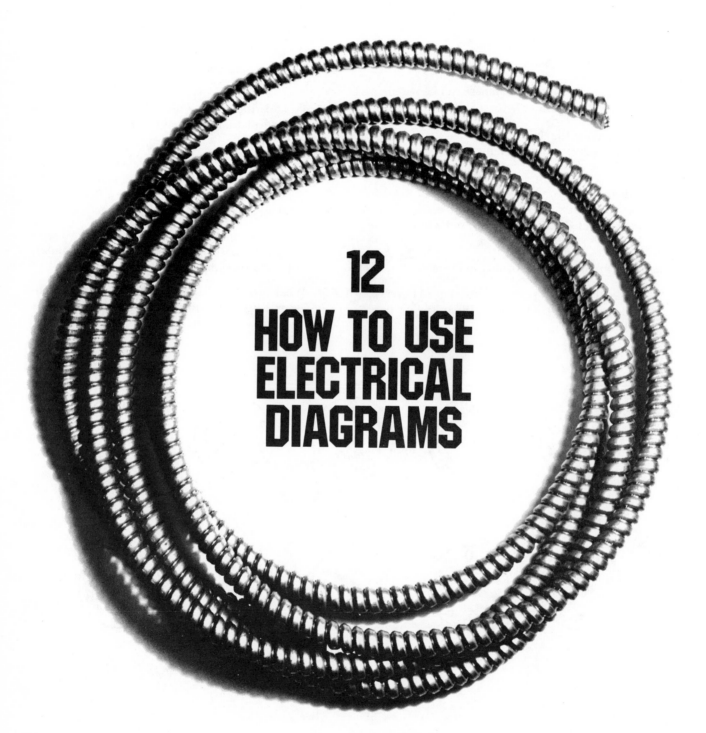

12
HOW TO USE ELECTRICAL DIAGRAMS

• INTRODUCTION •

Chapter 4 describes the use of architectural drawings and specifications to show how an electrical installation must be made. From time to time in the preceding chapters, schematic diagrams and electrical layouts are used to explain how devices work and how they are used. There is much more, however, that an electrician must know about using and understanding electrical diagrams.

Designers and electricians have learned through years of experience that the best way to provide information on electrical circuits and devices is through a combination of pictures and words. The pictures are electrical diagrams. The words are usually notes on the diagrams, but they may be in a separate document.

The diagrams make use of symbols to represent various electrical devices. Lines representing wires or cables show how these devices must be connected to work properly or to satisfy the builder's requirements. Electrical diagrams provide different kinds of information and different amounts of detail. A single diagram may cover a complete electrical installation or show only how one device operates. If you know the purpose of each diagram, you will know what information you can expect to get from it.

Electricians must be familiar with architectural symbols, as well as electrical symbols. The electrician must understand how a building will be constructed to know how cables must be routed and how devices must be installed.

This chapter provides an introduction to the kinds of drawings that electricians use. You will learn what kind of information each type of drawing provides. If you understand the purpose of each electrical diagram, you will know where to look for the information you need on the job.

• ARCHITECTURAL DRAWINGS •

Architectural drawings cover the complete construction of a building and are used by all the trades involved in the project. Two of the architectural drawings are of special interest to the electrical contractor.

Plot Plan

This is the basic drawing used to start construction (Fig. 12-1). It shows how the house will be situated on the building lot. The overall dimensions of the house are shown, as are adjacent roads, curb lines, and other property lines. Driveways, walkways, and sidewalks are also shown. This drawing is used by the electrical

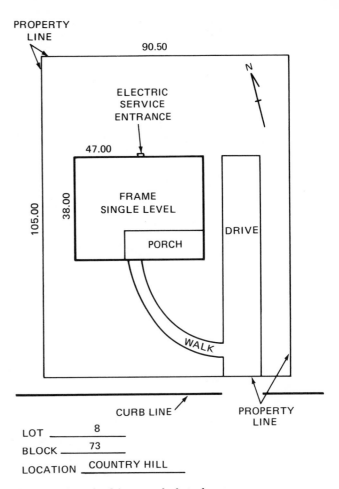

Figure 12-1. Architectural plot plan.

contractor and the utility company to determine where the service drop will be brought to the building and where the service entrance will be installed. All circuits in the building must be connected to the service panel for primary power and overcurrent protection. The service panel location, then, determines the starting point for every cable run in the building. If the project involves any outdoor underground wiring, the location of driveways and walks must be known to plan these runs. The plot plan also shows how the ground around the building will be graded. This information is needed to determine the depth that conduit or underground cabling must be buried.

Elevations and Sections

Drawings called elevations are made to show construction of interior and exterior walls (Fig. 12-2). Exterior elevations can be consulted to locate entrance lights, weatherproof receptacles, floodlights, and so forth. Interior elevations show the height of horizontal surfaces such as kitchen work surfaces. The kitchen elevation shows where appliance outlets are required above the work surface and where the receptacle for the refrigerator must be. Elevation drawings are

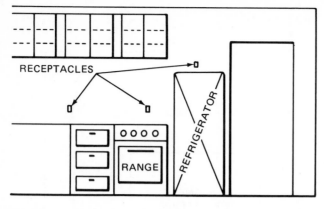

Figure 12-2. Interior elevation, kitchen.

sometimes supplemented with sectional drawings that show construction details (Fig. 12-3).

Electrical Floor Plan

Another architectural drawing used for the electrical installation is the electrical floor plan. This drawing or group of drawings (there is usually a separate drawing sheet for each living level) shows all electrical work to be done.

ANSI SYMBOLS • A standard set of symbols for electrical wiring and layout diagrams has been established by the American National Standards Institute, Inc. The symbols for architecture and building construction are covered by specification ANSI Y32.9. Figure 12-4 shows the symbols commonly used on residential electrical diagrams. A few symbols you

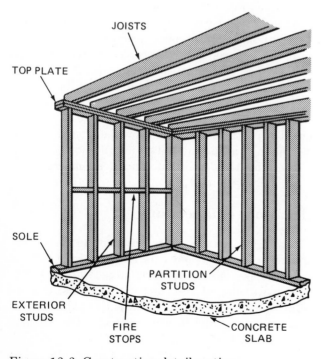

Figure 12-3. Construction detail section.

might expect to see, such as circuit breakers, fuses, and the watthour meter, are not included in ANSI Y32.9. Another specification, ANSI Y32.2, covers symbols for electrical and electronic diagrams. The ANSI policy is to include a particular symbol in only one specification. This is done to avoid the confusion that would occur if two specifications showed different symbols for the same device. The overcurrent protection devices and meters noted above, for example, do not ordinarily appear on architectural drawings. Architectural drawings show the location of the meter socket and the service panel. A symbol for the electrical *function* of these parts is needed on electrical diagrams These diagrams are described in the next section, Electrical Diagrams.

The architectural electrical floor plan shows where receptacles, switches, and fixtures are to be located and indicates what type of device is to be installed: weatherproof, ground-fault protected, split-wired, etc. (Fig. 12-5). Curved broken lines that run from switches to receptacles and fixtures show which items must be under switch control. These lines do *not* represent cable runs. The electrical floor plans for residences are, of course, two-dimensional drawings. The third dimension—the height at which items must be installed—is covered by notes on the drawing or by elevation drawings, or is described in the electrical specification. It is customary in most areas for wall receptacles to be 1 foot on center above the finished floor. Switches are usually 48 inches above the floor. In many cases height information is provided only if a different height is required.

When an experienced electrician or an electrical contractor looks at an electrical layout, a three-dimensional picture of the installation must be visualized in order to decide exactly how cable runs for each branch circuit should be made, taking into account the requirements of the NEC and local codes. Runs must be as short as possible, particularly for large appliances. Large appliances are likely to draw heavy current and must use large-size conductors. Short runs keep line loss low and the cost of materials to a minimum. The electrician or contractor must also be familiar with the construction details of the building. For example, wiring requirements are much different in concrete or cinder block construction than in wood frame buildings. Where wiring is protected by finished walls, nonmetallic cabling may be used. Wiring in unfinished areas—such as along floor joists in unfinished basements—may require conduit or armored cable.

When all these factors have been considered, the contractor marks up a set of electrical floor plans to show the order in which outlets should be wired, from the service panel to the end of the run (Fig. 12-6). While these drawings do not show the actual location of finished cabling, they are the basic drawings used to install the electrical system.

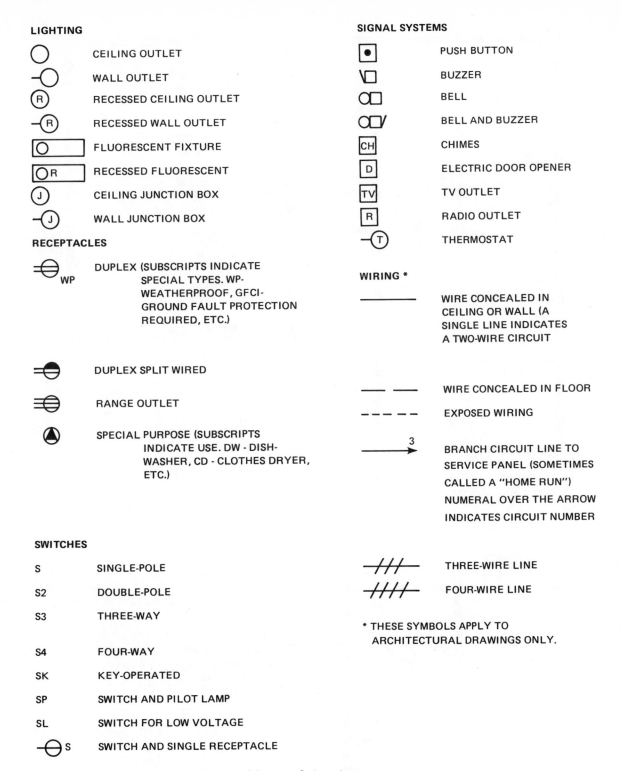

LIGHTING

○ CEILING OUTLET

─○ WALL OUTLET

Ⓡ RECESSED CEILING OUTLET

─Ⓡ RECESSED WALL OUTLET

FLUORESCENT FIXTURE

RECESSED FLUORESCENT

Ⓙ CEILING JUNCTION BOX

─Ⓙ WALL JUNCTION BOX

RECEPTACLES

DUPLEX (SUBSCRIPTS INDICATE SPECIAL TYPES. WP-WEATHERPROOF, GFCI-GROUND FAULT PROTECTION REQUIRED, ETC.)

DUPLEX SPLIT WIRED

RANGE OUTLET

SPECIAL PURPOSE (SUBSCRIPTS INDICATE USE. DW - DISH-WASHER, CD - CLOTHES DRYER, ETC.)

SWITCHES

S SINGLE-POLE

S2 DOUBLE-POLE

S3 THREE-WAY

S4 FOUR-WAY

SK KEY-OPERATED

SP SWITCH AND PILOT LAMP

SL SWITCH FOR LOW VOLTAGE

─○S SWITCH AND SINGLE RECEPTACLE

SIGNAL SYSTEMS

⊡ PUSH BUTTON

BUZZER

BELL

BELL AND BUZZER

CH CHIMES

D ELECTRIC DOOR OPENER

TV TV OUTLET

R RADIO OUTLET

─Ⓣ THERMOSTAT

WIRING *

───── WIRE CONCEALED IN CEILING OR WALL (A SINGLE LINE INDICATES A TWO-WIRE CIRCUIT

── ── WIRE CONCEALED IN FLOOR

─ ─ ─ ─ EXPOSED WIRING

$\xrightarrow{3}$ BRANCH CIRCUIT LINE TO SERVICE PANEL (SOMETIMES CALLED A "HOME RUN") NUMERAL OVER THE ARROW INDICATES CIRCUIT NUMBER

─///─ THREE-WIRE LINE

─////─ FOUR-WIRE LINE

* THESE SYMBOLS APPLY TO ARCHITECTURAL DRAWINGS ONLY.

Figure 12-4. Electrical symbols used on architectural drawings.

The scheme used to mark the wiring information on the electrical floor plan is not controlled by a specification. The contractor can use any system that will be understood. A few markings are widely used, however. A single solid line may be used to represent concealed two-wire cable. One or more broken line patterns may be used to represent exposed wiring or other special situations. An arrowhead at the end of a cable line indicates that that line must run to the service panel. When a cable run requires more than two conductors, short diagonal lines indicate how many conductors are needed. For example, when a receptacle or fixture is controlled by two or more switches, additional conductors are needed between the switch and the fixture

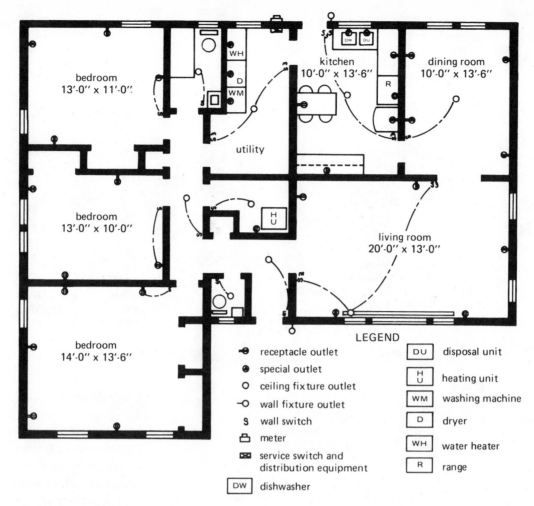

Figure 12-5. Electrical floor plan.

LEGEND

receptacle outlet	
special outlet	
ceiling fixture outlet	
wall fixture outlet	
S wall switch	
meter	
service switch and distribution equipment	
DW dishwasher	

DU	disposal unit
HU	heating unit
WM	washing machine
D	dryer
WH	water heater
R	range

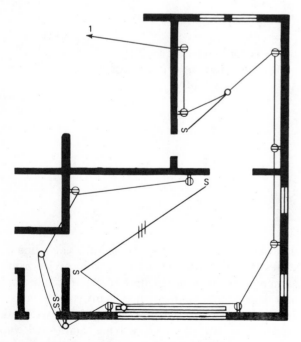

Figure 12-6. Floor plan marked for wiring.

or receptacle. This is shown by short cross strokes. These short diagonal lines also show the number of conductors in conduit.

Architectural drawings are always drawn to scale. This means that all dimensions on the drawings are proportional to the actual dimensions of the finished building. The scale used is always shown somewhere on the drawing. Typical scales are 1/8 or 1/4 inch equal to 1 foot. This means that every 1/8 or 1/4 inch on the drawing represents 1 foot on the actual building. This scale can often be used to make rough estimates of the cable lengths required. A word of caution here. It is frequently necessary to reproduce several copies of architectural drawings for use by the various trades. If, in the process of reproduction, the drawings are enlarged or reduced photographically, the scale can no longer be used. Many reproduction processes that make copies that appear to be the same size, actually reduce the drawing slightly in the process; others enlarge. When 1/8 inch represents a foot, even a slight change in drawing size can cause a large error in estimating cable length.

• ELECTRICAL DIAGRAMS •

The electrical floor plan provides the information needed to install cables, fixtures, conduit, electrical boxes, and receptacles. For other wiring tasks, electricians must be familiar with various types of electrical drawings. Power to many large appliances is supplied by connecting conductors directly to terminals on the appliance. Appliance wiring diagrams must be consulted to know where and how to make the connections. To wire low-voltage circuits such as intercoms, TV and FM antenna systems, remote control switching circuits, and security systems, schematic diagrams, block diagrams, and pictorial diagrams must be used. The paragraphs that follow describe the kinds of information provided by each type of diagram. A relatively simple doorbell circuit is used as an example. The principles apply, however, regardless of the complexity of the drawing.

Block Diagrams

The simplest form of electrical diagram is the block diagram. Each piece of a circuit that does a specific job is represented by a box or rectangle. Lines drawn between the boxes show the flow of current and the sequence in which things happen. Block diagrams do not show actual wiring.

The doorbell block diagram shows that the circuit consists of a transformer, a switch, and a bell unit. Functionally, the switch is between the transformer and the bell unit (Fig. 12-7). The block diagram also indicates that the transformer requires a connection to 120-volt power. The 120-volt input to the transformer and the line from the transformer to the switch, and to the bell unit is shown as a single line. This is customary on block diagrams because they show only the *relationship* of things, not electrical connections.

Block diagrams are most helpful when you are learning how a device or circuit works. They provide a simple, overall picture of how the main parts of the circuit or device are related to each other. Block diagrams are often used in service manuals and booklets to explain the general way in which something works. They are not primarily intended to be used for troubleshooting or repair; sometimes, however, you can use a block diagram in connection with other diagrams to narrow the source of trouble to one area. For example, if you can measure 10 to 12 volts at the

transformer output, but no voltage is present at the bell terminals when the doorbell button is pressed, the trouble is in the wiring or the switch. Other drawings would have to be used to make further checks.

Schematic Diagrams

Schematic diagrams show the electrical relationship of the devices on a circuit (Fig. 12-8). The schematic should provide enough information for the user to determine the conditions necessary for the circuit to operate. The schematic should also show what voltages must be present at various points in the circuit and what conditions must be met for the voltage to be present. Current flow may be shown or can be calculated from applied voltage and load data shown on the schematic. Schematic diagrams do *not* show the actual wiring runs or the physical appearance of devices.

From the doorbell schematic we can learn that the source power (120 volts, 60 hertz) is connected to the primary of a transformer. It is a step-down transformer and the secondary voltage is 10 to 12 volts. We can conclude from this that the doorbell will operate properly on any voltage between the limits shown. We can see that the bell circuit consists of two electromagnets in series. The front door push button and the rear door push button are in parallel. When either push button is pressed, the secondary voltage is applied to the bell electromagnets through a contact on the bell clapper arm. When the electromagnets are energized by the 10 to 12 volts, the clapper arm is pulled toward the magnets. This opens the contact in the clapper arm. Power to the electromagnets is interrupted. The clapper arm is spring-loaded and will return to the position shown on the diagram when not attracted by the magnets. This closes the contact on the clapper arm, energizing the magnet, and the cycle repeats. The clapper arm will continue to move back and forth, causing the bell to ring, as long as either push button is depressed.

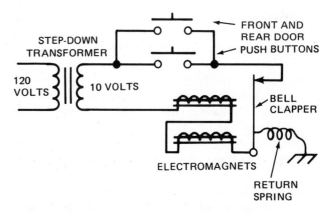

Figure 12-8. Doorbell schematic diagram.

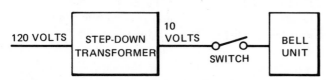

Figure 12-7. Doorbell block diagram.

Most of the circuit action described here should be evident by inspecting the diagram. One exception is the spring action of the clapper arm. The rule to remember for this sort of information is that mechanical and electromechanical devices such as push buttons, relays, and switches are normally shown in the off or unenergized position. Relay controls and controls such as the one on the clapper arm may be marked N.O. or N.C., meaning normally open or normally closed, if the condition is not evident from the drawing. The fact that the clapper arm is spring-loaded may be contained in a note and may not be shown on an electrical schematic. It could be guessed from the fact that the arm must return to the position shown in order for the bell to work.

Schematic diagrams, then, are troubleshooting aids. They show you how a circuit should function electrically and they show you what devices may be at fault if the circuit is not working properly. Remember that notes on schematic diagrams contain useful information that is not apparent from the drawing. Always read notes carefully. Remember, too, that schematics do not tell where the actual points are at which voltage can be measured. The schematic diagram tells you what the correct voltage should be. You must check a wiring diagram or a pictorial diagram to locate the points on which the voltmeter probes must be placed.

Wiring Diagrams

Wiring diagrams show you how circuits are actually wired (Fig. 12-9). They also show the points at which wires are connected and how the wire runs are actually made. We can see from the wiring diagram that the transformer primary power can be measured by opening an electrical box in which the transformer pigtail leads are connected to the power line conductors. It is also evident that only one wire runs from the transformer secondary to each push button. The other wire runs directly from the transformer to the bell. Wiring diagrams are the most useful drawings for general troubleshooting.

Pictorial Diagrams

These diagrams have about the same electrical information as wiring diagrams, but they show the actual appearance of the devices used in the circuit (Fig. 12-10). These diagrams are especially helpful in showing the layout and location of terminal strips and internal connections.

Other Diagrams

MECHANICAL SCHEMATICS • The schematic diagrams described previously were electrical schematics. Schematic diagrams can also show how me-

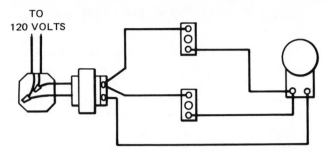

Figure 12-9. Doorbell wiring diagram.

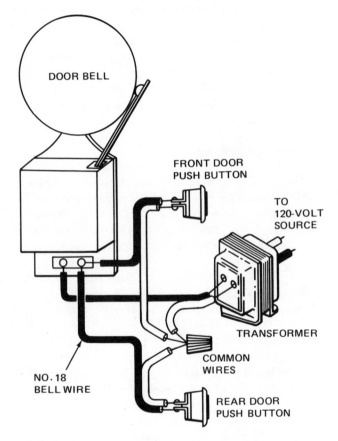

Figure 12-10. Doorbell pictorial diagram.

chanical devices work. Service data for motor-driven tools and appliances may include mechanical schematics. Generally these drawings use electrical symbols for the electrical parts of the diagram and mechanical symbols for other items. Mechanical symbols are highly simplified drawings of the things they represent. You will probably recognize most symbols without a guide.

EXPLODED VIEWS • Exploded views are also helpful in disassembling and reassembling mechanical items (Fig. 12-11). These drawings show the parts of a device as they actually look. All parts are shown to the same scale unless otherwise marked. The parts are positioned in the drawing as they would look if the device were taken apart and the parts were able to float

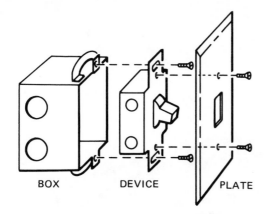

Figure 12-11. Exploded view.

in space. Dotted lines are sometimes used to show how parts must be moved to fit together. Exploded views usually show the complete disassembly of a device. If only partial disassembly is necessary, study the exploded view carefully to see which parts can remain assembled and which must be removed to get to the test or inspection point you want to reach.

CUTAWAY AND PHANTOM VIEWS • Two other types of diagrams can be used to troubleshoot mechanical devices. One is a cutaway view (Fig. 12-12). This drawing shows the internal workings of a device

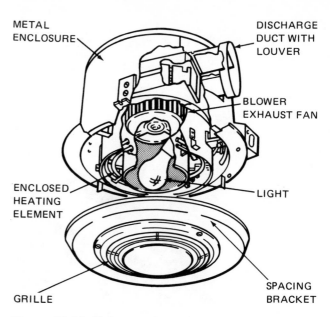

Figure 12-12. Cutaway view.

by showing it with outside housings and internal supporting parts cut away, so that the relationship of the operating parts can be seen. Another type of drawing, known as a phantom view, provides much the same kind of information by showing parts of a device as if you could see through them.

• REVIEW QUESTIONS •

1. What is the main piece of information the electrical contractor gets from the architectural plot plan?

2. What type of information can an electrician get from the architectural elevation drawings?

3. The figure below is part of an architectural electrical floor plan. What do the electrical symbols represent?

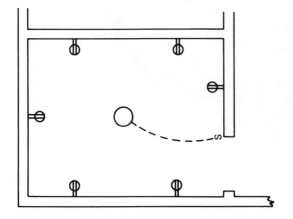

4. What kind of information does the electrical contractor add to the floor plan?

5. If an electrical floor plan has a scale of 1/8 inch = 1 foot, what is the actual distance between receptacles that are 1 1/4 inches apart on the drawing?

6. What type of information can you get from an electrical block diagram?

7. On schematic diagrams the abbreviations N.O. and N.C. are often included near symbols for switches and relays. What do the abbreviations mean?

8. Schematic diagrams often show the way a circuit operates under certain conditions. Where can you expect to find a description of these conditions?

9. What sort of information can you get from wiring diagrams and pictorial diagrams?

10. Four other types of diagrams are mentioned in this chapter. Name two.

13
WIRING BASIC
CIRCUITS

• INTRODUCTION •

This chapter and the ones which follow tell you how to wire and test the branch circuits used in residential electrical systems. Branch circuit wiring puts into practical use all the subjects covered in the preceding chapters. A knowledge of electrical theory (Chapters 1 to 3) helps you to understand why branch circuits are wired as they are. The rules that must be followed in branch circuit wiring are based on the NEC, local codes, and safe working habits (Chapter 4). You must be familiar with electrician's tools and test equipment (Chapter 5) in order to do the job. The materials you use are wire and cable (Chapter 6) or conduit (Chapter 7). An electrical box or equivalent enclosure must be installed at each outlet point in the branch circuit (Chapter 8). The devices installed in the electrical boxes are switches, receptacles, and fixtures (Chapter 9). Every branch circuit must have overcurrent protection at its source (Chapter 10). The source of power for all branch circuits is the service panel located at the building service entrance (Chapter 11). To install branch circuits you must use electrical diagrams (Chapter 12).

Branch circuit wiring is the principal job of electricians engaged in residential wiring. The houses that people live in are built in a wide range of sizes and thousands of designs. The branch circuits in this broad range of homes—and in many larger buildings as well—consist of four basic circuit types. The difference between large and small homes is mainly in the number of circuits required. This chapter describes the way the four basic circuits are wired, and covers the most frequently encountered special circuit variations. If you have been able to answer the review questions in Chapters 1 to 12, branch circuit wiring will be easy to understand and the wiring techniques described will make sense to you.

• SINGLE-OUTLET AND MULTIPLE-OUTLET CIRCUITS •

Electricians refer to the wiring of a branch circuit as a *run*. The run begins at the overcurrent protection device in the service panel (known as the source) and ends at the last outlet on the circuit (known as end-of-run). On single-appliance circuits (both 120-volt and 120/240-volt) there is, of course, nothing in between. Power goes directly from the service panel to the single outlet at the appliance (Fig. 13-1). The electrical connection is made by inserting the plug of the appliance into a receptacle or by connecting the power lines directly to terminals on the appliance. In these circuits (with one exception) switches to turn the appliance on and off are built into the appliance, and no special

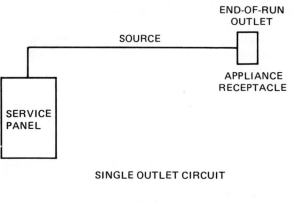

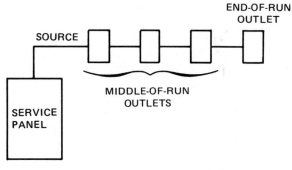

Figure 13-1. Single-outlet and multiple-outlet circuits.

wiring need be done for this purpose. The single exception is the wiring to heating units. Many local codes require that a remote switch (identified by a special faceplate) be included in these circuits.

General-purpose and appliance branch circuits provide power to two or more outlets. In addition to the source (the service panel) and an end-of-run (EOR) outlet, general-purpose circuits have outlets in between which are known as middle-of-run (MOR) outlets (Fig. 13-1). Middle-of-run outlets must be wired so that power is supplied to the device installed at the outlet and is also continued through the outlet to supply the rest of the run.

Appliance branch circuits generally have only two outlets. These circuits are intended for portable appliances with high wattage ratings. The number of outlets is limited to prevent overloading the circuit. Because the appliances have built-in switches designed to handle heavy current flow, no switch control of appliance outlets is needed.

General-purpose branch circuits are intended for lighting and small appliances. These circuits have many outlets and almost always include some switch control. There are four possible methods of wiring for single-switch control and several more methods of wiring for multiple-switch control. These switching circuits are described in the sections that follow.

Circuit Pictorial Diagrams

Pictorial diagrams are used in this chapter to show how basic circuits are wired with the most widely used types of cable. The symbols used in these diagrams are identified in Fig. 13-2. The color coding used in the pictorials is the coding that must be followed when wiring with cable. For simplicity, cable grounding wires are not shown. However, in addition to the wiring shown, grounding wires must be joined in each box and connected by jumper to the box (if it is metal) and to the green-tinted grounding screw on the switches and receptacles. The diagrams show how wiring and connections differ, depending upon the point in the circuit

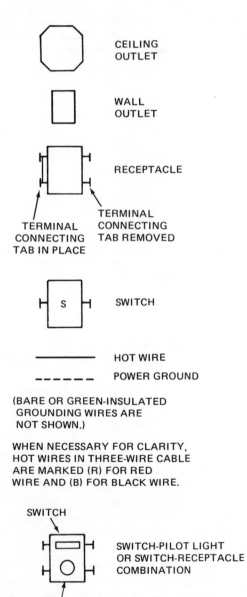

Figure 13-2. Symbols used on wiring pictorials.

at which source power is available or the type of switching being done. The relative positions of the outlets, switches, and fixtures in the pictorials is otherwise not important.

Standard wiring practice requires that only conductors with the same color insulation be joined (black to black, white to white, red to red, etc.). Cable wiring requires that some exceptions be made to this rule and the diagrams show these exceptions.

Differences in Cable and Conduit Wiring

The pictorial diagram in this chapter can be applied to conduit wiring, if three important wiring differences are kept in mind.

The first difference applies to the wiring of switch loops, the conductors between the source power and the switch. In Chapter 9 the description of switch uses states that switches are wired only in the hot wire. The hot wire is usually the black- or red-insulated wire. When wiring with cable, standard color code connections cannot always be maintained. Virtually all two-wire armored and nonmetallic cables have black- and white-insulated conductors. A red-insulated conductor is added in three-wire cables.

When a fixture must be controlled by a wall switch and source power is available at the fixture box, the hot wire must run from the fixture box to the switch and then back to the fixture box. The conductors that run from the fixture box to the switch are called a switch loop. Article 200-7 of the NEC describes how a switch loop must be wired when cable is used. The wiring must be done so that power *to* the switch is carried by the white-insulated conductor and power *from* the switch by the black-insulated conductor. This wiring requires that a white-insulated conductor and a black-insulated conductor be joined. This is the only situation in which that connection is permitted. This wiring is shown in Fig. 13-3. Note the wiring in the figure carefully. Standard white-to-white, black-to-black connections are maintained at the fixture. The black fixture wire is connected to the black switch-loop wire. The black and white switch-loop wires are connected to the switch. The white switch-loop conductor must be connected to the black hot wire to complete the circuit. At one time it was customary to mark the white conductor with black tape or black paint at both ends when used in a switch loop. This is not required by the NEC, but may be required by local codes.

When this situation is encountered in conduit wiring no exceptions need be made. Conductors with the correct color insulation can be fished through the conduit. Switch-loop wiring and conduit is also shown in Fig. 13-3.

The second difference between cable and conduit wiring can also be seen in Fig. 13-3. Note that the black wire from the source can be continued through the

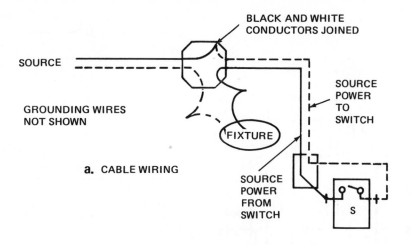

a. CABLE WIRING

SOURCE

GROUNDING WIRES
NOT SHOWN

BLACK AND WHITE
CONDUCTORS JOINED

FIXTURE

SOURCE
POWER
TO
SWITCH

SOURCE
POWER
FROM
SWITCH

S

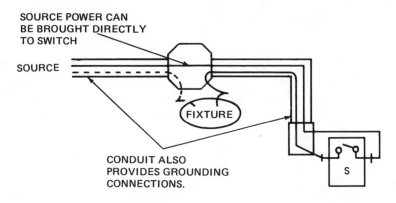

SOURCE POWER CAN
BE BROUGHT DIRECTLY
TO SWITCH

SOURCE

FIXTURE

CONDUIT ALSO
PROVIDES GROUNDING
CONNECTIONS.

S

b. CONDUIT WIRING

Figure 13-3. Cable and conduit wiring for switch loops.

ceiling outlet to the switch. Many wiring connections can be avoided in conduit installation.

The third difference has to do with the grounding conductors. Although grounding conductors are not shown on the cable pictorials, they would, of course, have to be joined throughout each circuit. If the conductors in the pictorials were enclosed in rigid, intermediate, or thin-walled (EMT) conduit, the wiring would be complete as shown. No separate grounding conductor would be required.

• SINGLE-SWITCH CIRCUITS •

There are four possible methods of wiring for single-switch control. The differences in switch wiring result from the differences in middle-of-run and end-of-run outlets and from the way the power lines are routed. A sample circuit which includes all four possibilities is shown in Fig. 13-4, which shows ceiling fixtures as the switched devices. Similar wiring can be used if receptacles are switched.

Basic Types

In the first fixture-switch combination, no. 1, the cable from the power source is routed to the ceiling box. Because this is a middle-of-run (MOR) outlet, power must continue through the outlet to the rest of the circuit. The switch must be wired to control the fixture mounted on the box, but not to interrupt power to the rest of the circuit. A single-pole single-throw (SPST) switch—such as the quiet switches described in Chapter 9—can be used in this circuit. Two wires must be routed from the ceiling box to the switch. Conductors to carry power to the rest of the circuit are also in this box. The connections shown in the figure allow the switch to control the fixture without affecting power to the rest of the circuit.

When source power is available at the switch outlet, standard color code connections can be maintained. Switch-and-fixture combination no. 2 in Fig. 13-4 illustrates this connection.

The power line having MOR combination no. 2 branches to feed two end-of-run (EOR) switch-and-fixture combinations. Combination no. 3 is essentially

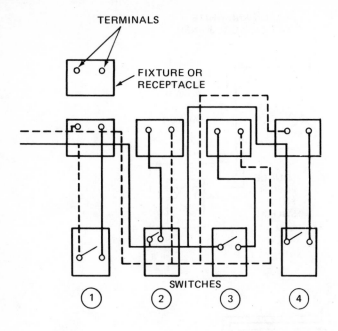

TERMINALS

FIXTURE OR RECEPTACLE

SWITCHES

① ② ③ ④

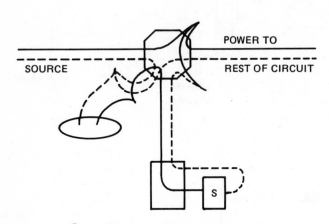

POWER TO

SOURCE

REST OF CIRCUIT

① MIDDLE-OF-RUN, POWER AT FIXTURE

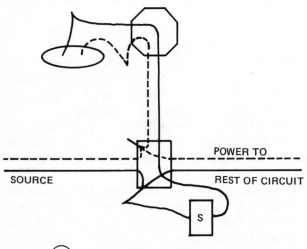

POWER TO

SOURCE

REST OF CIRCUIT

② MIDDLE-OF-RUN, POWER AT SWITCH

Figure 13-4. Single-switch circuits.

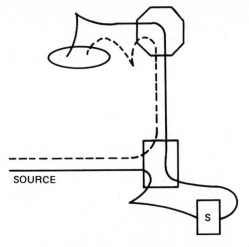

SOURCE

③ END-OF-RUN, POWER AT SWITCH

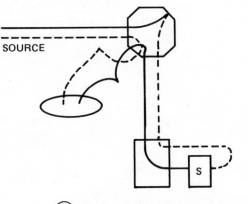

SOURCE

④ END-OF-RUN, POWER AT FIXTURE

Figure 13-4. *(Continued)* Single-switch circuits.

the same as no. 2, except that power does not continue beyond the switch outlet. Combination no. 4 employs a switch loop and is wired like combination no. 1 except that the run ends at the fixture.

Switch–Pilot Light Wiring

The wiring of switch and pilot light combinations is discussed briefly in Chapter 9. We will consider here the wiring variations that result from two possible locations of source power. The chief difference between pilot light switches and standard switches is that a power-ground conductor must be available at the outlet for connection to one of the pilot light terminals.

Figure 13-5 shows the wiring when source power is available at the switch–pilot light outlet. Two-wire cable can be used for the circuit and standard color code connections can be maintained. The figure shows a jumper between the brass-colored pilot light terminal and one switch terminal. Some switch-and-pilot light devices have an internal connection between these points and the jumper is not required.

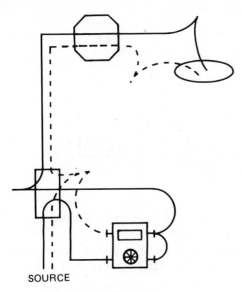

Figure 13-5. Switch–pilot light wiring, source at switch outlet.

When source power is available at the monitored fixture outlet, a three-wire cable is required for the switch-loop and pilot-light ground. This wiring is shown in Fig. 13-6.

Other Switch Combinations

Other devices combining switches with receptacles, timers, and dimmers are wired in much the same way as the switch–pilot light. Receptacles and timers require that a power-ground connection be available at the outlet. Figures 13-7 and 13-8 show the wiring for a switch receptacle with source power available at the switch outlet and at the controlled fixture outlet.

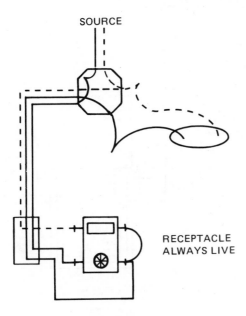

Figure 13-6. Switch-pilot light wiring, source at outlet.

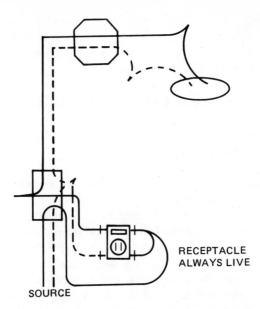

Figure 13-7. Switch-receptacle wiring, source at switch outlet.

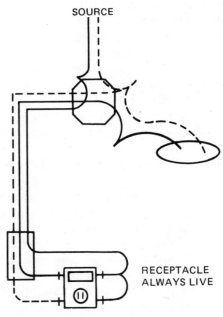

Figure 13-8. Switch-receptacle wiring, source at fixture outlet.

Timers can be wired the same way, the principal difference being that many timers have pigtail connections, rather than screw terminals. The power-ground connection is necessary at timers to operate the clock mechanism. Short-period timers (up to 1 hour) may have a spring-wound clock rather than an electric mechanism. These timers can be installed by connecting them into the hot wire, just as standard switches are.

Dimmer switches are available in many forms for incandescent and fluorescent fixtures. As described in

Chapter 9, some require a power-ground connection, and some can be wired as standard switches. When a power ground is needed, the wiring shown for the switch-receptacle can be used.

• MULTIPLE-SWITCH CIRCUITS •

Two-Switch Circuits

Another type of switching circuit is frequently used in residential and commercial wiring. This circuit employs two three-way switches and allows one receptacle or fixture to be controlled from either of two locations. You will recall from Chapter 9 that the so-called three-way switch is a single-pole double-throw switch with three terminals, one of which is common. In the standard toggle switch form the switch has no positions marked on or off. In both positions of the toggle, the common terminal is connected to one of the other two terminals. Figure 13-9 shows schematically how two of these switches can be wired to control one fixture. Note the following features of this circuit. The ground wire is connected directly to the fixture. The hot wire from the source is connected to the common terminal of one three-way switch. The other two terminals of the switch are directly connected to the equivalent terminals on the second three-way switch. The wires connecting these terminals are called *travelers*. The hot wire to the fixture is connected to the common terminal of the second switch. As shown, the fixture is off. Changing the position of either switch, in any order, will switch the fixture to the opposite condition.

The actual wiring for this circuit when source power is available at one switch outlet is shown in Fig. 13-10. A three-wire cable runs from one switch outlet to the other. A two-wire cable runs from the fixture to the second switch outlet. Standard color code connections can be maintained throughout this circuit. The same wiring can be used with conduit. Note, also, that both switch outlets contain five conductors, grounding wires if cable is being used, and the switch. The boxes installed at these locations must then be at least 3 inches

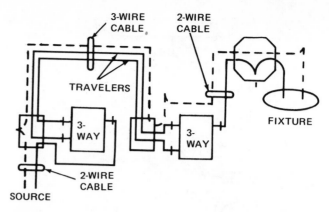

Figure 13-10. Two-switch control, source at switch.

× 2 inches × 2 3/4 inches deep to conform to the NEC requirements for maximum number of conductors per box size.

When source power is available at the fixture outlet, the wiring for two-switch control can be done with the same type and amount of cable, but the single switch-loop color code exception must be used to complete the circuit wiring (Fig. 13-11). Note that the black wire from the fixture is connected to the black wire in the two-wire switch-loop cable. At the switch outlet the black switch-loop wire is connected to the common terminal on the switch. The white wire in the two-wire cable is connected to the white wire in the three-wire cable. The red and black wires in the three-wire cable became the travelers. As in the single switch loop, the exception to the color code takes place in the connection of the white switch-loop wire to the black source wire and the connection of the white switch-loop wire to the common terminal of one three-way switch.

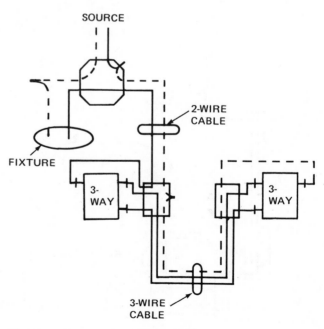

Figure 13-11. Two-switch control, source at fixture.

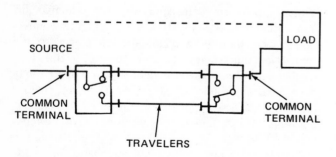

Figure 13-9. Two-switch control, schematic diagram.

Another version of the two-switch control circuit can be used when source power is available at the fixture outlet and the fixture is located between the switches. For this circuit, three wires are required from the source to each switch. If three-wire cable is used, the only exception to standard color code connections is the use of the white-insulated conductor as one of the travelers. The wiring for this circuit is shown in Fig. 13-12.

Three-Switch Circuits

One or more fixtures or receptacles can be controlled from three or more switches by adding four-way switches in the traveler portion of the two-switch circuits we have seen. The four-way toggle switch, you will recall from Chapter 9, has two unmarked positions and four terminals. In one position, the terminals are connected straight through. In the other position, the connections are interchanged (Fig. 13-13). A complete switching circuit is shown schematically in Fig. 13-14. Note that changing the position of any one of the three switches will change the circuit from off to on, or from on to off. Note also, any number of four-way switches can be added in the traveler wires and control will be possible from each. Three wiring arrangements for three-switch control are shown in Figs. 13-15 to 13-17.

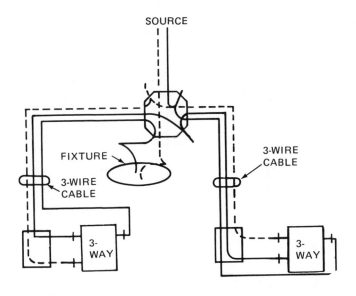

Figure 13-12. Two-switch control, source at central fixture.

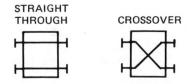

Figure 13-13. Four-way switch positions.

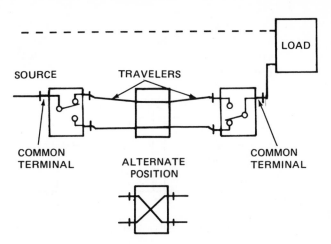

Figure 13-14. Three-switch control, schematic diagram.

The circuit shown in Fig. 13-15 is similar to the two-switch circuit shown in Fig. 13-10. Source power is available at one of the switch outlets. Power is routed through the other two switch outlets to the fixture being controlled. Three-wire cable must be used between the switches. Standard color coding can be maintained in this circuit.

The three-switch circuit shown in Fig. 13-16 is similar to the two-switch circuit shown in Fig. 13-11. Source power is available at the fixture outlet. Following the NEC requirement for switch-loop wiring, the black conductor in the two-wire cable is connected to the fixture. The white wire in the two-wire cable must then be connected to the black source wire. Three-wire cables are used between the switch outlets. The white wires are joined to carry the source hot wire to the three-way switch at the end of the loop. The red and black wires in the three-wire cables are the travelers. The four-way switch is connected to the red and black travelers.

Figure 13-17 is similar to Fig. 13-12. Source power is available at the fixture outlet, but the switches are located either side of the source. In this circuit the travelers must run through the source outlet and three-wire cable must be used for all wiring. Standard color code connections can be used in this circuit. Note that the white conductor in the three-wire cables is used as one of the travelers.

• SPLIT-RECEPTACLE WIRING •

The standard duplex receptacle used in house wiring has two brass-colored terminals on one side and two silver-colored terminals on the other side. The terminals on each side are joined by a strip of metal. A connection to one of the two brass terminals and to one of the two silver terminals supplies power to both upper

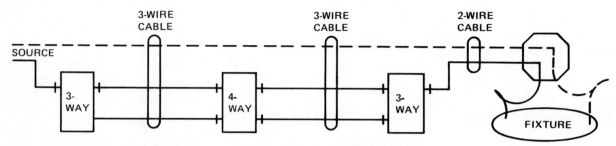

Figure 13-15. Three-switch control, source at switch.

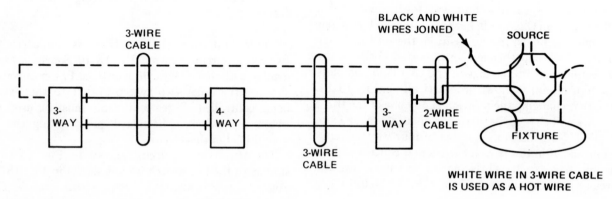

Figure 13-16. Three-switch control, source at fixture.

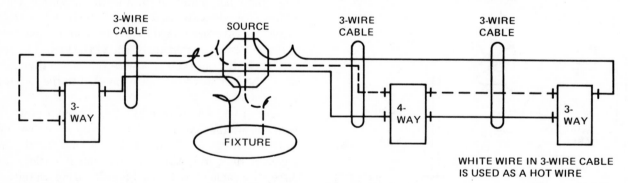

Figure 13-17. Three-switch control, source at central fixture.

and lower receptacles. As described in Chapter 9, the metal strip connecting the terminals can be broken off so that each receptacle can be wired separately. This is done to provide switch control of one half of a receptacle or to split the receptacle between two branch circuits.

Split-Switched Receptacles

Figure 13-18 shows the wiring for both a full-switched receptacle and a split-switched receptacle

when source power is at the receptacle. The metal strip between the brass terminals is broken off. A black insulated jumper is connected from the black source wire to one brass terminal. The return black conductor from the switch loop is connected to the other brass terminal. As shown in the figure, the lower part of the receptacle is switch-controlled. The upper part is always live. The wiring can be reversed, if desired, so that the lower part is always live.

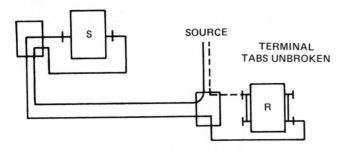

FULL-SWITCHED RECEPTACLE

SOURCE

TERMINAL
TABS UNBROKEN

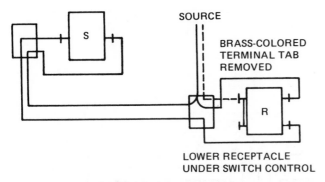

SOURCE

BRASS-COLORED
TERMINAL TAB
REMOVED

LOWER RECEPTACLE
UNDER SWITCH CONTROL

SPLIT-SWITCHED RECEPTACLE

Figure 13-18. Full-switched and split-switched receptacle, source at receptacle.

Figure 13-19 shows the wiring for a split receptacle when source power is available at the switch outlet. To wire this circuit, a three-wire cable must be used between the switch and receptacle outlets. Standard color coding can be maintained. As in the previous example, the lower half of the receptacle is switch-controlled.

Two-Circuit Receptacles

Residential electrical layouts can be designed so that receptacles in some areas are split, so that each half is

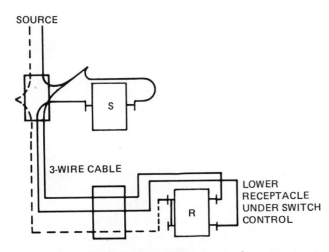

SOURCE

3-WIRE CABLE

LOWER
RECEPTACLE
UNDER SWITCH
CONTROL

Figure 13-19. Split-switched receptacle, source at switch.

powered from a different branch circuit. Figure 13-20 shows part of a floor plan for two-circuit receptacles, and how the wiring is done using three-wire cable. Splitting receptacles helps to balance the lighting load between the two utility hot wires, *if the two circuits used are powered from different service panel bus bars.* When wired in this way, the circuit breakers or fuses for each circuit should be "ganged" or locked together so that both circuits are automatically turned off together. This eliminates the possibility of shock from either circuit when working at split-receptacle outlets. In Chapter 11, in connection with load calculation, the current flow in the neutral conductor of three-wire circuits was explained. You will recall that if current flow in both hot wires is exactly equal, no current will flow in the neutral conductor. If current flow in the hot wires is unequal, current flow in the neutral conductor will be equal to the difference between the two hot-wire currents. It is important to remember that this rule on neutral conductor current flow is true in branch circuit wiring *only* when the two branch circuits are connected to opposite service panel buses.

In some cases—particularly when making additions or modifications to old work—it may be necessary to wire split receptacles to two branch circuits that are powered from the same service panel bus. When this type of split wiring must be done, three-wire cable cannot be used. Whenever conductors from two branch circuits powered by the same bus are present in a single outlet, it is important to keep the conductor pairs separate.

The reason for this separation of conductor pairs can be seen from Fig. 13-21a and b, which shows the wiring for a simple six-outlet circuit for six duplex receptacles to be split between branch circuits A and B. To simplify, we will assume that all branch circuit A receptacles have 200-watt loads for a total of 1200 watts. Assume that branch circuit B receptacles all have 250-watt loads, for a total of 1500 watts. When correctly wired, as in Fig. 13-21a, each conductor of branch circuit A has a current flow of 10 amperes and each conductor of branch circuit B has a flow of 12.5 amperes. Current flow in each circuit is well within the safe range for a 15-ampere circuit wired with no. 14 copper wire.

If wired incorrectly, as in Fig. 13-21b, current flow from both hot wires must be carried by one white conductor. Because both circuits are powered by the same bus, current in the two circuits will add. This causes the white conductor to carry 22.5 amperes. The circuit overcurrent protection device is in the black wires. Power will not be automatically cut off and the white conductor will become dangerously hot. Even if both hot wires were carrying less current, so that the white-wire current was 15 amperes, or less, this wiring error would be dangerous. In cables or conduit, the two

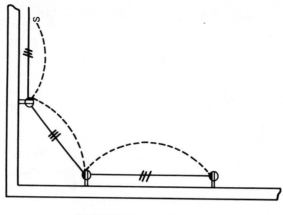

ELECTRICAL LAYOUT

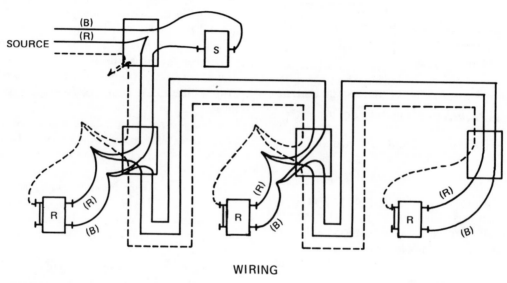

WIRING

Figure 13-20. Split receptacle floor plan.

circuit conductors, black and white, are parallel and close together. From the chapters on electricity, you will recall that current flow in a conductor creates a magnetic field around the conductor. The strength of the field is proportional to the amount of current. When current flow in both black and white wires is equal, the magnetic fields around each conductor are opposite at any instant and they cancel. When current flow in the wires is unequal, the wire carrying the larger current will induce a voltage in the other conductor that opposes the applied voltage. This has the effect of adding opposition to current flow, which can cause loss of power and overheating of conductors.

Figure 13-22 shows in detail how split receptacles are wired when both source circuits are powered by the same bus and when they are powered by different busses. Switch control can be added to this receptacle by using wire similar to that shown for single-circuit split-switched receptacles. Switched receptacle wiring for two-circuit receptacles powered from different busses is shown in Fig. 13-20. The wiring when both

circuits are powered by the same bus is shown in Fig. 13-23. Note that when both source circuits are powered by the same bus, special care must be taken to keep black and white conductors properly paired. Keep in mind, also, that the extra conductors may require the use of a larger box, as previously described.

• PLANNING BRANCH CIRCUIT WIRING •

Wiring runs for individual and multiwire branch circuits require little special planning. The location of an appliance determines the general location of the outlet. The exact location is chosen for ease of installation and connection to the appliance. It should not, however, be more than 6 feet from the appliance. Appliance branch circuit planning is also relatively simple. These circuits are installed in kitchens and workrooms. They are generally limited to two outlets

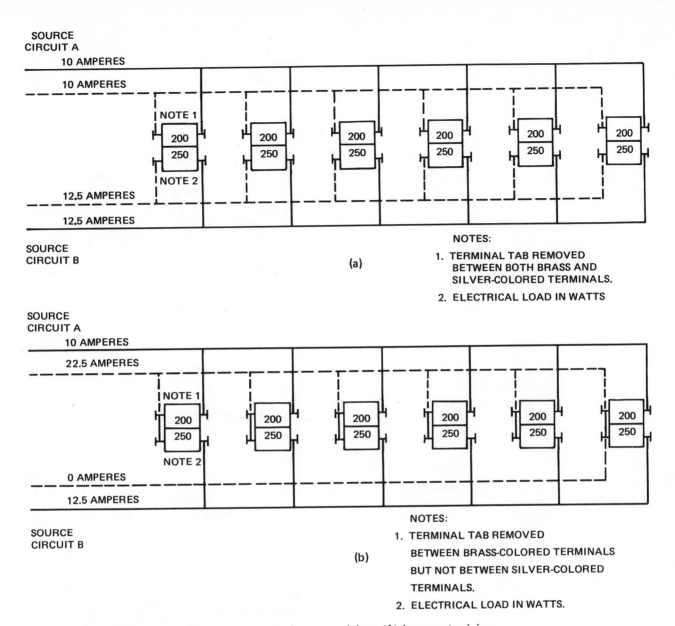

Figure 13-21. Split receptacle wiring. (*a*) Correct wiring; (*b*) incorrect wiring.

(although the NEC does not restrict the number). The outlets must be handy to work surfaces. The main decision to be made is in spacing the outlets so that receptacles are within reach of appliance cords. General-purpose branch circuit wiring, however, requires considerable analysis and planning. The wiring must conform with NEC and local code requirements; control of material and labor costs and balancing the load between the two utility hot wires must also be considered.

Location of Outlets

The NEC and most local codes specify the minimum number of outlets in terms of room size, but do not limit the number of outlets on one circuit. The NEC requires that no point along any wall shall be more than 6 feet from an outlet (Article 210-25(b)). One local code covers this same subject as follows. (Material is from the *Code Manual for the New York State Building Construction Code* and is quoted with permission.)

Receptacle outlets shall be provided in every kitchen, dining room, living room, parlor, library, den, sun room, recreation room, and bedroom. At least one receptacle outlet shall be provided for every multiple of 12 feet or major fraction thereof of the total distance around the room as measured horizontally along the wall at

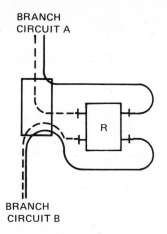

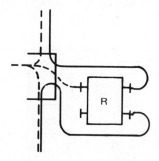

Figure 13-22. Wiring two-circuit receptacles.

the floor line. Receptacle outlets in the kitchen shall be supplied by not less than two small appliance branch circuits. At least one receptacle outlet shall be installed in the bathroom, for the laundry, and outdoors.

The difference between the two code requirements is an interesting example of how local codes and the NEC may differ. The NEC requirement is clear. No point on a wall shall be more than 6 feet from an outlet. The purpose of this requirement is to reduce the use of extension cords as much as possible. At first reading the local code appears to state the same requirement in somewhat different words. If receptacles are spaced 12 feet apart, no point on a wall will be more than 6 feet from a receptacle. However, this is not exactly what the local code says. The local code uses the room size to determine the *number* of receptacle outlets a room must have, but not *where* they must be. For example, compare the living room layouts in Fig. 13-24. The room is 20 × 13 feet. The total distance around the

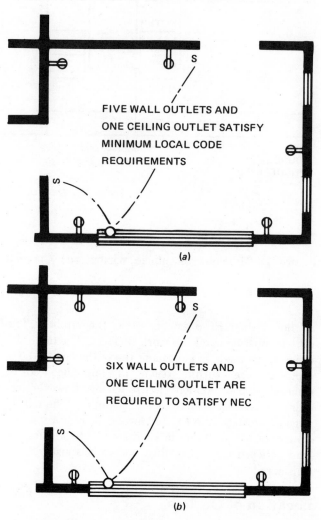

Figure 13-24. Living room electrical plans. (*a*) Local code; (*b*) NEC.

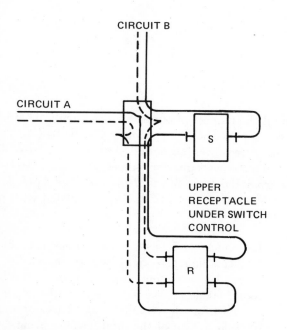

Figure 13-23. Wiring two-circuit switched receptacles.

room is 66 feet. If we must install one receptacle outlet for every 12 feet or major fraction of the distance around the room, we must install six receptacles (Fig. 13-24a). (A major fraction means one-half or more.) Note that the room does, in fact, have six outlets for receptacles. However, one of the outlets is a ceiling outlet, located over the living room window for special lighting. This leaves many areas of the living room wall more than 6 feet from an outlet, but the wiring shown meets the local code. This arrangement would *not* meet NEC requirements. To meet NEC requirements an additional outlet is needed and the outlets on the interior walls would have to be rearranged as shown in Fig. 13-24b.

Labor costs are kept low by simplifying installation as much as possible, thus reducing time on the job. Material costs are controlled by keeping cable runs as short as practical and combining functions wherever possible. For example, in an area where a junction box is required to join cables and a ceiling box is nearby, money can be saved by eliminating the junction box and installing a larger ceiling box to serve both purposes. Generally, it is desirable to group outlets on one circuit in one area of the house to keep cable runs short. Installation goes faster when as little wiring as possible is done in exterior walls. The insulation in exterior walls and the fire stops sometimes required slow down the installation of wiring. Of course, when scheduling permits, the wiring can be done before insulation is installed. The floor plans used as examples in this chapter show thirty-two general-purpose outlets; only eight of these are located on exterior walls.

Balancing the line load means keeping the load on each of the incoming hot wires as close to equal as possible. One way of doing this is to divide the receptacles among the branch circuits. The grouping has thirteen outlets in circuit no. 1, eleven in circuit no. 2, and eight on circuit no. 3. This is a reasonably balanced distribution. Note that both bathrooms are on circuit no. 2, even though the small bathroom is at the front of the house and the rest of circuit no. 2 is at the rear. This is still a good arrangement because it keeps the requirement for ground fault circuit interruption on one circuit.

Number of Outlets per Circuit

The NEC does not specify how many outlets a branch circuit can have. However, when computing a branch circuit load, outlets *not* used for lighting must be assumed to have a minimum load of 180 voltamperes (1.5 amperes at 120 volts for a resistive load. This must be taken into account, as noted below). Determining the number and location of outlets for a general-purpose branch circuit requires some calculations (wall space, for example) which can be used as

guidelines, but judgment and experience must also be included in the determination. For this reason, the architect, building contractor, electrical contractor, and, perhaps, the person who will occupy the house, all contribute ideas and suggestions to the electrical layout.

One of the general rules used to set a maximum limit on the number of outlets on a general-purpose branch circuit is to install no more than one outlet for each 1.5 amperes of overcurrent protection. This means that a 15-ampere circuit can have ten outlets and a 20-ampere circuit can have thirteen. This rule assumes an average load of 180 watts per outlet (1.5 amperes $\times$ 120 volts). When it is known that one outlet will have a greater load, this rule must be modified. For example, if a circuit will contain a ceiling fixture and the fixture is designed to hold six 60-watt lamps, this is equivalent to two average outlets. On a 15-ampere circuit only eight other outlets should be installed. It is important to understand that this method of determining the number of outlets per circuit must be used only for general-purpose lighting circuits when it is unlikely that all outlets will be fully loaded at any one time.

Another point to be considered is covered by NEC Article 220-2(a), which states that the continuous load on a branch circuit shall not exceed 80 percent of the circuit rating. The code defines a continuous load as one that is expected to continue for 3 hours or longer. A 120-volt, 15-ampere circuit has an 80 percent load of 1440 watts (15 amperes $\times$ 120 volts = 1800 watts $\times$ 80% = 1440). In locations where the lighting load is likely to reach this value for 3 hours or more, this limit must be taken into account. Normal room lighting rarely reaches the 80 percent value and probably never stays at that level for the 3-hour minimum. In some locations, however, this code restriction may require special wiring. For example, a luminous ceiling in a recreation room 12 $\times$ 15 feet or larger may have a fluorescent lamp load greater than 1440 watts. Recreation room uses are likely to mean the lights will be on for 3 hours or more. Outlets for the lighting in rooms such as this should be split between two branch circuits.

Another method of deciding how many outlets a circuit should have and where they should be installed requires that an analysis be made of how the various areas of the house are to be used and what the minimum lighting load will be. You will recall from the load calculation made in Chapter 11 that the local code, following the NEC value, specified that the lighting load should be figured at 3 watts per square foot of living area. A living room 13 $\times$ 20 feet, then, should have provision for at least 780 watts of illumination. Additional service should be added to this for such things as television sets, music systems, and other small appliances. If the average figure of 180 watts per outlet is used, five outlets would provide 900 watts, or slightly more than the minimum.

The electrician on the job sees the results of all these considerations. The actual wiring of a building is done from electrical layout drawings marked up to show how branch circuits will be wired. The markings show how outlets are grouped on circuits, and the order in which the outlets must be wired from the service panel to the end-of-run. This markup also shows how switches must be installed to control fixtures and receptacles. A single line is generally used to represent two conductors in cable or conduit. A short diagonal line across the single line means more than two conductors are required. A numeral near the short line indicates how many conductors are needed. In addition to the information about the electrical layout, the electrician must keep in mind the general rules for good wiring that must be followed when installing the circuits. If it is not noted on the drawing, the electrician must also figure the number of conductors that will have to be joined in each box and make certain the minimum size or larger box is installed in each location. For review, the important wiring rules and the rules for determining minimum box size are summarized in the following section.

• REVIEW OF WIRING BASICS •

General Rules

CONDUCTOR SIZE • Wire and cable conductors must be the proper size for the ampere rating of the circuit on which they are used. For residential wiring no. 14 copper is used on 15-ampere circuits, no. 12 copper on 20-ampere circuits. Large-appliance circuits require larger-size conductors. NEC Table 310-16 with notes covers most other circuit requirements.

CONDUCTOR MATERIAL • When conductors of aluminum or copper-clad aluminum are used, larger sizes are required than are specified for copper. Be sure you are using the right size for the conductor material.

WET AND DRY LOCATIONS • Conductors or conduit cables used must be suitable for the location in which they are installed. For cable and conduit used in residential wiring, check the UL listing or NEC Chapter 3 for details.

ADEQUATE SUPPORT • Cable or conduit must be adequately supported. The general rule for armored cable and nonmetallic cable calls for supports every 4.5 feet and within 12 inches of each outlet box. Intermediate metal conduit and EMT must be supported every 10 feet and within 3 feet of each outlet box. For rigid conduit, support requirements vary with diameter. See NEC Table 347-8.

NO SPLICES • All conductors must be continuous. Connections and splices can be made only in electrical boxes.

CABLE OR CONDUIT RUNS • When cable or conduit is installed in exposed locations, runs should be made to follow the line of studs and joists. This provides good support, prevents damage, and makes a neat appearance. Concealed runs need not follow studs and joists. The rule on concealed wiring is to make runs as short as possible.

PROTECT CABLES FROM DAMAGE • When cabling and thin-wall conduit are installed in a building, reasonable care must be taken to protect the conductors from damage. Holes bored in joists, beams, and rafters should be in the approximate center of the face of the structural member. Holes in studs should be not less than 1.25 inches from the stud edge. When notches are made for cable runs, the notch must be covered with a 1/16-inch steel plate. Cables run through metal studs or beams must be protected by bushings or grommets.

Rules for Determining Box Size

1. Count all current-carrying conductors that will enter the box, except fixture wires.
2. If the box has internal cable clamps, fixture stud, or hickey, add one to the above number (one maximum, not one for each item).
3. Add one to the above number if the box will contain a receptacle or switch.
4. Wires running continuously through the box (this can occur in conduit wiring) are counted as one wire.
5. Grounding wires (no matter how many) are counted as one wire.
6. Jumpers that begin and end in the box are not counted.

When the final number is determined, consult NEC Table 370-6(a) in the column appropriate to the conductor size to find the minimum box size that can be used.

• WIRING A SMALL HOUSE •

The one-family dwelling used for the load calculation in Chapter 11 is used in this chapter to illustrate some practical situations in branch circuit wiring. The house shown in the electrical layout is a single-level, moderate-sized house. The location of the service panel in the utility room suggests that the house does not have a full basement. The construction could be concrete block on a poured concrete slab, wood frame on concrete block crawl space, or wood frame on a poured concrete slab. For discussion purposes in this chapter, we will assume wood frame construction on a poured concrete slab foundation (Fig. 13-25). We will also assume that the service and appliances to be wired are the ones listed in the Chapter 11 load calculation. The power supplied by the utility company, as a result of

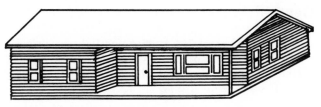

Figure 13-25. Wood frame house.

the load calculation, is three-wire 120/240-volt 150-ampere service. To serve the electrical needs of the house, the incoming power is divided into 12 branch circuits. As described in Chapter 11, the branch circuits consist of:

Three general-purpose, 15-ampere lighting and small appliance circuits
Two 20-ampere kitchen appliance circuits
Four 120-volt single outlet circuits
Three 240-volt large appliance circuits

The full service to be supplied is shown in Table 13-1.

From the plot plan (Chapter 12) we see that the service entrance is at the rear of the house, near the rear entrance (Fig. 13-26). A single-story house such as this would require a mast for connection of the service drop. Service entrance cable runs from the service head to the meter socket and then through the wall of the house to the service panel in the utility room. The grounding electrode system in this house could easily

TABLE 13-1. SERVICE REQUIREMENTS FOR TYPICAL HOME

Circuit Number	Area Served	Overcurrent Protection	Number of Outlets
1	Living room, dining room, entrance	15 A	13
2	Master bedroom, second bedroom, hall, furnace room	15 A	11
3	Kitchen ceiling light, rear entrance, outside receptacle, utility room, bathrooms, third bedroom	15 A	8
4	Spare		
5	Kitchen appliance	20 A	2
6	Kitchen appliance	20 A	2
7	Furnace blower	15 A	1
8 10 }	Dryer	Dual 15 A	1
9 11 }	Range	Dual 40 A	1
12 14 }	Water heater	Dual 20 A	1
13	Waste disposal unit	15 A	1
15	Washing machine	15 A	1
16	Dishwasher	20 A	1
17	Spare		
18	Spare		

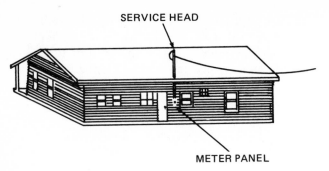

Figure 13-26. House service head.

consist of connection to a cold water pipe and, perhaps, a grounding rod. Plumbing lines are easily accessible in the utility room. The ground rod might be driven into the ground before the slab was poured. A long enough rod could be used to project through the slab so a cable and clamp could be installed. As an alternative method, the ground rod could be driven into the ground outside. In this case a grounding cable would have to be brought through the wall for connection to the service panel.

Single-level, slab foundation houses are usually wired by making use of the unfinished attic space for cable or conduit runs. Branch circuit cables run up from the service panel to the attic space and are then fanned out to the areas to be served. The cables or conduit enter the walls through holes bored in the wall top plate. Ceiling fixtures can, of course, be readily installed and wired from the attic space.

The paragraphs that follow describe one method of wiring the house we are using as an example. Keep in mind that the grouping of outlets on branch circuits can be done in many different ways and still conform to NEC and local code requirements. The examples described in this section represent only one approach to a wiring layout. Many others are possible.

General-Purpose Branch Circuit No. 1

This circuit contains thirteen outlets to provide power to nine receptacles, two ceiling fixtures, one entrance light, and one special lighting outlet. The floor plan layout for this circuit is shown in Fig. 13-27. This circuit will have 15 amperes overcurrent protection and can be wired with no. 14 copper wire. In addition to the outlets, the circuit requires three switch loops and a two-switch control for the special lighting outlet. As shown on the electrical layout, the special purpose outlet is typical of outlets for valance lighting. A valance is a short, decorative piece of cloth or wood

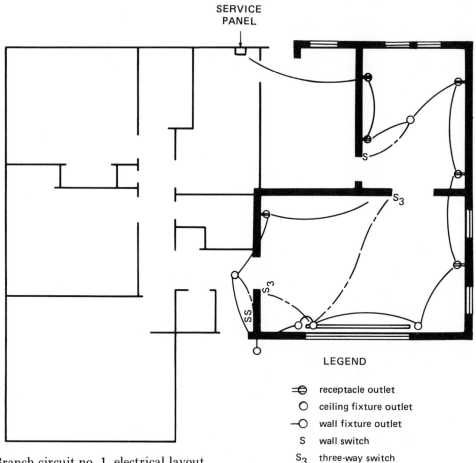

Figure 13-27. Branch circuit no. 1, electrical layout.

LEGEND

⊖ receptacle outlet

O ceiling fixture outlet

—O wall fixture outlet

S wall switch

S₃ three-way switch

that covers the top edge of window drapery. Lights placed behind the valance provide soft, low-level illumination. Fluorescent lamps and fixtures are often used in valances because they do not generate much heat and provide economical illumination. As shown in the layout, the valance extends aross the living room window (Fig. 13-28). The window is approximately 10 feet wide, so lighting could be done with two 4-foot, 40-watt fluorescent lamps. The outlet shown on the diagram could be recessed into the ceiling or mounted on the ceiling so as to be concealed by the valance. Electrical connections to the fixtures could be made by installing a receptacle in the outlet for cord and plug connection or running no. 14 wires from the fixture to the outlet box and connecting them to the source wires. Power to this outlet can be turned on and off from two locations. A simplified diagram of the complete branch circuit is shown in Fig. 13-29. A pictorial wiring diagram of the circuit is shown in Fig. 13-30. Note that this diagram simply combines as one circuit the receptacle wiring, switch loops, and two-switch control

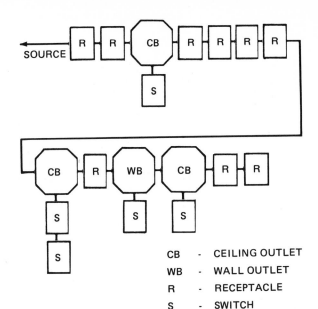

Figure 13-29. Branch circuit no. 1, simplified diagram.

that we have previously discussed separately. As noted earlier, many variations in branch circuit wiring are possible. For example, some designers might divide this branch circuit into two branch circuits so that a blown fuse or a tripped circuit breaker would not put so large an area in darkness.

General-Purpose Branch Circuit No. 2

This circuit provides power to two bedrooms, the furnace room, and one hall fixture. Like circuit no. 1, it has 15-ampere overcurrent protection. The circuit contains nine receptacles and two of the receptacles have single-switch control (Fig. 13-31). No new or unusual wiring is required for this circuit.

General-Purpose Branch Circuit No. 3

Because this circuit contains bathrooms and an outdoor receptacle, ground fault circuit interruption protection must be provided (Fig. 13-32). It was this consideration that suggested grouping both bathrooms on one circuit, even though this requires running the circuit from the back to the front of the house. You will recall from the discussion of GFCI's that three types are available. One type provides ground fault protection only for the outlet in which it is installed. Another type, known as a feed-through GFCI, protects not only the outlet in which it is installed, but also all outlets from the point of installation to the end of the run. The third type of GFCI is combined with a circuit breaker; it provides ground fault protection for all outlets on the circuit. Either of the last two types could be used in this circuit. The feed-through type is probably the most economical method to use here. The wiring with a feed-through GFCI is shown in Fig. 13-33.

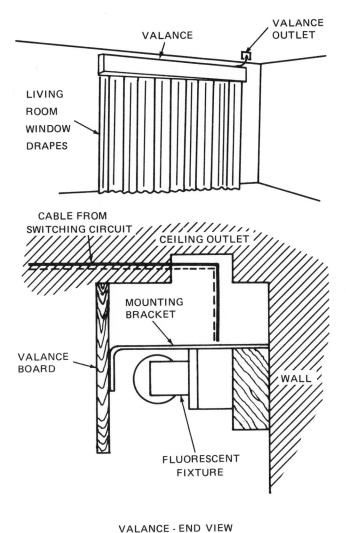

Figure 13-28. Valance lighting.

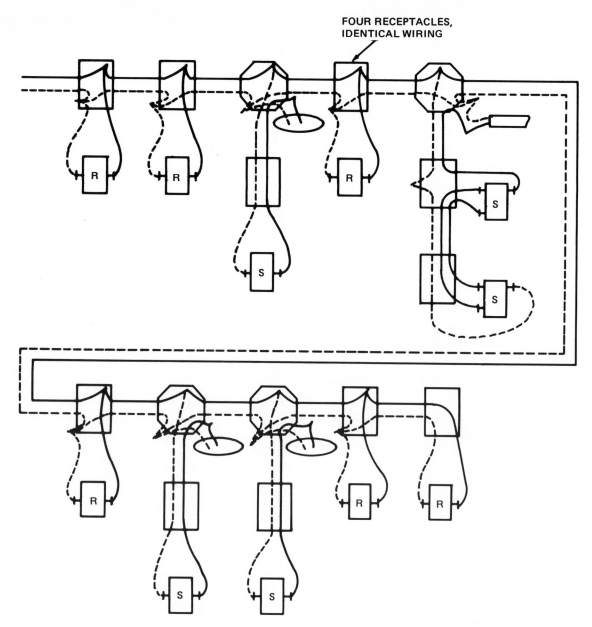

Figure 13-30. Branch circuit no. 1, pictorial wiring diagram.

Appliance Branch Circuits

Two appliance branch circuits, each supplying two outlets, are sufficient for a small kitchen. Both of these circuits must have 20-ampere overcurrent protection and must be wired with no. 12 wire, if copper is used. One circuit provides power for the refrigerator and one work surface appliance receptacle. The other circuit has two appliance receptacles (Fig. 13-34).

Individual Branch Circuits

Individual 120-volt branch circuits are usually installed for the units shown in Fig. 13-35. All of these units are relatively light loads when running, but all must handle motor starting surges. Typical nameplate wattages for these units are:

Heating unit blower motor 1/2 hp (450-600 watts)
Waste disposal unit 500-900 watts
Washing machine 500-800 watts
Dishwasher 1000-1500 watts

The first three appliances are suitable for 15-ampere overcurrent protection; the dishwasher requires 20-ampere protection. A wall switch is provided for the heater blower because some local codes require it. A special red faceplate is often used on the heater switch.

Multiwire Branch Circuits

These are three-wire circuits for units requiring 120/240-volt power (Fig. 13-36). All three units are located close to the service panel. This is desirable be-

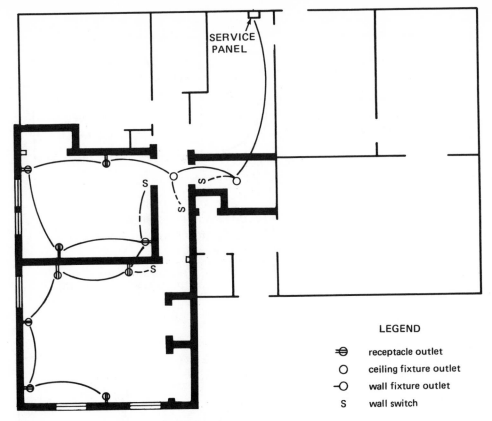

Figure 13-31. Branch circuit no. 2, electrical layout.

LEGEND

⊖ receptacle outlet

○ ceiling fixture outlet

─○ wall fixture outlet

S wall switch

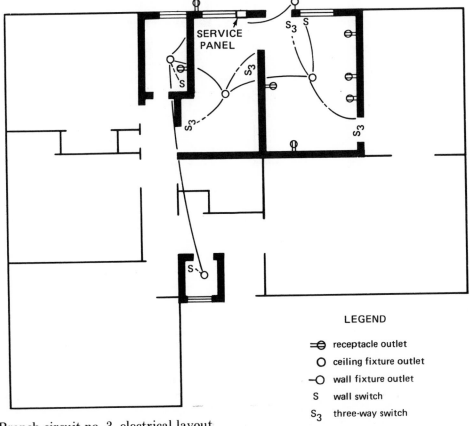

Figure 13-32. Branch circuit no. 3, electrical layout.

LEGEND

⊖ receptacle outlet

○ ceiling fixture outlet

─○ wall fixture outlet

S wall switch

S₃ three-way switch

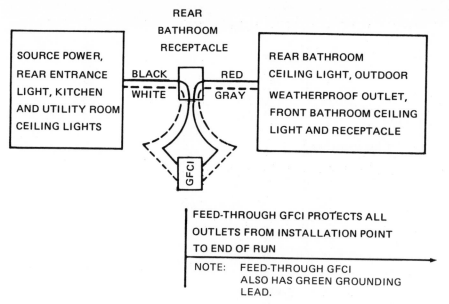

Figure 13-33. Branch circuit no. 3, GFCI protection.

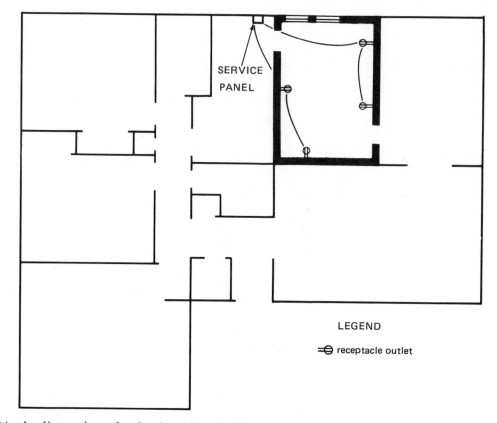

Figure 13-34. Appliance branch circuits, electrical layout.

cause it reduces line voltage drop. These units can draw heavy current, so line voltage drop must be considered when wiring the circuits. The dryer and water heater circuits as shown can be wired with no. 10 copper wire. Note that the NEC considers water heaters to be a continuous load. This means that the nameplate wattage rating must be increased by 25 percent when calculating wire size. As described in Chapter 11, electric ranges can be rated in accordance with a demand factor. This allows the range to be wired with smaller-size wire than the nameplate wattage would call for. The range in the example could be wired with no. 6 copper wire. When long wire runs cannot be avoided, larger-size conductors should be used.

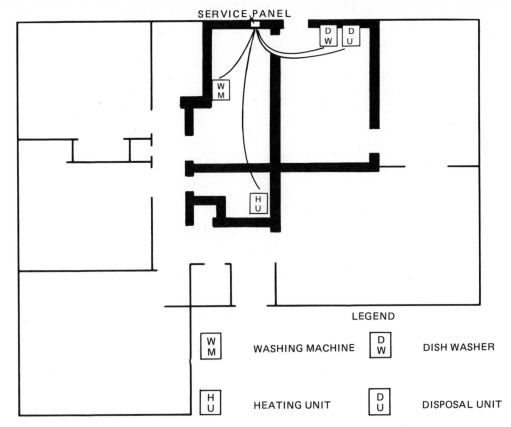

Figure 13-35. Individual branch circuits, electrical layout.

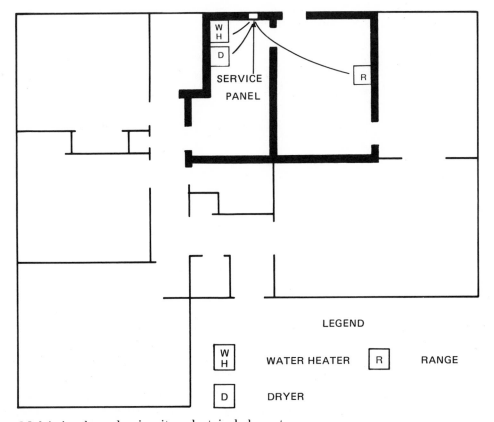

Figure 13-36. Multiwire branch circuits, electrical layout.

1. Branch circuit wiring is called a run. Where does the run start?

2. What do the phrases "middle-of-run" and "end-of-run" describe?

3. There are four kinds of branch circuits. What are they?

4. Name one important characteristic or use of each kind of branch circuit.

5. Standard color code wiring always joins black wires to black wires, white wires to white wires, and red wires to red wires. Switch wiring with cable requires an exception to this rule. Why?

6. Why is the exception in question 5 not required in conduit wiring?

7. When switches are combined with pilot lights, receptacles, or timers, both source power conductors must be available at the outlet. Why?

8. Below is a simplified diagram showing two switches controlling power to one load. What are the wires at A called? What type of switch is used for S1 and S2?

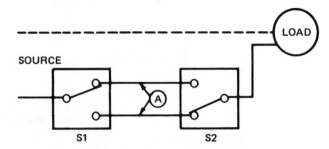

9. The circuit in question 8 can be changed to allow power to the load to be controlled from three switches. Where is the third switch added in the circuit? What type of switch is used?

10. Switches are sometimes wired to control power to one half of a duplex receptacle. What change must be made to the receptacle for this type of circuit?

11. What article in the NEC describes the permissible loads on branch circuits of various ampere ratings?

14
TESTING AND TROUBLESHOOTING BASIC CIRCUITS

• INTRODUCTION •

Part of the job of installing new wiring consists of testing the installation to make certain all connections have been made correctly. If the test indicates that there are errors in the wiring, troubleshooting procedures must be used to locate and correct them. This chapter tells you how to test and troubleshoot wiring in both new and old work.

For best results, testing must be done in a logical sequence. This approach not only saves time, but makes it easier to identify the source of trouble if a test indicates a fault. As noted in Chapter 5, many test instruments are available to make testing and troubleshooting easier and quicker. Simple voltage and continuity testers or "homemade" testers can be used for much of the testing. For more complete testing or to track down faults, outlet analyzers, voltage level meters, dial and pointer meters, or clamp-on meters may be used.

This chapter describes how to locate the most common faults that occur in residential wiring and how to test systematically to be sure all wiring is correct. Information is included on how incorrect test results are interpreted to identify the source of trouble. A special fault isolation procedure is included that can be used to locate hard-to-find troubles in branch circuits having numerous outlets and switches.

Planning and installing wiring are important, but the work isn't done until you have made sure the installed circuits are correctly and safely wired. Complete and careful testing assures a good wiring job.

• NEW WORK AND OLD WORK •

The procedures used to test wiring in new work and old work are basically the same. The first test is a low-voltage test of the new or added wiring alone, before switches, receptacles, and fixtures are installed. A second low-voltage test is made after all electrical devices are in place. The final check is a full-power test after the new wiring is connected to source power.

In the area of troubleshooting—finding and correcting wiring errors that are discovered during the tests—the procedures for new and old work require different approaches. This should be kept in mind as you study the test and troubleshooting procedures in the sections that follow.

Wiring in new work can be tested before walls and ceilings are enclosed. If faults are discovered, they can usually be corrected quickly and easily because all wiring is visible and accessible. An additional help on new work is the availability of electrical drawings. By referring to these drawings you can locate the best test points to use for fault isolation.

Troubleshooting additions and modifications to old work is more difficult because wiring is usually at least partly concealed. In addition, if you are adding outlets to an existing circuit, the switches, receptacles, fixtures, or wiring on the old part of the circuit can cause faults that appear to be due to the new work. Electrical drawings are rarely available when you are working on old wiring; much tracing of wires and cables must be done to identify test and tie-in points.

• SEQUENCE OF TESTING •

The sequence of testing is broadly divided into two parts: low-voltage testing and testing after source power is connected. Low-voltage testing allows you to check wiring with no danger to personnel. If wiring is not correct, no damage will be done to cabling or to the structure. Low-voltage testing is similar to continuity testing. Continuity tests are made to check connections and to make certain that there are no breaks in the wiring. Low-voltage tests also make these checks, and in addition they check complete circuits to determine that the wiring has been done correctly. Low-voltage testing can uncover many, but not all, errors in wiring and faults in material. To be completely certain wiring is done correctly and will be acceptable to an inspector, full-power tests must also be made.

• TEST EQUIPMENT •

Low-voltage Testers

Low-voltage tests can be made with a continuity tester, an ohmmeter, or a simple tester you can make yourself. Continuity testers and ohmmeters, although extremely useful for many tests, have two drawbacks when used for full-circuit testing. Both devices generally use one or two penlight batteries. This low voltage (1.5 to 3 volts) may not be sufficient to test long runs of cable containing many connections. Normal voltage drop along the line may produce the same test result as an open circuit. In addition, shorts in a circuit that will cause trouble under full power may not show up when only 1.5 to 3 volts are applied. Another shortcoming of continuity testers is that they provide only a visual indication of the result of the test. Quite often considerable leg work and time can be saved if you have a tester that gives an audible indication of the test result. For these reasons, electricians often construct a tester such as the one shown in Fig. 14-1. The tester can be as simple as the one shown, or it can be, packaged and wired like the one in Fig. 14-2. On large projects contractors sometimes use a movable storage battery power source that can be connected at some point at the service entrance to supply low-voltage power to a

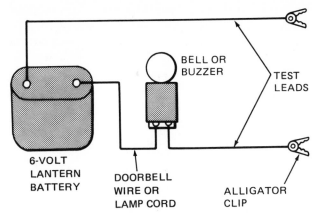

Figure 14-1. Simple battery-and-bell tester.

complete wiring installation. Many commercial versions of these testers are available at moderate cost.

Full-power Testers

Full-power tests can be made with a voltmeter, ammeter, or one of the special devices called analyzers. A voltmeter can do the same job as the analyzer, but it takes more time. For large jobs the time saved by the analyzer may mean an important saving. Ammeters—either the test lead type or the clamp-on type—are used primarily for troubleshooting special wiring—such as motor circuits—where current flow measurement may be needed.

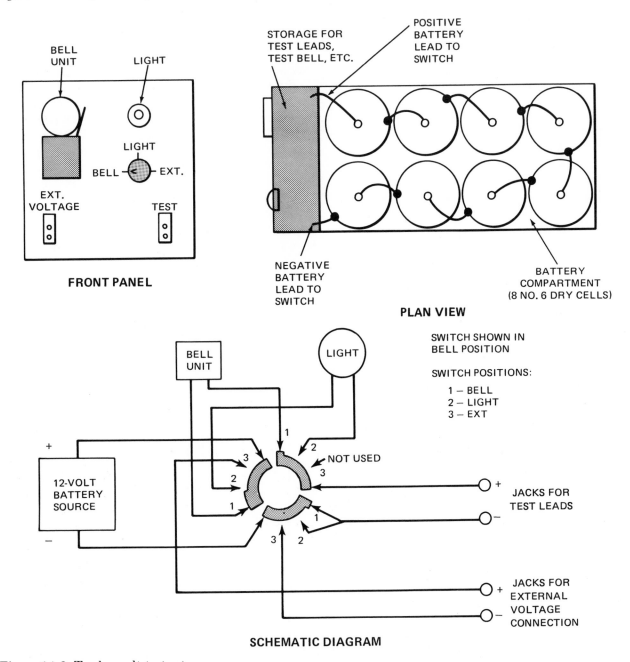

Figure 14-2. Twelve-volt test set.

• LOW-VOLTAGE TEST PROCEDURES •

The first low-voltage test should be made as soon as practicable after all boxes are installed and all wiring has been done from the service panel to the end of the run for all circuits, but before switches, receptacles, and fixtures are installed and, of course, *before* utility power lines are connected at the service entrance.

Wiring Test

This test procedure describes the tests using a homemade battery-and-bell tester. With minor variations the procedure can be used for any type of continuity tester. The bell tester rings when a continuous path for current flow exists between the test leads. The bell does not ring when there is no path for current flow between the test leads. The equivalent results for other testers are shown in Fig. 14-3.

Before any testing is done, all wiring from the service panel to each outlet should be checked visually. Make sure all permanent connections between conductors have been made. Conductors that will be connected to receptacles and fixtures should be pulled out of the outlet boxes and separated. Make sure hot wires are not touching power-ground wires, grounding wires, metal boxes, or other ground points, such as water pipes. In conduit systems, make sure conductors are not in contact with the conduit. At every point where a switch is to be installed, the wires to be connected to the switch should be connected together temporarily for the purpose of the low-voltage test. The easiest way to do this is to join the wires with solderless connectors (wire nuts). On multiple switch circuits where three-way switches will be installed, make the connection between the wire that will go to the common terminal

and to both of the travelers. Where four-way switches will be installed, join the travelers. This method of shorting three-way and four-way switches is shown in Fig. 14-4.

When all receptacle and fixture wires are separated and all switches are shorted, the wiring is in a condition that represents the final circuit when all switches are on, but no loads are present on the circuit. Figure 14-5 shows how Chapter 13 branch circuit no. 1 looks electrically when prepared for this test.

SERVICE PANEL TESTS • The first test should be made at the service panel.

Test Setup. Set all circuit breakers to off or remove all fuses. Connect a temporary test jumper between the red and the black hot-wire terminals. Connect one bell-tester lead to the shorted hot wires. Connect the other tester lead to the neutral busbar. This places 12 volts across each hot wire and power ground (Fig. 14-6). In other words, the low voltage is applied to the service panel in the same way the 120-volt power will be applied on the finished installation. The test voltage does not, however, represent the 240-volt power. With the red and black wire buses shorted, there is no potential difference across the 240-volt lines. A different test setup is made to check these circuits.

Test Procedure. The tester should not ring when connected between the hot wires and the neutral bus. If it does, a short exists in the service panel wiring. Because all circuit breakers are off or all fuses have been removed, a power ground or a grounding wire is probably in contact with one of the hot buses. A careful visual check should uncover the trouble. If necessary to help locate the trouble, remove the jumper between the hot-wire buses and connect the tester between just one

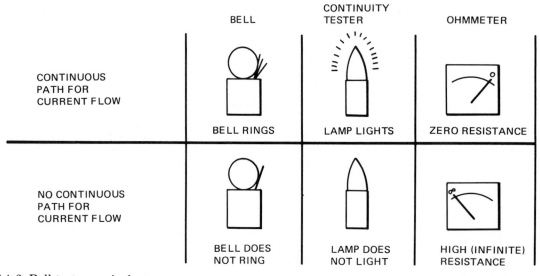

Figure 14-3. Bell tester equivalents.

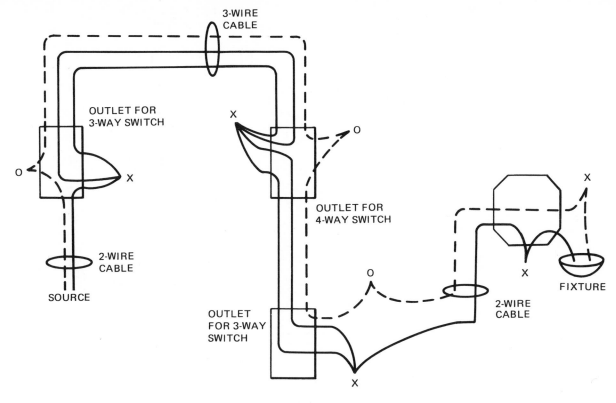

Figure 14-4. Temporary shorting of three-way and four-way switches.

hot-wire bus and the neutral bus. Make this test to determine which bus is shorted to ground. If the tester rings, the bus connected to the tester is the shorted one. If the tester does not ring, the short is on the other bus.

After the correct test result is obtained on the first service panel test, replace the jumper between the hot-wire buses, if it was removed. To continue the service panel test, turn on each circuit breaker or insert each fuse in order. This successively connects each branch circuit to the tester. The bell should not ring at any point in this test. If the bell does ring, a short or wiring error exists on the branch circuit that was connected, by circuit breaker or fuse, when the ringing started. Make a quick visual check of the outlets on the defective circuit to see if any of the leads pulled out of boxes have been disturbed. If the source of the short is not discovered by this check, recheck all wiring on the circuit from the service panel to the end-of-run to locate the short.

BRANCH CIRCUIT CONTINUITY TEST • If all circuits can be turned on with no fault indication, there is no short-circuit path in the complete wiring installation. However, this is not the end of the low-voltage check. An additional test must be made to establish that all circuit wiring is continuous, so that power will be applied to all outlets.

Test Setup. Remove the jumper between the hot-wire buses. Remove the bell from the tester. Connect one of the battery test leads directly to the neutral bus and connect the other battery test lead to a hot-wire bus (Fig. 14-7). Turn on all circuit breakers or insert all fuses. This connection supplies 12 volts to all the circuits connected to one hot-wire bus; it also supplies 12 volts to half of each 240-volt circuit. The presence of this voltage at each outlet supplied by the bus indicates that wiring from service panel to outlet is correct.

Test Procedure. The test procedure consists of connecting the bell to the black and white wires at each outlet. The bell should ring at every outlet. If the bell does not ring at any point, make a quick visual check for an open connection in wiring. If the outlet is controlled by a switch, make sure the switch wires are connected. If no open connection is visible and the bell does not ring, make a note of the outlet at which the trouble occurred. If a copy of the electrical layout drawing is available, that is a convenient place to make the notation.

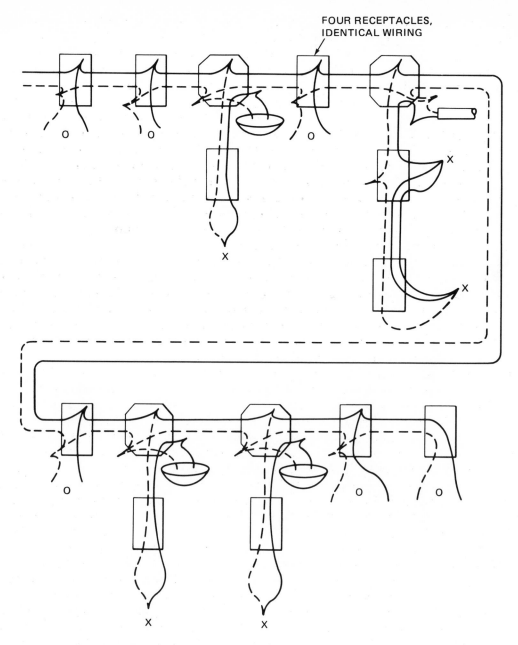

X — TEMPORARY TEST CONNECTION AT SWITCH OUTLETS

O — SPREAD WIRES AT RECEPTACLE OUTLETS

Figure 14-5. Branch circuit no. 1 prepared for low-voltage test.

Continue with the test until you have tested all outlets on that bus. The cause of trouble is often indicated by the pattern of final results. For example, if lack of power occurs only at one outlet, the trouble is localized to a small area of wiring. By removing connectors and making continuity tests of each conductor, the source of trouble can readily be found. If several outlets are without power, check the electrical layout drawing to find out which outlet is the first in the string that has no power. Make detailed continuity checks at that outlet. If the cause of trouble can be found at that outlet, the probability is quite high that

the trouble at the other outlets will also be corrected. With the connection used for this test, power at 240-volt outlets should be available only between one hot wire and neutral.

When all outlets on one bus have been tested, move the test lead at the service panel to the other bus and perform the same test at the remaining outlets. Power should be available at the other half of each 240-volt outlet for this part of the low-voltage test.

Careful inspection and testing will almost always solve problems that show up during the low-voltage tests. Wiring errors are the most probable cause of

234   Electrical Wiring Fundamentals

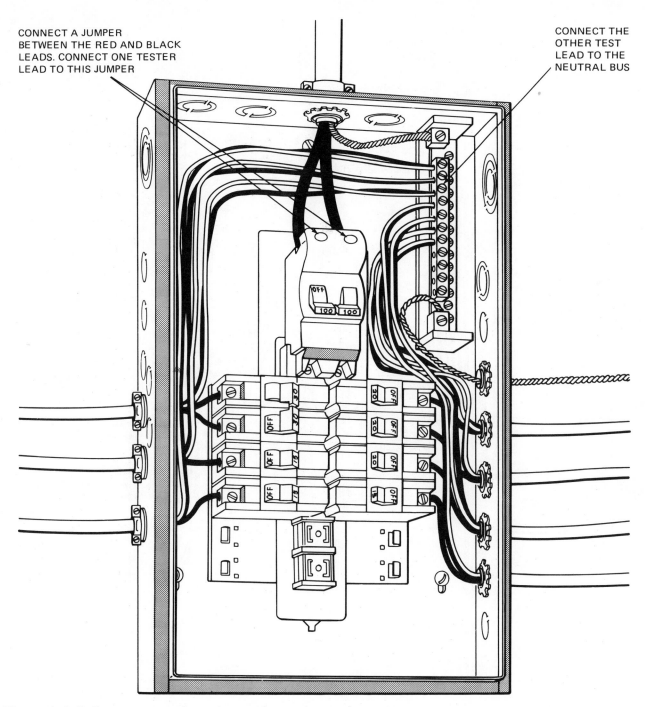

CONNECT A JUMPER
BETWEEN THE RED AND BLACK
LEADS. CONNECT ONE TESTER
LEAD TO THIS JUMPER

CONNECT THE
OTHER TEST
LEAD TO THE
NEUTRAL BUS

Figure 14-6. Bell tester connection points on the service panel.

trouble. Make sure proper connections are made. Refer to the electrical layout drawing as necessary. Make sure wires that are connected are connected firmly and that conductor surfaces are clean. On new work—with many trades working on the site at the same time—cabling can be accidentally damaged. A visual check will uncover damage to cables or conduit. Thin-walled electrical metal tubing (EMT) appears strong but is vulnerable to damage. Breaks in conductors in new wire and cable are extremely rare, but can happen. Continuity checks will reveal this problem.

TESTING ADDITIONS OR MODIFICATIONS TO OLD WORK • Wiring added to an existing electrical system can be tested in much the same way as wiring in new work. The differences in test procedure depend on the type of modification or addition that was made. If the work consists of adding a single outlet to an existing circuit, the low-voltage tests can be made before the power cable to the new outlet is connected to source power. If one or more complete branch circuits are added, the low-voltage tests can be made before the circuits are connected at the service panel. The test

Testing and Troubleshooting Basic Circuits **235**

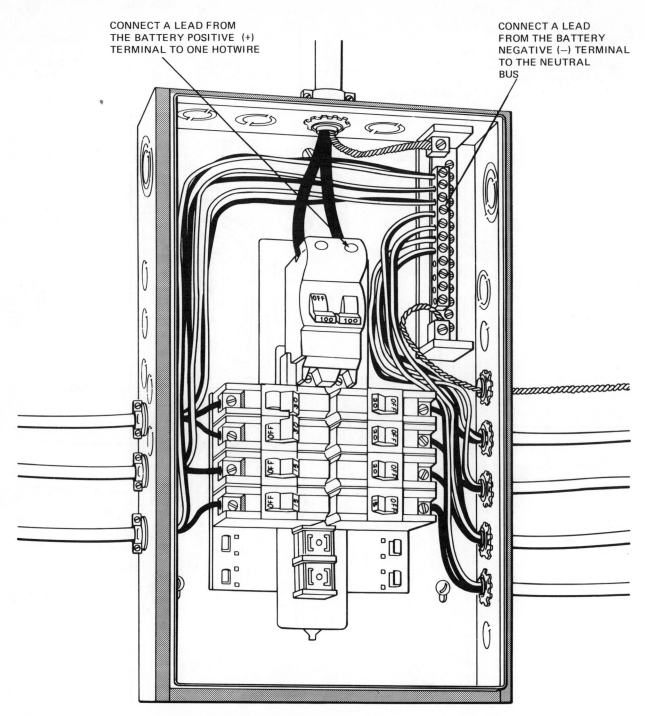

CONNECT A LEAD FROM
THE BATTERY POSITIVE (+)
TERMINAL TO ONE HOTWIRE

CONNECT A LEAD
FROM THE BATTERY
NEGATIVE (–) TERMINAL
TO THE NEUTRAL
BUS

Figure 14-7. Low-voltage branch circuit test connection points.

procedures given under Branch Circuit Continuity Test can be used to locate faults in old work.

Complete Circuit Test

When all wiring tests have been passed successfully, work on the installation can continue. All switches, receptacles, and fixtures can be installed. It is a good practice to make a brief low-voltage complete circuit test after the devices have been installed.

Turn off all circuit breakers or remove all fuses. Install a jumper between the service panel hot wires and connect the battery-and-bell unit, as it was connected for the initial service panel test. Turn on all branch circuit switches. At this point no lamps should be inserted in any fixture and nothing should be plugged into any receptacle. The condition of the circuits now is electrically the same as it was when switch leads were shorted together, and fixture and receptacle wires were pulled out of the outlet boxes. The difference at this point, of course, is that electrical connections have been made to all circuit devices.

With the battery-and-bell tester connected, turn on all circuit breakers or insert fuses in sequence. The bell should not ring at any time. If it does, a short exists in the circuit turned on when the ringing started. If the circuit passed the wiring test, the short must be due to an error in connecting the devices or to a defective device. Recheck all connections and, if necessary, test the switches, receptacles, and fixtures as described in Chapter 9.

Next, remove the bell from the tester and connect one of the battery test leads to the hot-wire bus and the other battery lead to the neutral bus. Add a couple of test leads to the bell and check for test voltage at each receptacle and fixture in the installation. At each receptacle, the bell should ring when the test leads are inserted in the slots. The bell should also ring when one test lead is inserted in the narrow slot and the other is inserted in the U-shaped grounding slot.

Incandescent lamp fixtures can be tested by touching one bell lead to the bottom center brass contact and the other to the screw shell of each lamp socket. If the bell does not ring, check the switch circuits controlling the fixture. Some fixtures may also have a switch on the fixture itself that can offset the test result.

Fluorescent fixtures cannot be tested with a low-voltage dc source because the fluorescent ballast operates only when 120-volt ac is supplied. Fluorescent fixtures are tested in the full-power test.

To check 240-volt outlets completely, the test battery must be separately connected between each service panel hot-wire bus and the neutral bus. The bell should ring when connected between only one of the outlet slots and the neutral slot for each bus connection. When the low-voltage test is complete, disconnect the test battery and remove the bus jumper from the service panel. If any other jumpers or temporary connections have been made to cover special situations, be sure to remove them before full power is applied.

• FULL-VOLTAGE TEST PROCEDURE •

A wiring installation that has passed all parts of the low-voltage test can safely be tested under full power. Keep in mind, however, that the full-voltage test is made because the wiring may still contain faults and errors. Some of these errors—an open grounding lead at one outlet, for example—could create a shock danger. Particular care should be taken when working on wiring that has not been fully tested. To repeat another precaution, make certain all temporary wiring done for the low-voltage test has been removed.

Before utility company personnel connect the service drop to the building and install a meter in the meter socket, the main service disconnect at the service entrance should be off. The main service disconnect may consist of a main disconnect switch, or from two to six circuit breakers or fuses. In addition to the main disconnect, all other circuit breakers should be off or all other fuses removed. When the utility company turns on power to the building, the first part of the full-voltage test can be performed. The procedure for this part of the test consists of turning on in sequence each disconnect from the main service disconnect through all branch circuits. Throughout this turn-on, no circuit breaker should trip to OFF (or no fuse should blow). If either of these indications of a short occurs, the cause must be found and corrected before continuing with the full-voltage test.

If no problems are encountered during the power turn-on, the next part of the test can be performed. This consists of checking for correct voltage, proper polarity, and grounding throughout each branch circuit. For this part of the test you can use a voltage tester, a dial-and-pointer voltmeter, a voltage level meter, or an outlet analyzer. The procedure below describes the use of a dial-and-pointer voltmeter. The equivalent indication for the other test units is shown in Fig. 14-8.

Three basic checks must be made at each receptacle. If correct results are obtained, nothing further need be done at that receptacle. If the first three checks do not show the correct results, the nature of the problem can be determined by what the tests do show. If you are using an outlet analyzer, the problem will be indicated directly on the analyzer.

Tests for 120-Volt Receptacles

Step 1. Insert the dial-and-pointer voltmeter test leads in the receptacle slots (Fig. 14-9). The meter should indicate 120 volts.

Step 2. Insert one test lead in the narrow slot and the other test lead in the U-shaped grounding opening. The meter should indicate 120 volts.

Step 3. Insert one test lead in the wide slot and the other test lead in the U-shaped grounding opening. The meter should indicate zero voltage.

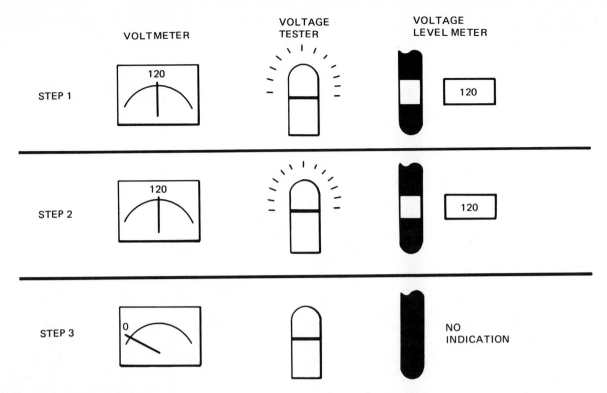

Figure 14-8. Voltmeter equivalents.

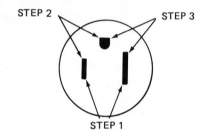

Figure 14-9. Testing 120-volt receptacles.

If incorrect results are obtained, the cause can be determined as follows.

Voltage Reading at Each Step			
1	2	3	Trouble
120	0	120	Polarity reversed. Hot-wire and power-ground connections reversed at some point.
120	0	0	Grounding wire open.
0	120	0	Power ground open.
0	0	0	Hot wire open.
0	120	120	Hot wire and grounding wire reversed.
0	0	120	Power ground hot. Hot wire open.

Tests for 240-Volt Receptacles

If a wiring error exists, 240 volts can be measured at any step. Be sure the voltmeter range switch is set high enough to measure this voltage safely.

Step 1. Insert one dial-and-pointer voltmeter test lead in the neutral (white-wire) slot (Fig. 14-10). Insert the other test lead in one hot-wire slot. The meter should indicate 120 volts.

Step 2. With one test lead still in the neutral slot, insert the other test lead in the other hot-wire slot. The meter should indicate 120 volts.

Step 3. Insert one test lead in one hot-wire slot. Insert the other test lead in the other hot-wire slot. The meter should indicate 240 volts.

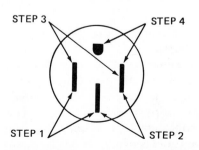

Figure 14-10. Testing 240-volt receptacles.

Step 4. Some 240-volt receptacles have an additional grounding slot. To check the grounding slot, insert one test lead in it and the other test lead in either hot-wire slot. The meter should indicate 120 volts.

If an incorrect reading is obtained at any step, the cause can be determined as follows.

Voltage Reading at Each Step				
1	2	3	4	Trouble
120	0	0	120	Hot wire open.
0	120	0	0	Hot wire open.
0	0	240	120	Neutral (white) open.
120	120	240	0	Grounding wire open.
240	120	120	120	Hot and white wires reversed.
120	240	120	0	Hot and white wires reversed.

• LOCATING CIRCUIT FAULTS •

Most circuit faults and wiring errors are easily found. A careful visual check and a review of the work done will often be all that is necessary to locate and correct a faulty circuit. Look for the simple causes of trouble first. On new work, wiring is visible and easily inspected, but it is also exposed to damage. Check cable and conduit between outlets for signs of damage. On old work, where cables and conduit are concealed, detection of damage is more difficult. When other possible sources of trouble have been checked and found correct, continuity tests between outlets may reveal broken or shorted conductors. Particularly in areas where renovation work has been done or is going on, cable damage can accidentally occur.

There are, from time to time, troubles that cannot be cleared by a simple inspection and recheck of wiring. These troubles are most efficiently tracked down by a systematic series of circuit tests. An example of this type of test follows.

Troubleshooting a Branch Circuit

The no. 1 branch lighting circuit in the wiring example in Chapter 13 is a good circuit to use to demonstrate systematic troubleshooting. Because of the number of outlets (thirteen) and switches (five) on this circuit, locating an unusual trouble would be a time-consuming job if a planned procedure were not used. Although the circuit we are using is an example of new

wiring, the same procedure can be used on circuits or circuit additions on old work.

The basic procedure consists of dividing the circuit or the new wiring as often as necessary and retesting after each division to see if the fault persists. As an example of the type of fault that can be difficult to locate, assume that branch circuit no. 1 cannot be permanently turned on. Each time the circuit breaker is set to ON, it remains on for about 1 minute and then trips to OFF. This indicates a short circuit. However, the fact that the breaker stays on for a minute indicates that the short is not direct. You will recall from Chapter 10 that most circuit breakers and fuses are designed to carry about 1 1/2 times their rated current flow for 1 minute before tripping to OFF or blowing. This indicates that the short in branch circuit no. 1 is not a direct path from the hot wire to power ground or some grounding point. A direct short would cause the breaker to trip or the fuse to blow immediately. The fault we are looking for is a path that limits current flow but allows a higher flow than the circuit rating of 15 amperes. Faults such as this are difficult to track down for two reasons. Power cannot be turned on long enough for voltage and polarity checks, and the resistance in the short path may prevent it from showing up as a fault on the low-voltage test.

The most practical approach, then, is to divide the circuit into parts and test each part. An examination of the electrical floor plan for circuit no. 1 (Fig. 14-11) shows that the outlet marked A is the midpoint in the circuit. The first step in the test is to disconnect the source power wires and the grounding wire at this point. Wire the outlet at A as if it were the end of the

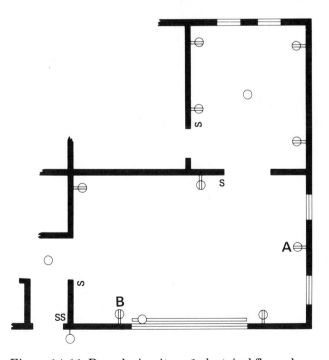

Figure 14-11. Branch circuit no. 1 electrical floor plan.

run. When this wiring change has been made, turn on the circuit. If power stays on, the circuit is good from the service panel to outlet A. The fault must be in the circuit between A and the end of the run. Of course, if the circuit breaker trips or the fuse blows as before, the fault is in the first half of the circuit. In either case, the next step is to divide the half containing the fault. If the fault is in the second half, turn off power and rewire outlet A as a middle-of-run outlet and open the source power wires at outlet B. Turn on power again to see if the fault persists. This procedure can be continued

until the fault is narrowed to a single outlet or a short run of wire. If necessary for further isolation of the source of trouble in a small area, the power and grounding wires can be connected one at a time. After each individual wire is connected, power is turned on. This procedure will show the short path. For example, if the short occurs when only the hot wire is connected, the short must be between the hot wire and some path to ground other than the power ground or grounding wires. This and other possibilities are shown in Fig. 14-12.

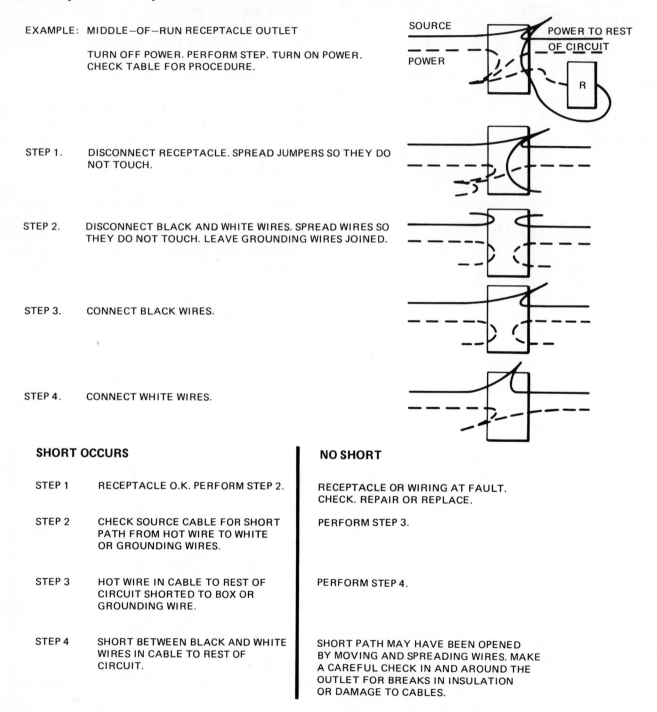

EXAMPLE: MIDDLE—OF—RUN RECEPTACLE OUTLET

TURN OFF POWER. PERFORM STEP. TURN ON POWER. CHECK TABLE FOR PROCEDURE.

STEP 1. DISCONNECT RECEPTACLE. SPREAD JUMPERS SO THEY DO NOT TOUCH.

STEP 2. DISCONNECT BLACK AND WHITE WIRES. SPREAD WIRES SO THEY DO NOT TOUCH. LEAVE GROUNDING WIRES JOINED.

STEP 3. CONNECT BLACK WIRES.

STEP 4. CONNECT WHITE WIRES.

SHORT OCCURS		NO SHORT
STEP 1	RECEPTACLE O.K. PERFORM STEP 2.	RECEPTACLE OR WIRING AT FAULT. CHECK. REPAIR OR REPLACE.
STEP 2	CHECK SOURCE CABLE FOR SHORT PATH FROM HOT WIRE TO WHITE OR GROUNDING WIRES.	PERFORM STEP 3.
STEP 3	HOT WIRE IN CABLE TO REST OF CIRCUIT SHORTED TO BOX OR GROUNDING WIRE.	PERFORM STEP 4.
STEP 4	SHORT BETWEEN BLACK AND WHITE WIRES IN CABLE TO REST OF CIRCUIT.	SHORT PATH MAY HAVE BEEN OPENED BY MOVING AND SPREADING WIRES. MAKE A CAREFUL CHECK IN AND AROUND THE OUTLET FOR BREAKS IN INSULATION OR DAMAGE TO CABLES.

Figure 14-12. Finding short by connecting wires individually.

Locating Ground Faults

When all or part of a circuit has ground fault protection, automatic shutoff will occur if the current flow in the power wires is not equal. This type of fault can also be located by using the previously described fault localization procedure. If ground fault protection for a complete circuit is provided at the service panel, the fault can be located by exactly the same process of dividing the circuit. If only a portion of a circuit has ground fault protection, the process can be simplified somewhat because only the outlets "downstream" from the GFCI need be checked. An example of this type of test is shown in Fig. 14-13 for branch circuit no. 3 of the Chapter 13 electrical layout.

Troubleshooting Fluorescent Fixtures

When fluorescent lamps do not light, always check the installation of the lamps in the holders. Lamps must be positioned properly in the holders to make electrical contact. Fluorescent lamps have marks (usually a small triangle) on the end which indicate proper position. Rotate the lamps until the mark is visible at the top center of the lamp holder.

Sometimes fluorescent fixtures produce a loud hum when turned on. This can be caused by a loose ballast. Open the fixture and check the ballast mounting screws. Tighten, if necessary. Excessive hum will occur if there is a mismatch between the ballast and the lamp type. The correct lamp type to be used is marked on the ballast.

Some hum is normal. All ballasts have a sound rating ranging from A (low) to F (high). Fixtures used in residences should have an A rating. A fixture intended for use in a store or commercial building would produce an annoyingly loud hum in a residence.

Some other common fluorescent problems, causes, and remedies are listed below.

Problem	Cause	Remedy
Lamp flickers or blinks on and off.	Newness (normal for a short period with some new lamps).	None required.
	Low line voltage.	Check with local utility company if low line voltage is suspected.
	Temperature below 50 degrees Fahrenheit in lamp location, or cold draft on lamp.	If condition is permanent, shield lamp from draft or install low-temperature ballast.
	Lamp-ballast mismatch.	Make sure lamp is type specified by fixture manufacturer.
	Lamp not properly seated in holders, or lamp pins bent.	Remove lamp. Check pins. If necessary, straighten pins with pliers. Clean pins with steel wool. Make sure lamp is properly seated in holders.
Short lamp life.	If lamp fails within a few hours, ballast is wrong for fixture or is incorrectly wired.	Correct the wiring or replace the ballast. (Correct wiring is marked on the ballast.)
	In some rapid-start and instant-start two-lamp fixtures, when one lamp burns out the other lamp will dim or fail.	Replace burned-out lamps immediately. Substitute a good lamp for each lamp in the fixture, one at a time. When burned-out lamp is replaced, both lamps will light.
Color variations in lamps of same type.	Some variation in color is normal.	None required.
	Significant age difference in lamps.	Substitute new lamps.
	Lamps operating at different temperatures.	Check for drafts. Equalize temperature as much as possible.

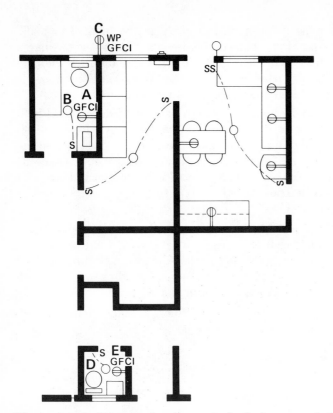

FOR CIRCUIT INFORMATION, REFER TO CHAPTER 13 (FIGURES 13–32 AND 13–33)

FEED THROUGH GFCI IS INSTALLED AT POINT A. POINTS B, C, D, AND E HAVE GFCI PROTECTION. DISCONNECT CIRCUIT AT POINT E. IF FAULT IS NO LONGER PRESENT, GROUND FAULT IS IN OUTLET E OR WIRING. INSPECT ALL AREAS CAREFULLY FOR LEAKAGE PATH FROM HOT WIRE TO GROUND. IF FAULT IS STILL PRESENT WITH E DISCONNECTED, E IS OK. RECONNECT WIRES AT E. DISCONNECT AT POINT D. CONTINUE "UPSTREAM" TO POINTS C, B, AND A UNTIL FAULT IS LOCATED.

Figure 14-13. Branch circuit no. 3 check for ground fault.

• REVIEW QUESTIONS •

1. What are the two main types of tests performed on new wiring?

2. Testing and troubleshooting wiring in new work and old work is essentially the same. There is one big difference, however. What is it?

3. What is the main advantage of the battery-and-bell tester over other low-voltage testers?

4. To prepare for the first low-voltage test, all receptacles and fixture leads are pulled from outlet boxes and separated so they do not touch each other or any metal surface. All switch leads are connected together. What circuit condition does this represent?

5. When the battery and bell are connected at the service panel, what does it mean that the bell rings when a circuit breaker is turned on?

6. When only the batteries are connected to the service panel, what does it mean that the bell rings when its leads are touched to the black and white wires at an outlet?

7. When a full-power test is being made, what do the following test results at a 120-volt receptacle indicate?

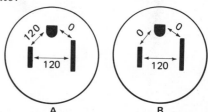

8. When a full-power test is being made, what is the correct voltage between each of the points listed below on the 240-volt receptacle diagram?

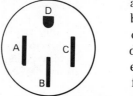

    a. A-B
    b. C-B
    c. A-C
    d. A-D
    e. C-D
    f. D-B

9. What basic test principle is used to locate troubles in the detailed branch circuit test procedure?

10. Under full-power, what is the most common fault if fluorescent lamps do not light?

15
WIRING IN FINISHED BUILDINGS

• INTRODUCTION •

Electricians refer to any wiring installation in a finished building as *old work*, whether the building is almost new or many years old. Regardless of the age of the building, electrical work becomes more difficult as soon as interior walls, ceilings, and floors are finished. There are so many different types of wiring jobs and so many building variations encountered in old work that it is impractical to try to cover specific situations. The approach used in this chapter is to discuss the most common individual problems in old work and describe ways in which they may be solved.

Most of the problems encountered are related to carpentry and masonry, rather than electrical wiring. The basic electrical rules for wiring new buildings apply—with only slight modification—to old work as well. In addition to a knowledge of electrical wiring, old work requires a basic knowledge of building construction. The best and easiest way to learn how buildings are put together is to study buildings under construction. If you can see a number of different types of buildings at various stages of construction, you will soon learn to look at a finished room and know what the structure behind the floor, ceiling, and walls looks like.

When planning electrical additions or modifications to a finished building, much thought must be given to where and how the new wire runs can be made. In many cases the electrical design must be made to fit the wire runs that are possible and practical in a particular building. Previous chapters cover the installation of wall and ceiling outlet boxes (Chapter 8) and the mounting of fixtures (Chapter 9). This chapter covers the installation of the cabling between the outlets.

• PLANNING •

Wiring in finished buildings can be as simple a job as adding another outlet to an existing circuit or as extensive as upgrading electrical service by installing a new service entrance and replacing old wiring. When wiring is done as part of a large renovation project—such as adding one or more rooms—most of the wiring can be done after exterior framing is complete and before the interior is finished. In this case the wiring can be done just as it is in new work. The special problems that electricians have when installing new wiring in finished buildings occur only when work must be done in areas where walls, ceilings, and floors are enclosed and finished. For these projects, additional planning is needed, taking into account the type of building construction, the purpose of the new wiring, and the location in which it will be installed.

Building Construction

Chapter 8 provides information on wood-frame construction and the techniques that can be used to mount electrical boxes in areas finished with plasterboard or plaster and lath. Figures 15-1 and 15-2 illustrate typical wood frame construction and identify some terms used in the building trades. Definitions of other common construction and architectural terms are given below.

Baseboard. A finishing board next to the floor.

Base Shoe. Molding in the angle between baseboard and floor.

Batt Insulation. Small blankets of mineral-fiber insulating material installed between studs and joists.

Beam. Any large piece of timber or other material used to support a load over an opening; a main horizontal timber supporting the floor of a building.

Bridging. Crisscross bracing installed between floor joists and studs.

Building paper. Heavy paper used as a sealer and insulator between rough and finished floors and under roofing material.

Button Board. Perforated panel mounted on studs to support plaster.

Casing. The finishing board that covers the joint between wall material and door and window frames.

Crawl Space. An area of a house just large enough to crawl through. The area may be under the living area in a house without a full basement or above the living area when the attic is too low to allow standing upright.

Diagonal Brace. A piece of lumber notched into framing at an angle, usually at corners, extending from top to bottom plate.

Finish Flooring. The top layer of flooring, laid over the subflooring.

Firestop. A piece of studding lumber installed horizontally between studs and halfway between floor and ceiling.

Furring. A strip of lumber nailed into masonry to support paneling or other wall finish.

Header. In doorways, the horizontal piece of lumber across the top of the opening.

Interior Finish. A term that describes the complete final interior appearance of a building. It includes the materials used and the manner of installation.

Jamb. The sidepost or lining of a door or window opening.

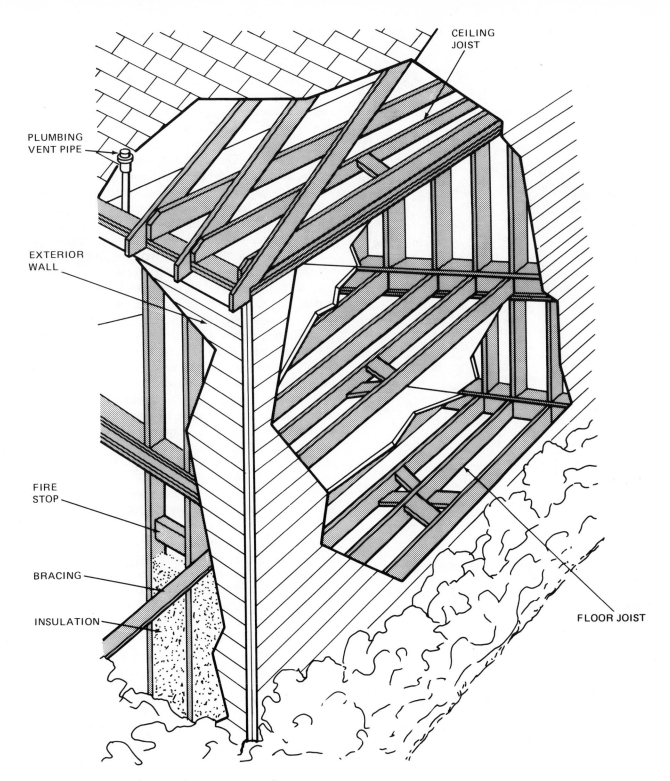

Figure 15-1. Typical wood frame construction.

Joists. The timbers that support the floor (or ceiling) of a building.

Moisture Barrier. Waterproof material that prevents or retards the passage of water vapor through an insulator. The barrier is placed on the warmer side.

Molding. Any ornamental wood strip used to finish an angle or a surface.

On-Center. A term that defines measurement points. It means that the measurement given is from the center of one structural member to the center of a corresponding member.

Plate. The horizontal timber across the top of a wall frame. It is supported by the wall studs and, in turn, supports joists of the floor above.

Wiring in Finished Buildings   **245**

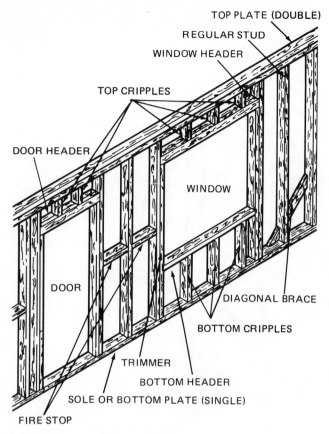

TOP PLATE (DOUBLE)
REGULAR STUD
WINDOW HEADER
TOP CRIPPLES
DOOR HEADER
WINDOW
DOOR
DIAGONAL BRACE
BOTTOM CRIPPLES
TRIMMER
BOTTOM HEADER
SOLE OR BOTTOM PLATE (SINGLE)
FIRE STOP

Figure 15-2. Outside wall framing.

Rafter.   A timber supporting a roof.

Ridge.   The highest part of a roof; the top board into which rafters are nailed.

Riser.   The vertical part of a step that supports the tread.

Rough Flooring.   The first rough lumber installed over joists.

Sash.   The framing in doors and windows in which panes of glass are set.

Sheathing.   The first boards or sheets of plywood nailed on studs and rafters to enclose the building.

Stop.   Molding or other object which prevents damage to adjacent surfaces between which movement can occur.

Purpose of Wiring

The intended use of the outlet or outlets to be added provides one indication of whether a new circuit is required or simply an addition to an existing one. In residential wiring the only circuits to which outlets can be added are 15- or 20-ampere general purpose circuits. Higher rated circuits of 30 to 50 amperes are intended for fixed major appliances such as ranges, wall ovens, and water heaters. These are single outlet circuits and no additional load should be placed on them.

If one or two outlets are needed for additional lighting or small appliances, these can generally be added to an existing general purpose branch circuit. You will recall from Chapter 13 that the NEC does not specify or limit the number of outlets on a general purpose circuit. The present electrical load on the circuit or circuits available in the location where the new outlets are required should be calculated to determine which circuit has the lowest normal load. (Mapping may be required to do this; mapping is explained later in this chapter.) To make this calculation, you simply add the wattages of all lamps, fixtures, and appliances plugged into receptacles on each circuit. The maximum load on a 15-ampere 120-volt branch circuit is 1800 watts; for a 20-ampere circuit it is 2400 watts.

There are a few NEC restrictions which must be kept in mind when adding outlets to a circuit. No outlet should be added to an existing circuit for a small appliance that is rated at more than 80 percent of the branch-circuit-permitted load. This means that a portable, cord-and-plug appliance rated at more than 1440 watts cannot be used on a 15-ampere circuit. For a 20-ampere circuit the maximum load would be 1920 watts. This is the maximum for a single portable appliance.

If fixed appliances (one or more) are to be permanently connected to the circuit, they must have a total rating of not more than 900 watts for a 15-ampere circuit and 1200 watts for a 20-ampere circuit. Of course, you must include in these restrictions any appliances already on the circuit. Appliances exceeding the branch-circuit limits must be put on individual circuits. If more than one circuit has sufficient capacity to handle the new outlets, consider where the connection to source power can be made on each circuit, then select the circuit that can most easily be joined.

If the new wiring is required for a single large appliance, whether 120- or 240-volt, wiring must be done from the service panel to the location of the new outlet. You must determine that space exists at the service panel for another single (120-volt) or dual (240-volt) circuit breaker or fuse. If space for another circuit is not available, the service panel can be modified as explained later in this chapter. The electric service supplied to the house may not be adequate for a new circuit. A complete load calculation may be necessary to determine if larger capacity service is needed. The procedure for making load calculations is described in Chapter 11.

Keep in mind that if the new wiring includes bathroom or outdoor receptacles, ground fault circuit interruption protection is required. GFCI protection is a relatively new requirement of the NEC. Bathroom and outdoor wiring in older houses may not have GFCI units, but they should be added when new wiring is installed.

Location of Wiring

If the type of cable that is installed in the house is suitable for the new location, the same type of cable should be used for the added outlets. For example, if a house is wired with nonmetallic cable such as type NMC, it will be easier to continue with NMC for the new wiring. Wherever the new cable is added to old outlets, the internal cable clamps will probably be the type used for nonmetallic cable. You can make use of these for the added cable.

There are situations in which new wiring cannot be done in the same manner as the original wiring. One instance in which this is true is where the house has been wired with rigid or thin-walled (EMT) conduit. If the new wiring must be installed under finished walls and ceilings, nonflexible types of conduit cannot be used without ripping out the interior finish of the room. Either cable or flexible conduit must be used if wall and ceiling repair is to be kept to a minimum. If any part of the new wiring is in an unfinished location, rigid conduit or EMT can be used in that area. If the existing conduit is large enough, additional conductors can be fished through for connection to the new wiring.

Consideration must also be given to the possibility of dampness or moisture in new wiring areas. In old work it may be necessary to install new wiring wholly or partially underground or in locations where water may be present. When this is the case, cable such as type UF, which is suitable for wet locations, must be used. For example, if a wall has many interior obstructions or is expensively paneled, it may be less expensive to bore through it to the outside, make the cable run on the outside of the building, and then bring the cable back through the wall where needed.

• SPECIAL PROCEDURES •

Make Maximum Use of Unfinished Areas

It is easier to install new wiring in unfinished areas where studs, joists, and beams are exposed. Wiring for first-floor outlets can usually be installed along floor joists, working from the basement area. The wiring can be brought to a point below the wall in which the outlet is to be installed. A hole can then be drilled through the flooring and the wall sole plate. After the wall opening is made, the wiring can be fished up through the hole.

A reverse procedure can be used to install second-floor outlets when the space above the second floor is an unfinished attic or crawl space. In this area the wiring can be installed along ceiling beams and brought to a point above the wall in which the outlet is to be added. A hole can be drilled down through the wall top plate. The wiring can then be brought through the hole to the outlet opening.

Use Interior Walls Whenever Possible

Whenever possible, add new outlets to interior walls. The space between wall surfaces in interior walls is usually empty and free of obstructions. The only things you are likely to encounter are existing wiring and gas or water pipes. Exterior walls, on the other hand, are difficult to work with. In wood frame construction, as defined under building construction, exterior walls often have horizontal sections of 2 × 4 lumber between studs and midway between floor and ceiling. These are called firestops; they are installed to slow down the spread of a fire. In some construction diagonal bracing is notched into the studs at corners. It requires more time and involves more wall repair when you have to run cabling past these obstructions.

Most exterior walls contain some type of insulation that limits the space available for new wiring. Exterior walls in masonry (or concrete block) construction are difficult to break into, may not have clear interior space for cable runs, and require extensive repair work after they have been opened.

Standard Access Areas

It is customary in wood frame construction to leave some open space around the plumbing vent pipe. The vent pipe is near other plumbing lines and generally joins the sewer line in the basement near the point where the sewer line leaves the building. The pipe then runs vertically to the roof. It extends above the roof level and is weathersealed to the roofing material. It is the open space around the pipe that can be useful to an electrician. For example, if a separate circuit is being installed for a second floor air-conditioning unit, the new cable must run from the service panel to the second floor. Service panels are usually located in the basement. The cable run is much simplified if the vent pipe area is used. The cable can be run from the service panel along or through floor joists to the vent pipe opening. The cable can then be pulled up through the opening to the attic. In the attic the cable can be routed to a point above the wall in which the air-conditioner outlet will be installed, and then down to the outlet.

Special Outlet Boxes

Two types of electrical boxes are designed especially for use in old work. One type is similar to a standard rectangular wall box used for receptacles or switches, but has two beveled surfaces at the rear (Fig. 15-3).

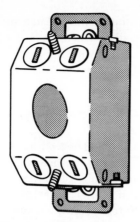

Figure 15-3. Wall box designed for old work.

These surfaces have knockouts so that cable can be brought into the box at that point. The angle of the surface allows these boxes to be pushed into standard size wall openings after cables have been connected (Fig. 15-4). (The various methods of securing boxes in wall openings are covered in Chapter 8.)

The NEC permits ceiling boxes as shallow as 1/2 inch to be used if fixtures are mounted on them. Boxes as shallow as 15/16 inch can be used for switches and receptacles. These sizes are often useful in old work. The shallow ceiling boxes (Fig. 15-5) can be mounted on the surface, rather than recessed into the ceiling. Cable can be fished through a small opening in the ceiling and secured in the box by a cable clamp. Electrical connection can be made to the fixture and the

fixture can then be mounted on the box. The box is completely or largely hidden when the fixture is in place. A similar procedure can be used with the shallow wall boxes. After the switch or receptacle is wired and mounted in the box, the box can be concealed by mounting a special faceplate (Fig. 15-6) on it.

Problems with Old Wire and Cable

The insulation on wire and cable dries out after long periods of time. This drying out shows up as brittleness. When making connections to old wiring, handle the wire carefully to avoid cracking the insulation. If cracking cannot be avoided, the wire or cable must be replaced.

• MAPPING OLD CIRCUITS •

Old work often requires adding outlets to existing circuits as well as installing completely new ones. If no electrical layout drawing is available (and usually it is not), it is difficult to decide which branch circuit to add to and where to make the connection to source power. One way to solve this problem is to make your own electrical layout drawing by "mapping" the outlets in the building. This procedure really involves working backward from the standard procedures in new work.

Start by making a floor plan of the complete building, or only part of it if your work will be confined to one area. The floor plan need not be as complete as an electrical layout drawing, but it should be neat and accurate. Mark on the drawing where outlets are located in each room. Use standard symbols; if your drawing must be used by someone else, it will be easier to understand.

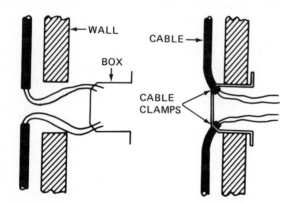

Figure 15-4. How to use the wall box.

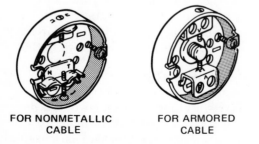

FOR NONMETALLIC CABLE   FOR ARMORED CABLE

Figure 15-5. Shallow ceiling box.

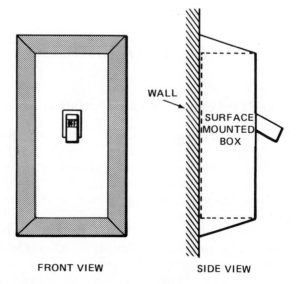

FRONT VIEW   SIDE VIEW

Figure 15-6. Faceplate for surface wall box.

At the service panel, number each circuit breaker or fuse. The usual numbering system runs from left to right, top to bottom. Adhesive tape or stick-on labels are easiest to use. Use a piece of tape or a label large enough to allow notes to be added as you go along.

Next, turn on all ceiling and wall fixtures and any lamps plugged into receptacles. At the service panel, turn off the no. 1 circuit breaker or remove the no. 1 fuse. Make notes on your layout of which lights are now out. On your floor plan, mark the number of the turned-off circuit next to each outlet on the circuit that is now without power. Use a test light to check unused receptacles. Continue checking until you have found all receptacles controlled by the circuit breaker you turned off, or the fuse you removed. Operate all switches to find out which fixtures and receptacles they control. Don't overlook switches on fixtures and appliances.

Now turn on the no. 1 circuit breaker or replace the no. 1 fuse. Turn off circuit breaker no. 2 or remove fuse no. 2 and note the fixtures and receptacles on that circuit. If you need a complete picture of the total electrical installation, repeat this procedure for all the lighting circuits on the service panel. Appliance circuits, whether 120- or 240-volt, can be checked by using a voltmeter at the receptacles, or by momentarily turning on the appliances plugged into them. Remember to include outdoor lights, receptacles, and fixtures, as well as receptacles in such places as utility rooms and furnace rooms.

When you finish you should have a number by each outlet on your floor plan (Fig. 15-7). This will tell you which circuit or circuits are available in the area where new wiring is to be installed. If more than one circuit is available, you can add up the wattages of all lights and appliances on each circuit to find which has the most unused capacity. You can use your circuit map as a starting point in determining where you can join the new wiring to source power, but you must also make some checks of the wiring in the available outlets to find the best place to join the circuit.

Identifying Wires in Boxes

Mapping tells you which outlets are on each branch circuit. If you are adding one or two more outlets for lighting or small appliances, you can use one of the existing outlets as the source of power for the added circuits. Usually the connection for source power can be made at more than one outlet. To decide which outlet to use you must remove the faceplate from wall outlets, or the fixture from ceiling outlets, and inspect and identify the wires in the box.

Particularly in old work—when other modifications may have been made—take special care when inspecting the wires in a box. Sometimes a box may contain source wires for more than one circuit. Use your voltage tester to check, before you touch bare conductors. To identify the source wires in a box, first turn off power at the service panel. Pull the conductors and switch or receptacle out of the box. Remove the solderless connectors and fan out the wires so they do not touch. Turn on power at the service panel. Use your voltage tester to check black and white pairs until you find the live conductors (Fig. 15-8). It will be helpful later if you tag or mark these wires. To identify the remaining wires, *again turn off power at the service panel.* You can use your continuity tester to check the remaining wires. Switch-loop wires can be identified by checking your map to see which switch controls the outlet. Connect your continuity tester to a black and white pair of wires, and operate the switch. If your continuity tester goes on and off when the switch is operated, the tester must be connected to the switch loop. If the outlet is in the middle of a run, you may want to identify the wires that continue power to the rest of the circuit. To do this, first connect your continuity tester to the brass and silver terminals on the next receptacle on the circuit. Then, at the outlet you are checking, touch the bare conductors in pairs of black and white wires. When you find the pair that causes the tester to light, you have found the cable carrying power to the rest of the circuit.

Connecting New Outlets to Old

Outlets containing unswitched receptacles are usually the best place to make the connection to source power. However, any outlet where source power is available can be used. Some of the typical tie-in points are described below.

CONNECTING NEW CABLE TO END-OF-RUN RECEPTACLE • This connection (Fig. 15-9) is the easiest to make. The outlet shown is not under switch control. The new circuit is connected to the spare screw terminals at the outlet.

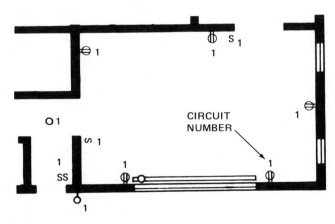

Figure 15-7. Completed circuit map (partial).

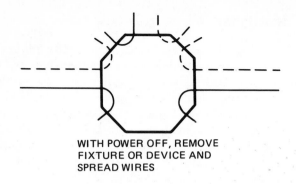

WITH POWER OFF, REMOVE FIXTURE OR DEVICE AND SPREAD WIRES

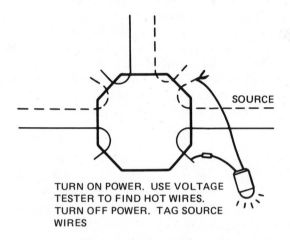

SOURCE

TURN ON POWER. USE VOLTAGE TESTER TO FIND HOT WIRES. TURN OFF POWER. TAG SOURCE WIRES

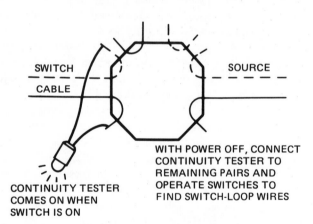

SWITCH CABLE

SOURCE

WITH POWER OFF, CONNECT CONTINUITY TESTER TO REMAINING PAIRS AND OPERATE SWITCHES TO FIND SWITCH-LOOP WIRES

CONTINUITY TESTER COMES ON WHEN SWITCH IS ON

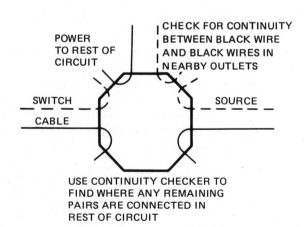

POWER TO REST OF CIRCUIT

CHECK FOR CONTINUITY BETWEEN BLACK WIRE AND BLACK WIRES IN NEARBY OUTLETS

SWITCH CABLE

SOURCE

USE CONTINUITY CHECKER TO FIND WHERE ANY REMAINING PAIRS ARE CONNECTED IN REST OF CIRCUIT

Figure 15-8. Identifying source wires in a box.

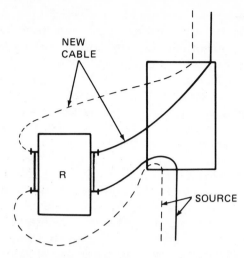

NEW CABLE

R

SOURCE

Figure 15-9. Connecting new cable to end-of-run receptacle.

CONNECTING NEW CABLE TO END-OF-RUN CEILING FIXTURE WITH LOOP-SWITCH CONTROL • When the new circuit is added (Fig. 15-10), two of the connections will join three wires, rather than two. Be sure to use large-size solderless connectors (wire nuts). As shown, the new circuit will not be controlled by the loop switch. If control by the switch is desired, connect the new-circuit black wire with the switch cable and fixture wire at point A.

CONNECTING NEW CABLE TO MIDDLE-OF-RUN CEILING FIXTURE WITH LOOP-SWITCH CONTROL • As with the end-of-run ceiling fixture with loop-switch control, if you want to control the new circuit from the present ceiling light switch (Fig. 15-11), connect the new-circuit black wire at point A. Assuming that the box shown also contains a mounting stud for the ceiling fixture, the box would have to be at least 2 1/8 inches deep to accommodate the additional connections. This point must be kept in mind when selecting a tie-in point. The box must be large enough to handle two additional conductors.

CONNECTING NEW CABLE TO MIDDLE-OF-RUN SWITCH • To provide switch control of the new outlets, connect the new-circuit black wire, the

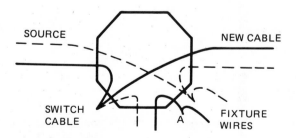

SOURCE

NEW CABLE

SWITCH CABLE

A

FIXTURE WIRES

Figure 15-10. Connecting new cable to a ceiling fixture with loop-switch control.

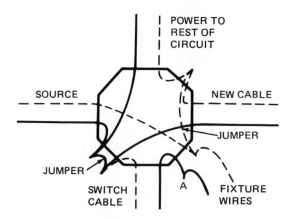

Figure 15-11. Connecting new cable to middle-of-run ceiling fixture with loop-switch control.

black wire to the fixture, and a jumper to the switch as shown in the lower part of Fig. 15-12. If the connections are made as shown in the upper part of Fig. 15-12, the new outlets will be unswitched.

CONNECTING NEW CABLE TO MIDDLE-OF-RUN RECEPTACLE • Screw-type terminals should never be used to join wires. Only one wire should be connected to each terminal. Make additional connections with jumper wires, as shown in Fig. 15-13, or by using spare terminals.

CONNECTING NEW CABLE TO JUNCTION BOX • This junction box (Fig. 15-14) contains wiring for only one circuit. Remember, junction boxes may contain wiring for more than one circuit.

CONNECTING NEW CABLE TO RECEPTACLE WITH LOOP-SWITCH CONTROL • If the new-circuit black wire is connected to the spare brass-colored terminal on the receptacle (Fig. 15-15), the new circuit will be controlled by the same switch that controls the receptacle.

• TYPICAL OLD-WORK JOBS •

Service Panel Modifications

Renovations and modifications often require changes to the service panel. New circuits must be connected to source power at the service panel and overcurrent protection must be installed for them. Whenever major modifications are made, or outlets are added for major appliances, a load calculation (Chapter 11) should be made to determine if the existing service is adequate.

If no space is available in the service panel for a circuit breaker or fuse for the new circuit, and if the

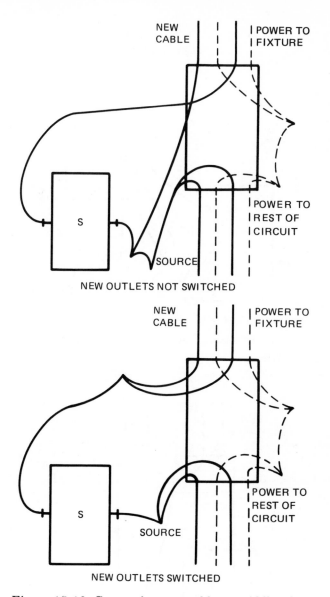

Figure 15-12. Connecting new cable to middle-of-run switch.

load calculation indicates that higher-rated service is required, there are two possible ways to solve the problem: a new, larger service panel can be installed, or a second panel can be added in parallel with the original panel. If the present service is adequate, and only additional space is needed to install more overcurrent protection devices, an auxiliary panel can be installed in series with the original panel.

The main considerations in deciding which option to choose are conformance to NEC requirements and cost. It is almost always less expensive to install a parallel panel or an auxiliary panel, but two NEC requirements must be considered before a final decision is made. The first NEC requirement is that it must be possible to turn off all power in the building by turning off a maximum of six switches. The second requirement is that the grounding system must be maintained.

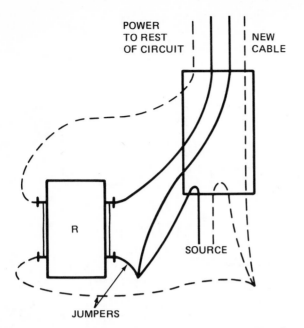

Figure 15-13. Connecting new cable to middle-of-run receptacle.

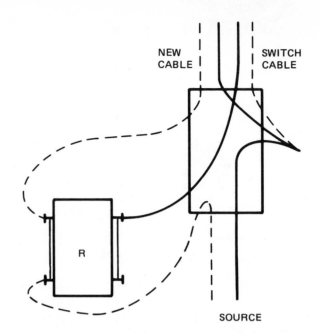

Figure 15-15. Connecting new cable to receptacle with loop-switch control.

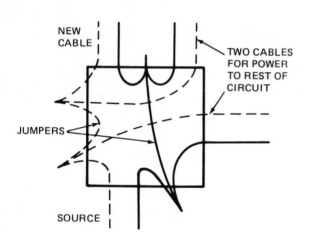

Figure 15-14. Connecting new cable to junction box.

When a second service panel is installed in parallel with the old one, the incoming power lines from the meter feed both panels directly. In this arrangement power must be turned off at both panels to meet the Code requirement for a means of disconnecting utility power. If both the old panel and the one to be installed have main breakers or fused main switches, full disconnect can be done within the six-switch limit. If the old panel did not have a main disconnect and the NEC disconnect requirement was met by turning off six circuit breakers or removing six fuses, a parallel panel cannot be added. A new service panel must be installed. It is also important that space be available to mount the parallel panel close to the old panel, so that the full power disconnect can be done as quickly as if all

switches were on a single panel. It is also desirable that the grounding connections to the new panel be kept as short as possible. Long grounding wires are more susceptible to damage than short ones. There is also the possibility that resistance can be added to the line, especially if any intermediate connections are made. To be effective, all points in a grounding circuit should have zero resistance to ground.

If the requirement for six or fewer main disconnects can be met and a suitable location is available, a parallel panel is an inexpensive and efficient way to expand service (Fig. 15-16). All requirements for internal grounding and connection to the grounding system must be met for the parallel panel, just as they are for the original panel (Chapter 11).

When an auxiliary panel (sometimes called a subpanel) can provide all the additional capacity needed, the installation is somewhat simpler and the location is more flexible. The auxiliary panel is in series with the main panel and is connected to the grounding system in the same way all other parts of the electrical system are connected, that is, by the grounding wire in the source cable, or by rigid conduit or EMT itself if the source conductors are enclosed in conduit. Source power for the auxiliary panel comes from the main panel; it is protected by an overcurrent device with a rating high enough to provide power for all the branch circuits on the auxiliary panel. For example, a dual 30-ampere circuit breaker in the main panel could provide a three-wire source to an auxiliary panel. At the auxiliary panel each hot wire could feed two 15-ampere branch circuits (Fig. 15-17).

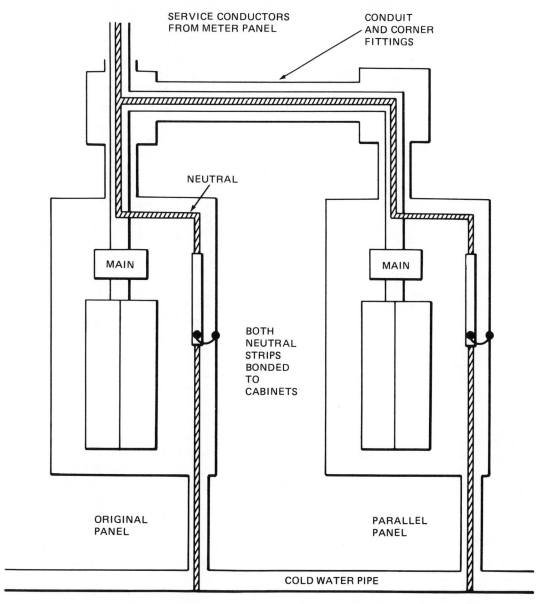

SERVICE CONDUCTORS FROM METER PANEL

CONDUIT AND CORNER FITTINGS

NEUTRAL

MAIN

MAIN

BOTH NEUTRAL STRIPS BONDED TO CABINETS

ORIGINAL PANEL

PARALLEL PANEL

COLD WATER PIPE

Figure 15-16. Increasing service by using a parallel panel.

Cable Runs in Basement and Attic

The NEC requirements for cable runs in basement and attic are the same for old and new work. In new work cable runs are usually planned so that the installation can be done in the quickest and easiest manner, using the least amount of cable. In old work, these considerations are secondary. Cable runs must be made so that cables are brought into walls at points that will keep the work in the finished area of the house to a minimum. Regardless of how a cable run is made in a basement or attic, it must still be done in accordance with the following rules. The general NEC rule is that cable installed in exposed areas must be protected from damage. All the specific rules are based on this requirement.

1. Cable running parallel to a joist or beam can be secured to the side of the joist or beam (Fig. 15-18).
2. In basements, cable running parallel to joists can be secured to the bottom edge of the joist (Fig. 15-19), unless the basement ceiling is to be finished. Cable on the bottom edge would interfere with the installation of ceiling material. In this case the cable may be secured to the side of the joist.

The rules for cable installation in attics make a distinction between attics considered accessible and those not accessible. The difference between the two is not clearly defined, but generally an attic space is accessible if any kind of stairway leads to it from the floor

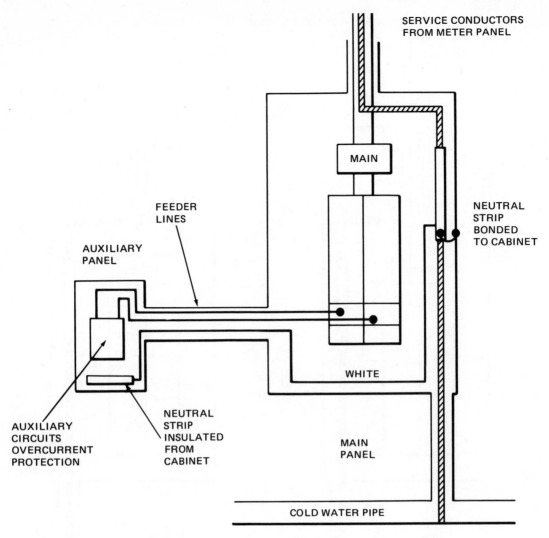

Figure 15-17. Adding circuits by using an auxiliary service panel.

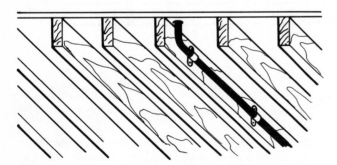

Figure 15-18. Cable run on side of joist.

below. This includes stairways that fold up and are concealed when not in use. Attics and crawl spaces that can be reached only through openings in the ceiling of the floor below are usually considered not accessible.

3. Cable can be secured to the top edge of beams in non-accessible attics in all areas more than 6 feet from the opening to the attic.

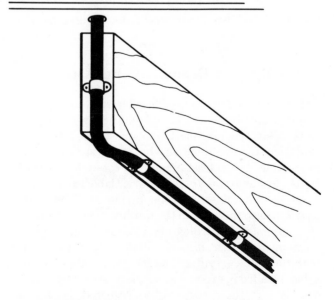

Figure 15-19. Cable run along bottom of joist.

4. Cable runs at right angles to joists or beams can be made in either basement or attic by boring holes in the approximate center of the beams and running cables through the holes (Fig. 15-20). This is not permitted if, at any point in the run, there is a space greater than 4 1/2 feet between supports.

5. In accessible attics, cables can be run at right angles to beams across the top edges only if protected by guard strips as thick as the cable is high (Fig. 15-21).

6. In basements, cables run at right angles to the joists and below them must be secured to running boards (Fig. 15-22). An exception is made if the cable contains at least three no. 8 conductors or two no. 6 conductors. These sizes and larger are considered strong enough to be strung along the bottom edge of joists without the additional support of running boards.

Installing New Outlets in Finished Areas

A typical situation in old work is the installation of new switch-controlled ceiling fixtures. In the example described below, the new ceiling fixture is on the first floor of a building in which the second floor is completely finished. The usual cable run in this case must be made from a switch or receptacle outlet on the wall, up to the ceiling level, and then between the second floor joists to the new fixture location. We will assume that an unswitched, end-of-run receptacle is available near the location of the fixture switch. We will also assume that an investigation, perhaps including mapping, has shown that the receptacle circuit can handle the additional load of the fixture. Electrically, this circuit will be connected as shown in Fig. 15-23. One procedure that can be used to make this installation is described in the following paragraphs.

Access holes must be cut in the wall and at the point where wall and ceiling meet (Fig. 15-24). (These are patched when the job is done.) Top plates and sometimes studs must be notched to accept the cable. The cable will run laterally across two studs (Fig. 15-25), up through the wall to the switch, from the switch to the ceiling, and between joists to the ceiling box opening.

Step 1. To start, cut access holes where the wall and ceiling meet and at each stud. Through the access

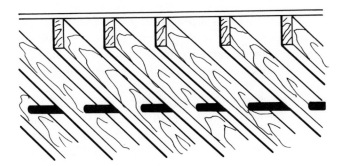

Figure 15-20. Cable run through joists.

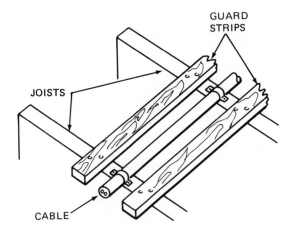

Figure 15-21. Cable run protected by guard strips.

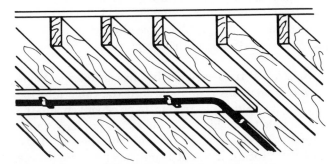

Figure 15-22. Cable on running boards.

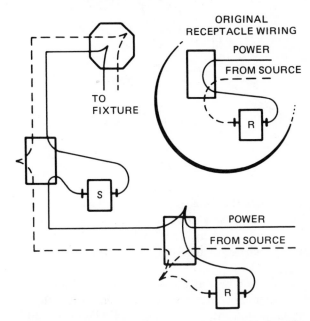

Figure 15-23. Circuit for new switch-controlled ceiling fixture.

Figure 15-24. Cutting access hole.

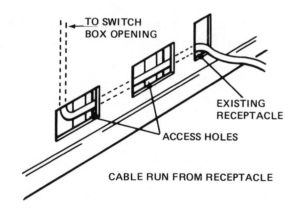

Figure 15-25. Lateral cable run.

holes, notch the studs so you can run the cable laterally across them. Use a saber saw to make two horizontal cuts, about 1 inch apart, in each stud. The cuts should be 1/2 inch deep. Chisel out the wood between the cuts.

The access hole at the point where wall and ceiling meet should be large enough to extend below the wall top plate and into the ceiling itself. The best way to make the holes in the wall and ceiling depends on the material. For plasterboard, the hole can be cut with a

utility knife, or a pilot hole can be drilled and the material cut out with a keyhole saw. Plaster walls with wood lath can be similarly cut. If the plaster is over metal lath, a saber saw with a fine-toothed cutting blade can be used.

Step 2. When you have cut the top-plate hole and the corresponding lower access hole, select a spot for mounting the switch. The switch box can be mounted on a stud or between studs. The various methods for mounting boxes are described in Chapter 8. Whichever method you choose, use the switch box as a template. Mark the outline on the wall. Drill a pilot hole and cut the opening. Use a similar procedure to cut the ceiling box opening. Do not mount either box until the cable has been fished through.

Step 3. Feed a fish tape (or a length of stiff wire) into the switch box opening and up to the top-plate opening. Connect cable to the fish tape at the switch box. Pull the cable through to the top-plate opening.

Step 4. Feed the fish tape into the ceiling box opening and over to the top-plate hole. Pull the tape out and attach it to the end of the cable (Fig. 15-26). Reel in the fish tape to pull the cable through. Have your helper (if you have one) feed the cable into the switch box opening to prevent kinking.

Step 5. Pull a foot of cable out through the ceiling box opening. Disconnect the fish tape. Cut the cable at the switch box, leaving a foot of cable at that point.

Step 6. Feed the cut end of the cable into the switch box opening and down toward the lower access hole. Fish tape is not required for this type of run.

Step 7. At the service panel, turn off power to the receptacle where the new circuit will be connected.

Step 8. Remove the faceplate and the receptacle mounting screws and pull the receptacle out of the box. Remove a knockout from the receptacle box. On the cable that was fed from the access hole to the switch box opening in step 6, measure the amount needed for the lateral run to the receptacle. Add a foot to this measurement and cut the cable. Feed the cut cable behind the wall toward the receptacle.

Step 9. If the receptacle box has an internal cable clamp, loosen this clamp and feed the cable into the knockout and under the clamp. (You can reach into the access hole to do this.) If the box does not have an internal cable clamp, slide a clamp onto the end of the cable and secure it a foot from the end. Remove the locknut from the clamp. Feed the cable into the

CEILING BOX OPENING

SWITCH BOX OPENING

TO
EXISTING
RECEPTACLE

Figure 15-26. Installing cable.

knockout until you can push the threaded end of the clamp into the box. Put the locknut over the end of the cable extending out of the box. Thread it on the end of the clamp and tighten it securely.

Step 10. Use cable staples to fasten the cable into the notches in the studs. Use a cable staple to fasten the cable at the top-plate opening also.

Step 11. Install a ceiling box and a switch box, using one of the techniques described in Chapter 8. To complete the job, mount and wire the ceiling fixture and the wall switch.

The foot of cable that was left at each opening is a working length during installation when a bit of extra cable length is often useful. Cable should never be pulled tight when it is installed. Cut off the excess cable at each opening. Four to six inches is sufficient conductor length for electrical connections.

Step 12. Make the electrical connections for this circuit as shown in Fig. 15-23.

When wiring is complete and has been tested, the temporary access holes must be repaired. The repair procedure depends on the wall material and the way it is finished. Detailed information on repair of finished interiors is beyond the scope of this book.

Drilling Cable Openings between Finished Floors

To bring cable into walls and ceilings and to run cable between floors, you must drill openings through wall and floor material. The best situation is when the wall is directly above or below the point from which the cable will enter. It is a simple matter to drill directly through the top or bottom plate (Fig. 15-27). In some locations you may have to drill through as much as 5 or 6 inches of material. Use a paddle bit or a standard bit, long enough to clear this much material.

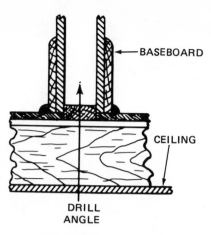

Figure 15-27. Drilling into interior walls.

To make cable openings in outside walls, or between walls and ceilings, an extension bit must be used because of the distance the drill must penetrate. The method of drilling into outside walls is shown in Fig. 15-28. Figure 15-29 shows one way of running cable from a ceiling opening to a wall opening.

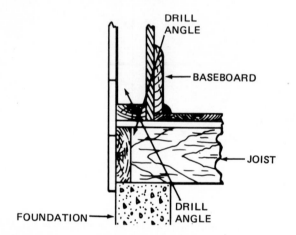

Figure 15-28. Drilling into exterior walls.

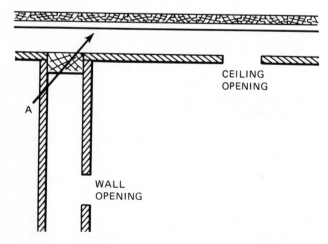

Figure 15-29. Drilling an opening between ceiling and wall.

Step 1. After drilling a hole indicated by the arrow, feed a fish tape into the opening at point A and up into the ceiling, pulling it out at the ceiling opening.

Step 2. Attach the cable to the end of the tape and pull the cable back through the ceiling and out opening A. Detach the cable from the fish tape.

Step 3. At the ceiling opening, pull the cable back slightly until the end of the cable is within the wall behind opening A.

Step 4. Push the cable at the ceiling opening to feed it down until it can be pulled out at the wall opening. The opening at A will be patched when the job is finished.

Concealing Cable in Finished Areas

It is sometimes practical to recess cable into the wall material or to conceal it behind trim for part of a run. This is often done when a cable must be installed between two or more wall outlets. For example, to install a switch which will control a wall fixture, a run such as that shown in Fig. 15-30 can be made.

Step 1. Cut wall openings for the switch and the fixture.

Step 2. Remove the baseboard and make an opening large enough for the cable behind the baseboard and beneath each outlet opening.

Step 3. Cut a trough into the wall material between the cable openings.

Step 4. Feed the cable into the switch opening and fish it out through the hole below. Run the cable in the trough to the other access cable hole.

Step 5. Feed a fish tape into the fixture outlet opening and pull it out of the access hole below. Attach cable to the top and pull it up and out of the fixture outlet opening.

Step 6. If the diameter of the cable is greater than the thickness of the wall material, cut notches into the studs between the openings.

Step 7. Replace the baseboard to conceal the cable.

In many finished interiors the baseboard trim is continued around doorways. If there were a doorway between the two outlets in the example above, the cable run could be continued by removing the trim around the doorway, running the cable inside the door frame, and replacing the trim.

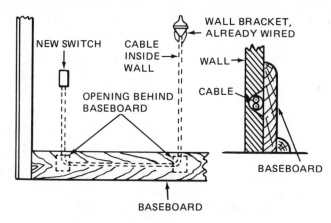

Figure 15-30. Cable run from switch to wall fixture.

Notching Studs

Notching may be used in any situation where it is not practical to bore holes through studs or beams. Make the notch as small as possible to avoid weakening the timber. If a notch in a stud is not protected, the NEC requires that a steel plate 1/16 inch thick be installed over the notch. In the example preceding, the baseboard would provide enough protection and the plates would not be required. In any unprotected area the cover plates must be used. These plates are also required if a hole is drilled through a stud less than 1 1/4 inch from either edge.

Knob-and-Tube Wiring

In houses that are quite old you may find a type of wiring known as "knob and tube" (Fig. 15-31). This wiring uses individual conductors supported by porcelain insulators. One type of insulator was made in two pieces with a nail through the center. The wires were placed in grooves in the porcelain and held in place when the nail was driven into a stud or joist. When the wires were fed through studs or joists, porcelain tubes were inserted in drilled holes and the conductors were fed through the tubes.

The NEC does permit this type of wiring for extensions of existing installations, but the materials involved are so difficult to find today that it is more practical to use modern cabling for additions. If the new cable must be joined to knob-and-tube conductors, the splice should be in an outlet box (Fig. 15-32). If necessary, a junction box can be installed. Solderless connectors can be used to join the conductors, as in other wiring.

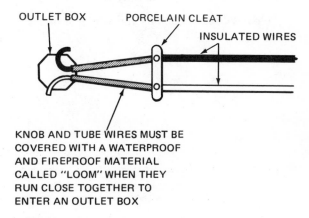

Figure 15-32. Joining cable to knob-and-tube wiring.

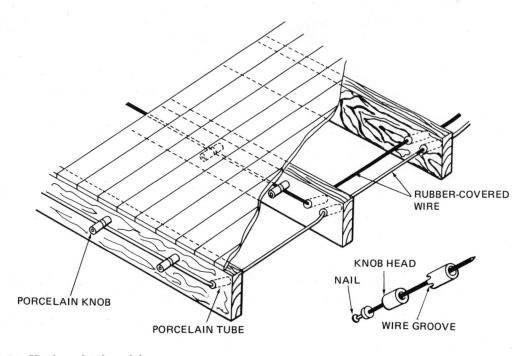

Figure 15-31. Knob-and-tube wiring.

1. Planning is an important part of installations in old work. There are three things that must be considered in your work plan. Name them.

2. In old work it is often necessary to decide if an existing general purpose branch circuit has enough unused capacity to handle additional outlets. What points must be considered in making this decision?

3. What limits the use of rigid conduit and electrical metal tubing (EMT) in old work?

4. In old work interior walls are generally easier to work in than exterior walls. Why?

5. Two types of electrical boxes are particularly useful in old work. What are they?

6. What is circuit mapping and why is it done?

7. When selecting an existing outlet to supply power to new wiring, the outlet chosen should meet three conditions. What are they?

8. One way to increase the capacity of a service panel is to install an auxiliary panel. How does this increase capacity?

9. Another way to increase the capacity of a service panel is to install a panel in parallel with the old one. How does this increase capacity?

10. The NEC specifies a number of requirements for cable installations in basements and attics. All the requirements are related to one safeguard. What is it?

11. What article in the NEC describes how conductors shall be protected against physical damage?

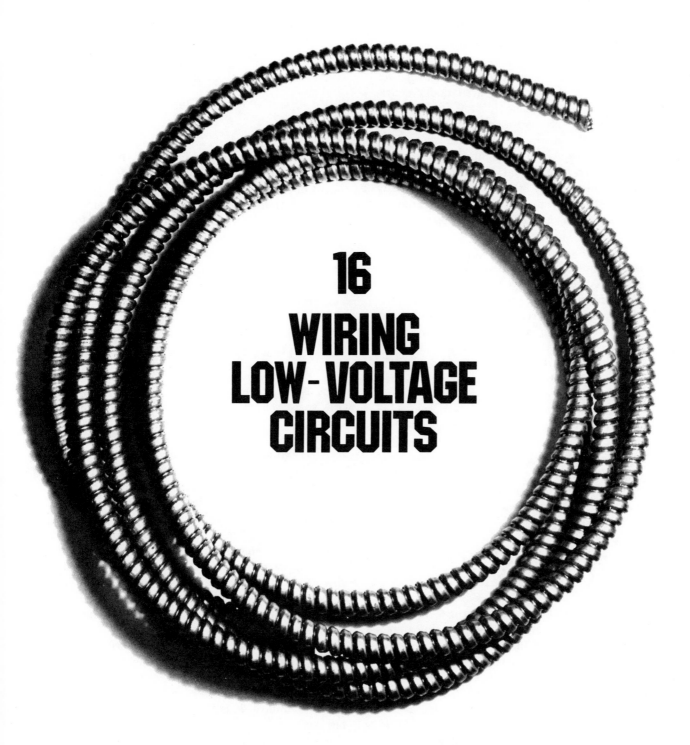

16
WIRING
LOW-VOLTAGE
CIRCUITS

• INTRODUCTION •

In addition to the branch circuit wiring discussed in previous chapters, almost all residential electrical systems include low-voltage circuits. These circuits employ voltages of 30 volts or less to control higher voltages or to perform work directly. The use of low voltages allows wiring to be done with less expensive materials and with fewer NEC requirements and restrictions. Residential low-voltage circuits use power sources that limit current flow—even under short-circuit conditions—to reasonably safe levels. Low-voltage wiring, then, presents a greatly reduced shock and fire hazard. For this reason the NEC allows low-voltage wiring to be done using light, flexible conductors with thin layers of insulation. Low-voltage wire runs can be made quickly and inexpensively in many areas where 120/240-volt wiring would be far more costly to install. Devices designed for use on low-voltage circuits are similarly low in cost and simple to install.

The most common low-voltage circuit is the doorbell or chime. However, low-voltage wiring has many other applications. Most heating and air-conditioning systems use low-voltage control circuits. Fire and burglar alarms generally operate on low voltage. Low-voltage circuits control automatic garage door openers, attic ventilating fans, and outdoor lighting. Low-voltage wiring is also used for television and FM antennas, intercoms, and remote speakers for music systems.

This chapter describes the general rules that apply to the most widely used low-voltage circuits and the NEC requirements that the electrician must be familiar with. Devices and materials especially designed for low-voltage use are also covered. A sound knowledge of this chapter will give you the information you need to install low-voltage circuits that will provide satisfactory, trouble-free service.

• LOW-VOLTAGE TRANSFORMERS •

Typical low-voltage circuits operate at 6, 10, 18, or 24 volts. The low voltage is obtained by using a step-down transformer whose primary winding is connected to a 120-volt source and whose secondary winding provides the low voltage (Fig. 16-1). For residential use, low-voltage circuits are rated at 100 volt-amperes or less. Typical low-voltage transformers for 24-volt systems are rated at 40 volt-amperes continuous load, and 75 volt-amperes momentary load. Low-voltage transformers are rated in volt-amperes, rather than watts; this is done because most devices used on low-voltage circuits—such as relays, solenoids, and transformers—are inductive. You will recall from Chapter 3,

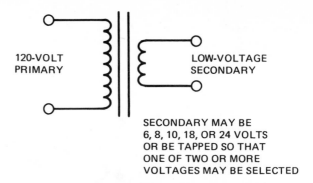

SECONDARY MAY BE
6, 8, 10, 18, OR 24 VOLTS
OR BE TAPPED SO THAT
ONE OF TWO OR MORE
VOLTAGES MAY BE SELECTED

Figure 16-1. Low-voltage transformer, schematic diagram.

that when a circuit is largely inductive, that is, when it contains electromagnetic devices, voltage and current are out of phase. When voltage and current are out of phase, volt-amperes measure circuit load more accurately than watts. Thus, in the typical transformer mentioned above, continuous current flow rating at 24 volts is about 1.7 amperes and the momentary rating is about 3 amperes. Low-voltage transformers are designed to limit current flow to safe values even when short circuits occur.

• LOW-VOLTAGE WIRING •

The NEC and most local codes allow low-voltage wires to be routed in any place and in any manner necessary to get the job done with only a few restrictions. The NEC prohibits running low-voltage wires and 120/240-volt wires in the same cable or conduit. Low-voltage and full-voltage wires cannot be present in the same outlet unless the full-voltage wires are brought into the outlet solely to provide power for a low-voltage circuit.

Indoors, low-voltage wires must be kept at least 2 inches from full-voltage wires and cables. Outdoors, the low-voltage wires (this includes antenna wires) and full-voltage wires must be at least 2 feet apart. When both low-voltage and full-voltage wires are securely fastened, most codes allow closer spacing. The NEC minimum is 4 inches.

Connections and splices of low-voltage wires must be properly insulated with tape or other material, but they do not have to be made in electrical boxes. There is no universal color code for low-voltage wire as there is for full voltage. It is helpful, however, to establish your own color code for each low-voltage circuit. The color code speeds up connecting wires and reduces the possibility of incorrect connections. Low-voltage wire is available in many colors and combinations of colors in cables. A form of low-voltage circuit known as remote-control wiring is described later in this chapter.

Color coding is widely used for these circuits. To prevent confusion, do not use remote-control colors in other circuits.

Low-voltage transformers—like all power transformers—are efficient and durable devices. The main cause of failure in low-voltage transformers is overheating. It is essential that low-voltage transformers be installed in areas when there is a free flow of air. Do not install low-voltage transformers in enclosed areas or under insulation.

Standard low-voltage transformers have pigtail connections on the primary winding and screw-type terminals on the secondary. The connection between the pigtail leads and the 120-volt wires must be made in an electrical box. Low-voltage transformers are designed to mount directly on electrical boxes or adjacent to them. Low-voltage transformers are available permanently attached to covers that fit 4-inch-square boxes (Fig. 16-2). The pigtail primary leads are brought into the box through a hole in the cover. Other transformers are designed to be mounted in box knockouts (Fig. 16-3). Preassembled combination units, including a transformer and outlet box, are also available.

Wire Types

The low-voltage wiring from the transformer secondary to the devices on the circuit can be made with any one of several types of wire made for low-voltage circuits. Two commonly used types of low-voltage wire are described below. Other types of wire especially made for television, FM, and music systems are described later in this chapter.

BELL WIRE • The most common type of low-voltage wire is known as bell wire because of its wide use for doorbell circuits (Fig. 16-4). This wire is available as a single conductor with plastic insulation

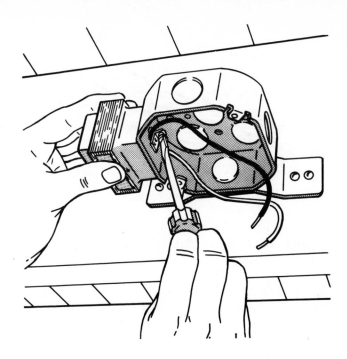

Figure 16-3. Low-voltage transformer for knockout mounting.

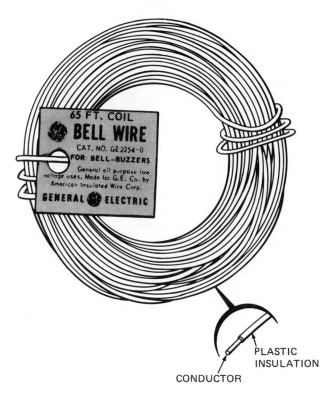

Figure 16-4. Low-voltage bell wire.

and as a two-conductor combination, called a "twisted pair." The conductors are generally copper and are made in no. 18 and no. 22 sizes. Multiple-conductor twisted wire containing two, three, or four separately insulated stranded conductors (Fig. 16-5) is used in low-voltage remote-control circuits, described later in

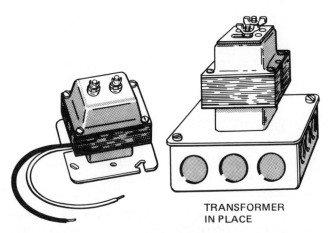

Figure 16-2. Low-voltage transformer for box mounting.

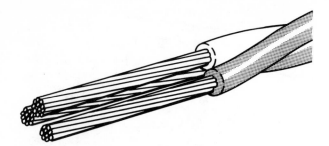

Figure 16-5. Twisted indoor wire.

this chapter. Plastic insulation is made in a wide variety of colors to allow color coding numerous lines of a circuit.

LAMP CORD • This two-conductor wire with plastic insulation is commonly used for lamps and small appliances, but is also suitable for many low-voltage applications. When used for low-voltage wiring, this wire can be routed through floors and walls. Lamp cord must never be used this way on full-voltage circuits.

Several techniques are available for making connections in low-voltage wiring. Keep in mind that some of these techniques are approved *only* for low-voltage wiring.

Connectors

Connectors that can be quickly and securely crimped onto conductors are available for use on both full-voltage and low-voltage wires. These connectors are particularly useful in low-voltage wiring where connectors approved for full-voltage use are too bulky or expensive to use. Crimp connectors can be used on TV antenna wire, as well as low-voltage power wire. Two connector types are available: spade terminals for connection to screw terminals, and male/female disconnects which can be used both as disconnects and for splices (Fig. 16-6). Remember, in full-voltage wiring, splices must be made in electrical boxes. The connector sleeves are color coded to show the wire sizes they are designed to fit.

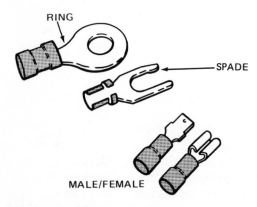

Figure 16-6. Connector types.

Sleeve Color	For Wire Size No.
Red	22 - 18
Blue	16 - 14
Yellow	12 - 10

To attach a connector, simply strip 1/4 inch of insulation from the end of the wire. Slip a connector over the exposed conductor and crimp the sleeve of the connector to secure it to the wire.

Soldering

Soldering full-voltage conductors is covered in Chapter 6. Soldering low-voltage wire is the same in principle, but somewhat different tools and techniques can be used. The smaller size conductors used in low-voltage wiring can be soldered with a pencil iron or a soldering gun. Soldering guns provide higher heat and are generally quicker and easier to use. The basic rules are listed below.

1. Heat the wire until it is hot enough to melt the solder. Do not apply the iron directly to the solder.
2. Use solder with rosin-core flux, suitable for electrical work.
3. Tin the tip of the gun by applying a thin coat of solder to it.

Step 1. Make sure all surfaces to be soldered are clean. Make a good mechanical connection first to take up any strain on the joint.

Step 2. Heat the joint until solder flows freely into the joined wires. A good joint will have a smooth, bright finish. A dull or grainy surface means a poor joint.

Step 3. Reheat the joint until the solder has the correct appearance. Use high enough heat to heat the joint quickly so that solder will flow before the wire becomes hot enough to melt the plastic insulation.

When making a solder connection to a plug or a terminal point, the job will go faster and be done better if the conductors are coated with a thin layer of solder before making the actual connection.

Step 4. Apply the bared conductor to the end of the iron.

Step 5. Touch the tip of the solder to the wire. When the solder melts, let it cover the end of the conductor.

Step 6. When the wire has cooled, make the mechanical connection to the terminal or insert the tinned lead into the plug. Heat the joint and apply a bit more solder to complete the job.

Heat-Shrinkable Tubing

Soldered connections in low-voltage wiring can be quickly and easily covered by using a special type of plastic sleeving known as heat-shrinkable tubing. This tubing is available in various lengths and diameters for use with different wire sizes. For most low-voltage wire, the 1/4-inch diameter size will do. This tubing can be used, for example, to cover a splice in a two-conductor low-voltage cable.

Step 1. Before joining the wires, slip a 6-inch length of heat-shrinkable tubing over the end of the cable.

Step 2. To make a neat connection and to avoid shorts between conductors, cut the conductors to different lengths so that the two soldered joints will not be opposite one another.

Step 3. Twist the conductors together to make a smooth joint, then solder each connection.

Step 4. When the joints have cooled, slide the heat-shrinkable tubing over the two joints. Heat the tubing by moving a lighted match rapidly back and forth under it. The tubing will shrink and form a tight seal around the joints (Fig. 16-7).

Routing Low-Voltage Wires

The same techniques used to route 120-volt wires can be used for low-voltage wires. Openings can be drilled between floors and into walls, just as they are for standard power wiring. Of course, the openings need not be as large. Low-voltage wires are small enough to be concealed below and behind moldings when there is not sufficient room for 120-volt cables. Low-voltage wires can be routed under carpeting, if carefully done. The wires should be put under carpeting only in areas having little or no traffic, to avoid wear and abrasion which could damage the thin plastic insulation and

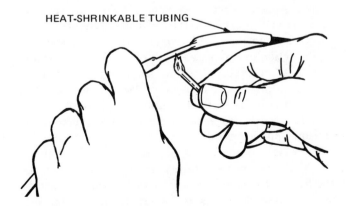

HEAT-SHRINKABLE TUBING

Figure 16-7. Heat-shrinkable tubing.

cause shorts or breaks in the wires. Keep in mind that wires for low-voltage devices will spark when a short occurs. This spark can ignite flammable material. Low-voltage wires can be concealed and protected under wall-to-wall carpeting by the following procedure.

Step 1. Plan the wire run so that it follows the wall line.

Step 2. Peel back the carpeting from the baseboard. The carpeting is held in place by carpet lath. Carpet lath is thin, flat strips of wood through which many tacks have been driven such that the pointed ends of the tacks project through the lath about 1/4 inch. The laths are nailed to the floor with the points up. The tacks stick into and hold the carpet.

Step 3. The thickness of the lath is about equal to the diameter of most low-voltage wire. Staple the low-voltage wire to the floor, close to the lath.

Step 4. Put the carpet back in place and press it down with your foot, so that the tack points will grip the carpet. The thickness of the lath will protect the wire from abrasion.

The variety of colors of low-voltage wire insulation allows the wire to be run in exposed locations without being conspicuous. For example, low-voltage wire with white insulation can be stapled to white baseboard and trim. If possible, for the least conspicuous installation, run the wire in grooves or ridges in molding or in the angle where baseboard and floor meet.

Low-voltage wire with heat-resistant insulation can be fished through heating and air-conditioning ducts. This method is particularly useful when adding low-voltage control circuits to a finished, hot air heating system. More information on heating and air-conditioning control circuits is provided later in this chapter.

Low-voltage wiring is also available with weather-resistant insulation. These wires can be run in areas exposed to the weather. To maintain a neat exterior appearance, the wire should be concealed as much as possible. For example, low-voltage wires can be stapled under the edge of a row of shingles, to the bottom edge of trim boards, and under porches. Plan low-voltage wire runs to avoid the area of the service entrance.

Installing Doorbells and Chimes

Every residential electrical installation includes at least one low-voltage circuit for the doorbell or chimes. These circuits consist of a transformer, a signaling unit, and one or more pushbutton switches. The simplest type of doorbell circuit is shown in Fig. 16-8. One wire from the transformer secondary runs directly to the bell unit. The other wire from the secondary runs to the bell button and then to the bell unit. When the bell button is pressed, the transformer secondary voltage is applied to the bell unit and the bell rings. For a simple bell unit, a transformer with a 6-volt secondary would be satisfactory.

Chimes are somewhat more complicated and require more power than bell units, using a transformer with a 16- to 18-volt secondary, but the basic circuit operation is the same. Chimes are rung by applying voltage to one or more solenoid units so that the solenoid plunger strikes the chimes. Most chime units have provision for both front- and rear-door signals. For this type of operation the units have three terminals. As in the simple circuit, one wire from the transformer secondary goes directly to a terminal on the chime unit marked "COMMON" (Fig. 16-9). Two wires must then

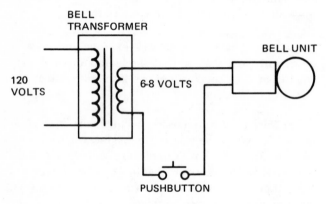

Figure 16-8. Simple doorbell circuit.

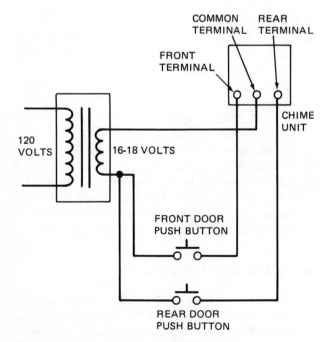

Figure 16-9. Two-button chime circuit.

be connected to the remaining transformer secondary terminal. One wire is routed to each push button and then to the chime terminal marked "FRONT" or "REAR."

Burglar Alarms

Low-voltage devices and wiring are used in many burglar alarms. In addition to transformers, relays, and rectifiers, burglar alarms have some type of *sensor*. Sensors are switches that open or close when movement occurs in a door or window. The use of low voltage for the alarm system allows the sensors to be small and inconspicuous. One type of sensor consists of two metal contacts, one mounted on a door, the other on the door frame. With the door closed, the contacts touch. When the door is opened, the contacts are separated, opening a circuit. Similar contacts can be mounted at the hinge side of a door and on the door frame such that when the door is opened, the contacts touch, closing a circuit. These sensors can be combined with metallic tape that can be applied to glass areas. If the glass is cut or broken, the tape is broken, opening a circuit.

There are basically two types of alarm systems. In one type the sensors are wired in parallel, in the other they are wired in series. In the parallel system the sensors are normally open (Fig. 16-10). When a door or window is opened, the movement closes one of the sensors, completing a low-voltage path to close the contacts on a relay. The relay contacts can be wired to turn on a low-voltage or a 120-volt circuit to sound an alarm, turn on lights, or activate an automatic alarm to police. In this system the alarm circuit is normally open so there is no current flow until a sensor is closed. Relays used in alarm systems are designed to latch in the alarm position. Even if the sensor is opened again, the alarm will stay on until turned off by a manually operated switch.

The series system uses sensors that are normally closed (Fig. 16-11). Current flows from the low-voltage transformer through the sensors to keep the relay energized. As long as the relay is energized, the alarm circuit is open. When movement of a door or window opens a sensor and breaks the low-voltage circuit to the relay, the movable relay contact is then moved by spring action to complete the alarm circuit. The series system is "on" all the time until a sensor is opened. The current drawn by the relay is not large, however. The series circuit has the advantage of automatically sounding the alarm if any accidental break occurs in the low-voltage circuit.

Both types of alarms usually have a test button to check the system periodically. In the parallel circuit, the test button completes the circuit to the relay, just as the sensors do. In the series circuit, the test button opens the circuit to the relay.

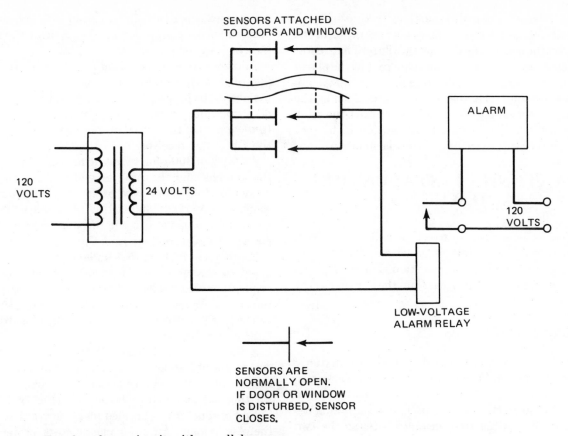

Figure 16-10. Burglar alarm circuit with parallel sensors.

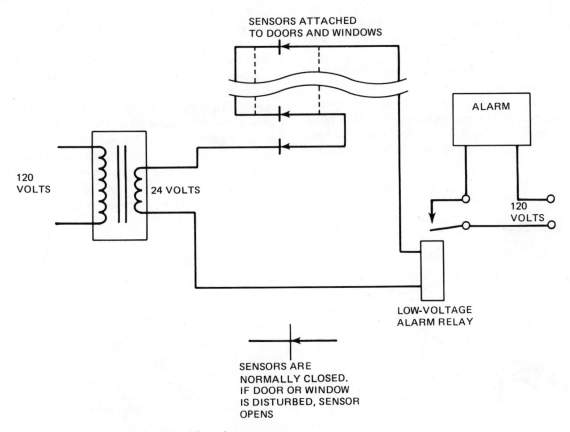

Figure 16-11. Burglar alarm circuit with series sensors.

Most alarm systems include a backup battery-powered voltage supply that takes over operation of the system in the event of utility power failure. This part of the system is usually a self-contained part of the alarm unit and requires no special wiring. The backup system includes a second relay that is always energized by the primary power supply. When the supply fails, the relay is deenergized. The relay contacts switch in battery power to the sensor circuit and the alarm circuit.

• ANTENNA INSTALLATION AND WIRING •

Antenna connections for television and FM receivers are another form of low-voltage wiring. This type of wiring involves no power connection. The wiring carries the television or FM signal from the antenna to the receiver.

Antenna Wire

Special types of wire are used for these installations. The most common form of TV and FM wire is known as *twin lead* (Fig. 16-12). This wire consists of two conductors in plastic insulation. The insulation is designed to maintain a fixed distance between the two leads. In one form of twin lead the wire is flat, in another form the twin leads are enclosed on opposite sides of a plastic tube. The flat lead is primarily for indoor use; heavier versions are made for outdoor use. The round twin lead is weather-resistant and can be used both indoors and out.

Both of these types of twin leads are susceptible to electrical interference. Particularly in heavily populated areas, this may seriously downgrade TV and FM reception. In areas where interference may be a problem, a special shielded type of twin lead should be used (Fig. 16-13). In shielded twin lead, the two conductors

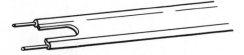

Figure 16-12. Antenna twin lead.

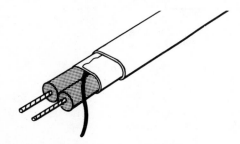

Figure 16-13. Shielded twin lead.

are encased in plastic foam, wrapped with foil, and then covered with a plastic outer cover. Shielded twin lead may be used indoors or out.

All three types of twin lead are often referred to as *300-ohm* wire. Some brands have this designation molded in the plastic. The term "300-ohm" in this case does not refer to resistance, but to a characteristic (*impedance*) of the wire that matches it to the input of the TV or FM receiver.

All types of 300-ohm twin lead are connected at both the antenna and the receiver end by attaching each conductor to a screw-type or spring-loaded terminal, sometimes marked "300 OHM." Any sharp-edged instrument can be used to cut through and remove the plastic insulation on twin lead. Cut the insulation at an angle to avoid nicking the conductors. Scrape the exposed conductor until the metal is shiny. Bend the conductors into hooks and secure them to the screw terminals. Connections can also be made by attaching connectors to the conductors. This subject is discussed later in the chapter.

A third means of TV and FM antenna connections uses a type of lead known as *coaxial*. This type of wire consists of a solid center conductor surrounded by insulating material such as plastic foam (Fig. 16-14). The second conductor is a braided metal sleeve that covers the plastic foam. An outer plastic cover protects and seals the wire. The braided metal sleeve acts as a shield to block interference from reaching the center conductor. This type of lead is known as *75-ohm* wire; sometimes it is called *RG-59 coax*. This type of conductor requires a special connector designed to fit into a jack on the TV or FM receiver. The jack for this connection is usually marked "75 OHM." Coaxial cable must have one of these connectors at each end. The connectors are known as F-type. They keep the center conductor and the outer braid from touching and they provide a good electrical contact from the cable to the receiver at one end and the antenna at the other. The procedure for installing F-type connectors is as follows:

Step 1. Use a wire stripper to remove about 3/4 inch of the outer cable covering to expose the braided shield.

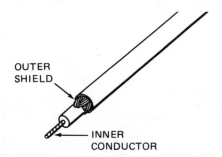

OUTER SHIELD

INNER CONDUCTOR

Figure 16-14. Coaxial antenna cable.

Step 2. Loosen the braid and fold it back over the outer covering of the cable. Trim the braid to about 1/4 inch.

Step 3. Remove about 3/8 inch of the inner insulation. The no. 16 opening on the stripper can be used to make this cut.

Step 4. Remove the collar from the F-type connector. Slide the collar over the end of the cable, past the folded braid.

Step 5. The tapered end of the F-type connector is designed to slip under the braid (Fig. 16-15). Push it into the cable end until the tapered part of the connector is covered by the outer insulation and the folded braid is against the back of the connector.

Step 6. Now slide the collar up to the back of the connector and crimp it with long-nose pliers to secure it to the cable (Fig. 16-16).

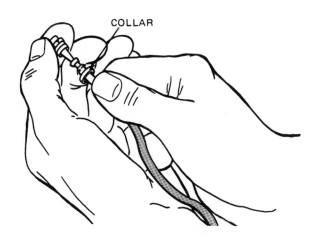

Figure 16-15. Adding an F-type connector.

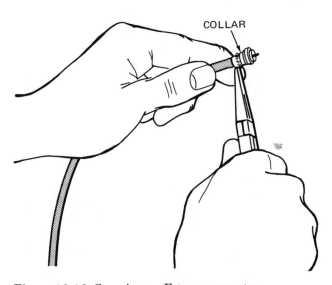

Figure 16-16. Securing an F-type connector.

Antenna Location

TV or FM antennas are usually mounted by strapping supports to a chimney or by mounting brackets on a side wall. When selecting an antenna location, you must take into account the requirements of the NEC, the guidelines for a good antenna system, and the location and structure of the building.

The NEC (Article 810) generally specifies that TV antennas and lead-in wire be kept as far away as practical from power lines, particularly the service entrance. Antenna and lead-in conductors must not cross over electric light or power circuits. In addition, they should be so placed that accidental contact is unlikely. Outdoors the minimum distance allowed between antenna conductors and conductors for 120/240 volts is 2 feet. If both antenna and power conductors are securely supported, so that no significant movement is possible, the clearance between them can be 4 inches. Indoors, the clearance between antenna and power conductors can be 2 inches.

The guidelines for a good antenna system are the following.

1. The antenna should be mounted in as high a location as practical.
2. The antenna should be so placed that the conductor run from antenna to receiver is as short as possible.
3. The antenna and lead-in wire should be kept as far as possible from sources of interference.

Of course, it is highly unlikely that all of these conditions can be fully satisfied in any one installation, so the best possible compromise must be found. For example, the highest mounting point on the house may be the farthest point from the receiver. In this case, the distance of the house from the transmitting station could be a deciding factor. This is covered in greater detail below. The main source of interference with TV or FM signals is any device that radiates electromagnetic energy. Such devices include x-ray and diathermy equipment, some amateur or citizens band radio equipment (particularly if not properly maintained), unshielded internal combustion engines (tractors or power generating equipment, for example), or high-voltage transmission lines. When interference from any of these sources is frequent or constant, an antenna location should be chosen as far as possible from the source. Most TV and FM antenna are highly directional. That is, they pick up signals most efficiently when they are pointed at the source (Fig. 16-17). This characteristic can be used to reduce interference by choosing a location for the antenna that places the source of interference at right angles to the antenna's strongest pickup direction.

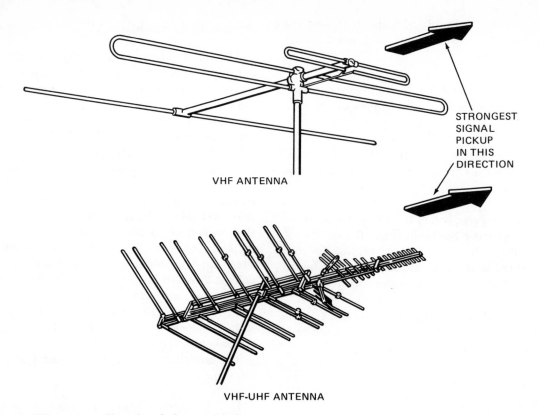

VHF ANTENNA

STRONGEST
SIGNAL
PICKUP
IN THIS
DIRECTION

VHF-UHF ANTENNA

Figure 16-17. TV antenna directional characteristics.

The location and structure of the building can also affect the choice of antenna location. Buildings in city or suburban areas close to transmitting stations can have good reception from almost any antenna location. In these areas, the antenna can be located wherever it is convenient (with perhaps some adjustment to reduce "multipath" interference). Buildings in rural areas far from transmitting stations need as much antenna height as possible for good reception. Antenna height should be a prime consideration when the building is a long distance from the transmitter.

It was noted above that for close-in locations the antenna can be located wherever convenient. This can, in some cases, include attic mounting. Both FM and VHF (channels 2 through 13) TV antennas will provide good reception from attic locations in strong signal areas. Where UHF (channels 14 to 83) TV reception is important, however, attic locations are not satisfactory.

Wall mounting of antennas is difficult in buildings of brick or masonry construction. Use a chimney mount, if at all possible. In wood frame construction either type of mounting can be used. The general procedure for each type of mount is described below.

• ANTENNA MOUNTING •

TV and FM antennas are made in a wide range of sizes. Close-in locations can use simple and fairly light

antennas and obtain good reception. Remote locations require more elaborate, and therefore heavier, antennas. The weight of the antenna must be considered when planning the mounting. Keep in mind that, in addition to the weight of the antenna, high winds and ice will put a strain on the antenna mount.

Chimney Mount

This is an easy and effective way to mount an antenna. If the structure is old, inspect the chimney carefully to be sure it can stand the antenna load. If mortar is loose or has fallen out in many areas, the chimney should be repaired before mounting an antenna on it.

The chimney mount consists of metal straps that encircle the chimney and hold mounting brackets for the antenna. The mounting brackets are designed to fit on the corners of the chimney (Fig. 16-18). The corner location keeps the bracket from shifting when subjected to high winds.

Step 1. Assemble the eyebolt with the strap attached to the bracket (Fig. 16-19). Secure the eyebolt with a few turns of the eyebolt mounting nut.

Step 2. At the chimney, select the corner on which the antenna will be mounted. Hold the bracket in place and loop the strap around the chimney. Locate the bracket about 1 foot above the roof.

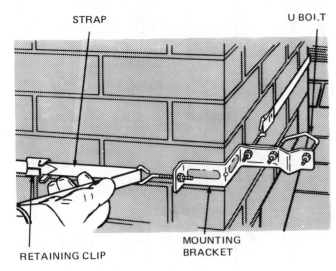

Figure 16-18. Chimney antenna mount.

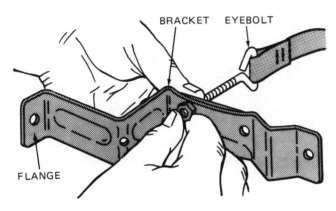

Figure 16-19. Assembling bracket and strap.

Step 3. Slip a retaining clip over the free end of the strap. Feed the strap through the second eyebolt and pull it tight.

Step 4. Feed the free end of the strap back through the retaining clip. Hammer the end of the retaining clip down to hold the strap.

Step 5. Cut off the extra strap and tighten the nuts on the eyebolts to secure the bracket firmly to the chimney.

Step 6. Use the same procedure to install another corner bracket higher up on the chimney. As a general rule, the distance between the straps should be at least one-third the length of the antenna mounting mast.

Step 7. The mast is held in place by U-bolts mounted on the corner bracket. Slide the mast through both U-bolts (Fig. 16-20). Tighten the nuts on the U-bolts to secure the mast to the bracket.

Figure 16-20. Mounting the antenna.

Wall Mount

Wall mounting is done by fastening brackets into the sidewall of the house and then securing the mast by tightening U-bolts in the bracket, just as is done in the chimney mount. The wall mount brackets come in various sizes so that the antenna can be mounted far enough out from the wall to clear the roof overhang.

Step 1. When the mounting location has been chosen, mark a plumbline on the wall so that the mast will be straight (Fig. 16-21).

Step 2. Center a bracket over the plumbline and mark the mounting holes. Use a small drill (1/8-inch diameter) to check the mounting location.

Step 3. If the drill penetrates the wall easily, the area is clear. At that point use a bolt, nut, and washer to secure the bracket. Use a larger drill (usually 1/4 inch) to drill a bolt hole. Feed the bolt through the bracket and into the wall. Have a helper—working in the attic or crawl space behind the wall—put a washer and nut on the bolt and tighten it securely.

Step 4. If the drill does not penetrate readily, there is a stud behind the mounting hole. Choose another location so that a bolt, washer, and nut can be used, or

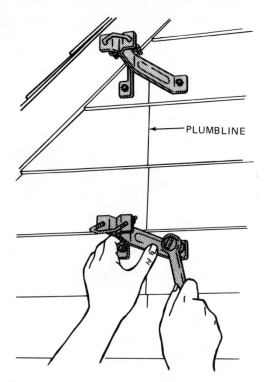

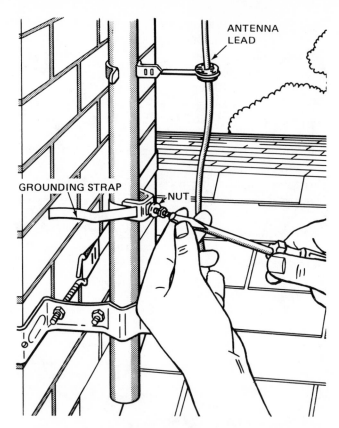

Figure 16-21. Antenna wall mounts.

use a lag screw. The lag screw can be fastened in place from the outside, by using a socket or box wrench.

Step 5. Repeat the procedure for the second bracket. As noted for the chimney mount, the space between the brackets should be at least one-third the length of the antenna mast. When the brackets are in place, secure the mast to the brackets with U-bolts.

Grounding

The NEC requires that the antenna mast be grounded and a discharge path be provided for the antenna lead-in. The antenna mast is grounded by installing a grounding strap or clamp (Fig. 16-22) on the antenna mast and attaching a grounding wire (no. 10 copper or equivalent) to it. The grounding wire is then brought down to ground level and attached to a grounding rod or cold water pipe. The grounding wire should be kept separate from the antenna lead.

The antenna lead-in must be provided with a discharge path to ground for high-voltage surges that may occur. Such surges can occur, for example, if a high-power line breaks loose in a storm and touches the antenna. The antenna lead-in cannot be directly grounded. Such a connection would greatly weaken or completely eliminate the TV or FM signal. Instead a direct grounding path is provided as close as possible to the lead-in insulation. If a high-voltage surge occurs, it will cause an arc-over from the lead-in conductor to the adjacent grounding path. The arc-over effectively

Figure 16-22. Installing a ground clamp on the mast.

grounds the surge. The device that provides the discharge path should be located as close as practical to a ground rod or other good ground. The device should be installed outside, before the antenna lead enters the building. There are two types of discharge devices, one for coaxial lead-in, and another for flat or round twin lead. The coaxial grounding block requires that the coaxial lead be cut and F-type connectors installed on each end of the cable. The grounding block is mounted on the wall of the building. The two connectors are screwed into the mating terminals on the grounding block (Fig. 16-23). The grounding wire (no. 10 copper

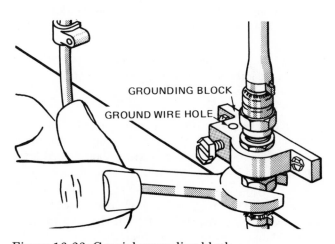

Figure 16-23. Coaxial grounding block.

or larger) or grounding rod is then inserted in an opening on the grounding block. A setscrew on the grounding block secures the wire or rod and assures a good ground connection. The grounding wire should be equivalent to no. 10 copper wire, or larger.

The discharge devices for round or flat twin lead do not require the lead to be cut. The device is attached to the wall near the point where the antenna lead enters the building. A cover is removed from the device and the antenna lead is pushed into a channel running through the device (Fig. 16-24). The cover is then replaced and a grounding wire (no. 10 copper, or larger) is attached to the discharge unit. A screw terminal is provided for the grounding connection.

Antenna lead can be secured to the building with special coaxial clips that can be nailed into the sidewall. Standoff insulators that can be nailed or screwed into the wall can also be used. Remember to keep the mast grounding wire separate from the antenna lead. Secure each one at least once every 4 feet. At the point where wires enter the building, allow enough slack for a drip loop. Be sure to caulk all openings.

When coaxial lead-in is used, special devices may be required to make the electrical connection at the antenna and the receiver. These devices are called *matching transformers*. An antenna or a receiver designed for 300-ohm twin lead will not work well with 75-ohm coaxial lead unless this matching transformer is installed. The transformers match the impedance of the antenna lead-in to the impedance of the TV receiver input at one end, and to the antenna impedance at the other end. Maximum transfer of energy occurs between circuits when their impedances are matched.

Impedance is expressed in ohms; it is the total opposition to current flow in an ac circuit. Total opposition means the sum of the resistance and the reactance. (Resistance and reactance are described in Chapters 2 and 3.)

The transformer for the receiver connection is a small rectangular box with one or two lengths of twin lead at one end and a coaxial connector at the other (Fig. 16-25). Connect the twin lead to the appropriate screw terminals on the antenna or the receiver. One length of twin lead is marked VHF, the other UHF. Connect them to the appropriate terminals on the receiver. Attach the coaxial lead to the coaxial connector. The matching transformer for the antenna connection is usually a circular device having a coaxial terminal on one end and a length of twin lead on the other (Fig. 16-26). Connect the twin lead to the screw terminals on the antenna. Connect the coaxial lead-in connector to the coaxial terminal on the device. Because these matching transformers are exposed to the weather, some means must be provided to seal the device after making the electrical connection. Directions for making the weathertight seal are provided with the device.

• REMOTE CONTROL •

Low voltage can be used to control full voltage. This action is known as remote control because the low-voltage switches and other devices can be some distance from the full-voltage light or motor that is being controlled. Low-voltage remote control provides great convenience to the user at relatively low cost. To an electrician low-voltage remote control provides more flexibility in installation than full-voltage wiring.

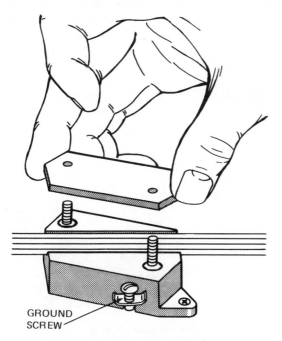

GROUND SCREW

Figure 16-24. Twin lead discharge block.

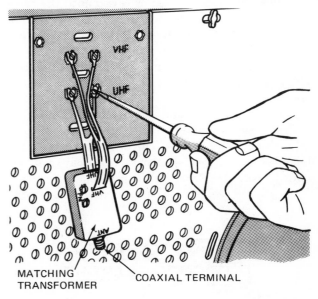

MATCHING TRANSFORMER   COAXIAL TERMINAL

Figure 16-25. Receiver and matching-transformer connection.

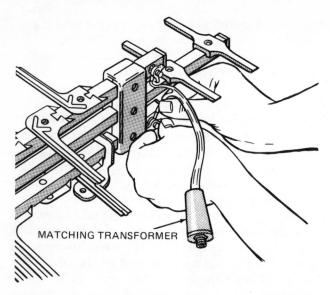

Figure 16-26. Antenna and matching-transformer connection.

Low-voltage wire and devices can be installed quickly and easily in either new or old work.

Relays

The key device in low-voltage remote control is a special relay (Fig. 16-27). This relay consists of an armature, two coils, a latching mechanism, and a set of single-pole single-throw switch contacts. When either of the two coils is energized, the armature moves by electromagnetic action. The armature is linked to the switch contacts, causing the switch to open or close depending on which coil is energized. When the armature is moved by a magnetic field in either coil, it is latched in position. The switch contacts linked to the armature then remain open or closed until the other coil is energized, causing the armature to move and be latched in the other position. Note that the latching action of the relay means that a short pulse of low voltage is all that is needed to change the switch from off to on. In practice, the SPST switch contacts close or open the full-voltage circuit. In this way, a short pulse of low voltage turns a full-voltage circuit on or off.

Relays for low-voltage remote control are designed to work with and fit into full-voltage hardware. The relay consists of a cylinder mounted on a small rectangular base (Fig. 16-28). The base contains terminals

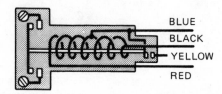

Figure 16-27. Low-voltage relay, schematic diagram.

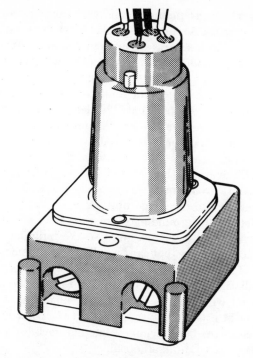

Figure 16-28. Low-voltage relay.

secured by setscrews for the full-voltage hot-wire connection. These connections are equivalent to the hot-wire connection to a standard wall toggle switch. The relay armature and coils are contained in the cylinder. Three pigtail leads at the top of the cylinder provide the low-voltage connections. The relay can easily be mounted in any electrical ceiling or wall box at least 1 1/2 inches deep. The relay is mounted by removing one of the cable knockouts and inserting the cylinder through the knockout so that the cylinder projects out of the box (Fig. 16-29). Slight pressure on the rectangular end causes the relay to snap into the box opening. This holds the relay in place.

Switches

Low-voltage switches are small rocker-type switches that make contact between a center terminal and one

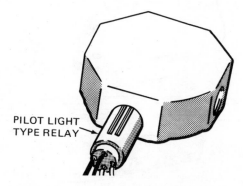

Figure 16-29. Low-voltage relay mounted in electrical box.

of two other terminals when either the ON or OFF section is pressed (Fig. 16-30). When pressure is removed, the switch returns to the neutral position. The momentary contact supplies the low-voltage pulses that cause the relay to open or close the switch contacts.

Low-voltage switches (Fig. 16-31) can be mounted on wall surfaces or recessed into the wall. Electrical boxes are not required when making connections in low-voltage circuits. For surface mounting, low-voltage wires can be brought through a small hole in the wall material and connected to the switch leads. The connection can be made by soldering and taping, or by solderless connectors. The switch can then be snapped into a small mounting frame (Fig. 16-32). The frame is attached to the wall with small screws.

Another method of mounting requires a wall cutout. This method is particularly well suited to mounting clusters of two or three switches. The hardware required consists of two mounting clamps, a switch mounting strap, and the switches. Make a cutout in the wall approximately 1/4 inch larger than the mounting strap. Install a bracket at each end of the opening (Fig. 16-33). Tighten the brackets enough to hold them in

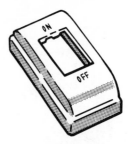

Figure 16-32. Surface frame for low-voltage switch.

MOUNT IN USE

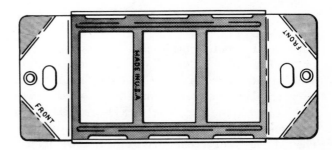

Figure 16-33. Recessed mount for low-voltage switch.

place, but loose enough so they can be moved. Hold the mounting strap in place and slide the brackets until the mounting holes in the brackets line up with the mounting holes in the switch strap. Tighten the brackets securely in this position. Make the electrical connections to the switches. Snap the switches into the mounting strap. Secure the strap to the brackets.

Rectifiers

Most low-voltage relays operate on 24 volts. The source can be any low-voltage transformer with a 24-volt output terminal. Unlike doorbell and chime circuits, the 24-volt transformer output is not applied directly to the relay. A combination device containing a terminal board and rectifier is usually mounted near the transformer (Fig. 16-34). The 24-volt output is connected through the rectifier to terminals on the board. You will recall from Chapter 5 that a rectifier allows current to flow in only one direction. This action changes the full-wave ac output of the transformer to half-wave form. This is done to protect the relay from damage that would result from prolonged application

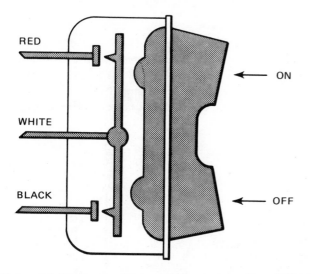

Figure 16-30. Low-voltage switch, cross-sectional diagram.

RED

WHITE

BLACK

ON

OFF

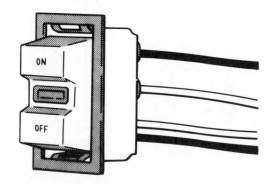

Figure 16-31. Low-voltage switch.

ON

OFF

Wiring Low-Voltage Circuits   **275**

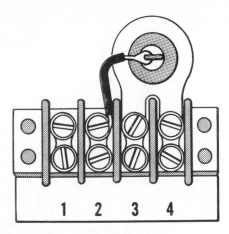

Figure 16-34. Low-voltage rectifier and terminal board.

of full-wave voltage. Such prolonged application could be caused by a defective or shorted switch or a short in the wiring.

Remote-Control Circuits

A typical remote-control circuit is shown in Fig. 16-35. The color coding shown is widely used in remote-control circuits. Power from the 24-volt source is connected to the center terminal on the switch and to the blue lead on the relay. The black and red relay leads are connected to the two momentary switch contacts. The red relay lead is connected to the red switch lead. This is the ON terminal. The black leads control the

OFF circuit. The two black hot wires that complete the 120-volt circuit are connected to the setscrew terminals on the base of the relay.

If a second switch is required for the same circuit, it can be added simply by running three wires from the existing switch to a second switch. Additional switches can be added in the same way. Pressing the ON portion of any switch will close the relay and pressing the OFF portion of any switch will open the relay. The actions of the switches in no way interfere with one another. If the ON portion of any switch is pressed when the relay is already on, no change occurs and the relay remains on. Note that adding switches to a remote-control circuit provides the same control as the three-way and four-way switches described in Chapter 9, but the circuit is simpler and the materials less expensive.

Low-voltage relays are available with an added set of switch contacts to provide a pilot light indication at the control switch. Pilot light circuits require an additional wire between the relay and the switch. This wire, usually color-coded yellow, applies 24 volts to the pilot lamp in the switch when the relay contacts are closed.

Heating and Air-Conditioning Control Circuits

In old work you may find heating and air-conditioning control circuits that operate on full voltage. In these circuits 120 volts is routed to the thermostat and the bimetallic element in the thermostat actually turns on the 120-volt power to the heating or

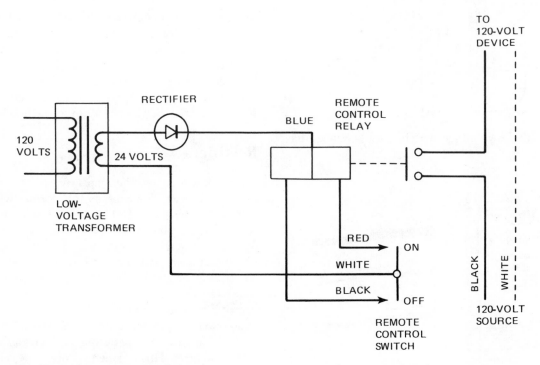

Figure 16-35. Typical remote-control circuit.

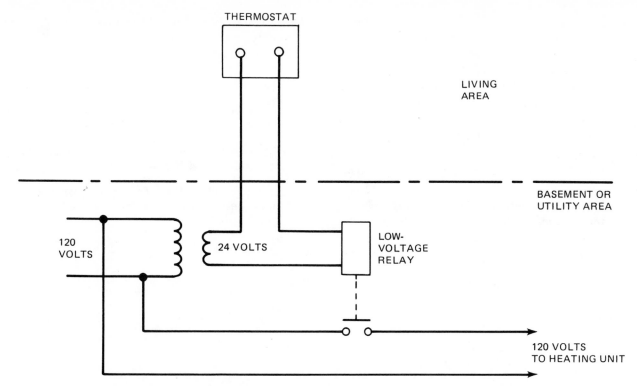

Figure 16-36. Low-voltage heating control.

air-conditioning unit. The use of full voltage throughout the circuit required that standard nonmetallic cable, armored cable, or conduit be used for all wiring. Switching of full voltage in the thermostat caused arcing and burning of contacts. For these reasons, modern heating and air-conditioning control circuits operate on low voltage. The operation of these circuits is basically the same as the switching circuit previously described. However, in place of an on-off switch, the circuit is controlled by a thermostat or a combination of thermostat and timer. The relay that turns on full voltage to the furnace or air conditioner is the same as that described earlier, but is built into the unit (Fig. 16-36).

• REVIEW QUESTIONS •

1. Transformers supply power for low-voltage circuits. What type of transformer is used?

2. What is the minimum distance that must be maintained between low-voltage wires and full-voltage wires indoors?

3. Outdoors low-voltage wires and antenna wires must be at least 2 feet from full-voltage wires under some conditions, but can be as close as 4 inches under other conditions. What are the conditions that determine which spacing is required?

4. What is the main cause of low-voltage transformer failure?

5. What is the main difference between a simple doorbell or buzzer circuit and one for chimes?

6. There are two types of burglar alarm systems. In one type the sensors are wired in parallel; in the other they are wired in series. What difference does this make in circuit operation?

7. In TV and FM antenna installations what advantage does shielded twin lead have over standard twin lead?

8. What characteristic does a TV antenna have that can be used to reduce interference from outside signals?

9. What are the two most common outdoor mounting places for TV antennas?

10. Two things must be done to protect buildings and their occupants from damage and injury resulting from lightning striking a TV antenna mast or from high-voltage lines touching the antenna. What are the two things?

11. Name two advantages of remote-control wiring.

12. What is the function of the low-voltage relay in a remote-control circuit?

13. Remote-control switches make electrical contact only as long as they are pressed. How does this short pulse at low voltage keep the full voltage turned on or off until the opposite half of the switch is pressed?

14. Why is a rectifier used in remote-control circuits?

15. What is the most common application of remote control in most homes?

16. What article in the NEC specifies the permitted location of outdoor TV antenna lead-in conductors?

GLOSSARY

Alternating Current (ac) An electric current that reverses direction at regular intervals, and whose magnitude varies continuously. The voltage causing this current flow varies continuously, and periodically reverses polarity.

Ampacity The maximum safe current flow for each size of wire and type of insulation; the ampere capacity of the wire.

Ampere The basic unit of current, or electron flow.

ANSI Symbols The symbols specified by the American National Standards Institute for use on electrical wiring and layout diagrams.

Appliance Branch Circuit Household branch circuits designed for areas such as kitchens and workshops, where many appliances may be used.

Armature A device consisting of a shaft and many turns of wire, usually in the form of a drum. In a generator, the armature connects to the external circuit to provide the output voltage. In a motor, the armature connects to the electrical source that drives the motor.

Back-off Scale An ohmmeter scale on which the high and low readings are opposite to the voltage scales.

Ballast A special transformer used in fluorescent fixtures to provide an initial high-voltage "kick" and to limit current flow after the lamp is illuminated.

Bar Hanger The name used for several types of ceiling fixture supports that are mounted between beams.

Battery A grouping of cells. If the cells are grouped in series, the battery voltage is equal to the sum of the cell voltages. If the cells are grouped in parallel, the battery will have the same voltage as one cell, but will have more current capacity.

Battery and Bell Tester A simple test instrument consisting of a doorbell or buzzer and dry cell batteries. The tester is used for low-voltage circuit testing.

Bell Wire A type of wire used on low-voltage circuits. It may be stranded or solid, nos. 18 or 22 size.

Block Diagram An electrical diagram on which each part of a circuit that does a specific job is represented by a rectangle. Lines between the boxes show current flow.

Box Ears Adjustable brackets on electrical boxes. These brackets can be adjusted to provide flush mounting for the box.

Branch Circuit An individual circuit that consists of conductors running from an individual fuse or circuit breaker on the service panel to one or more outlets for receptacles, switches, or fixtures where power is used.

Bridging A building term that describes the crisscross bracing installed between floor joists and studs.

Brushes Stationary pieces of graphite that are mounted so they press against a rotating commutator. The brushes, usually in pairs, provide a path for current flow between the stationary and rotating parts of a motor.

BX This is one manufacturer's trade name for cable having an outer shield of spiral steel or aluminum. The term is often used to mean any brand of armored cable.

Cable A grouping of two or more insulated conductors enclosed in a heavy outer covering.

Capacitance The ability of a capacitor to hold a charge.

Capacitive Reactance The opposition offered to the flow of an alternating current by capacitance, expressed in ohms.

Cell A combination of two electrodes and a chemical solution, which produces direct-current (dc) voltage.

Circuit A combination of a power source, conductors, a means of controlling power (a switch), and a load. A circuit must exist in order for electricity to do useful work.

Circuit Breaker A mechanical device which turns off power to a circuit in the event of excessive current flow.

Circular Mil The area of a circle whose diameter is 1 mil.

Clamp-on Meter A meter that reads current flow in amperes when a clamp on the meter is closed around a conductor. No direct wire connection is made between the meter and the conductor.

Commutator A device on the armature of a motor or generator. The commutator consists of metal segments insulated from each other by mica. Each commutator segment is connected to a different group of windings on the armature. As the armature rotates, the commutator segments come in contact with the brushes so that current flow to the windings is switched on and off.

Conductor A material or device that readily passes or conducts electrical current.

Continuity In electrical work, a term indicating that there is a continuous path for current flow.

Coulomb A unit of electric charge; the flow of 6,250,000,000,000,000,000 electrons past a point in one second.

Crawl Space A building term that describes any area of a house large enough to crawl through, but not large enough to stand up in.

Current Electron flow, measured in amperes.

Current Flow The movement of free electrons that occurs in a conductor when a difference in electrical potential (a voltage) exists between two points.

Demand Factor   The amount of lighting or other load that would actually be in use at any one period of time compared to the maximum load possible.

Demand Load   A value for appliance loads; similar to demand factor.

Dielectric   An insulator; the insulating material between the plates of a capacitor.

Direct Current (dc)   An electric current that flows in one direction. The voltage that causes the current flow remains constant in amount.

Electrical Floor Plan   One of the architectural drawings made for building construction. This drawing shows where electrical switches, receptacles, and fixtures must be installed.

Electrical Metal Tubing (EMT)   A form of conduit having walls less than one-half the thickness of rigid conduit.

Electromagnetic Field   The magnetic field which an electric current produces around any conductor through which it flows.

Electrostatic Field   A field created when a voltage difference exists between two points.

End-of-Run (EOR)   The last outlet on any branch circuit.

Farad   The unit of capacitance.

Frequency   The number of cycles per second (hertz) in any form of wave motion such as alternating current.

Furring   A building term that applies to a strip of lumber nailed into masonry to support paneling or other wall finish.

Fuse   A device that contains a piece of metal which melts and opens the circuit when current flow exceeds the fuse rating.

Ganging   A means of making a larger metal box from two or more small boxes by removing side panels and joining the boxes to form an enclosure.

Generator   A machine which converts mechanical energy into electrical energy.

Ground Fault Circuit Interrupter (GFCI)   A device which monitors the current flow in each conductor in a circuit. If the current is greater in one conductor than the other by a preset amount, power to the circuit is automatically turned off.

Ground Rod   A rod, usually copper or copper-coated, that is driven into the ground to serve as a system grounding electrode.

Header   A building term that describes the horizontal piece of lumber across the top of a door or window opening.

Henry   The basic unit of inductance.

Hertz   A unit of frequency equal to one cycle per second; formerly known as "cycles per second" (cps).

Hickey   An apparatus used to bend rigid or intermediate conduit. The term is also applied to a device used in wall and ceiling fixtures to provide an opening for wiring.

Hotbox   A special device which heats nonmetallic conduit electrically, softening it so that it can be bent as desired.

Hot Wire   Any ungrounded wire in an electrical system. When power is applied, current can flow between any hot wire and any grounded point.

Impedance   The total opposition to current flow in an ac circuit. It may consist of resistance and reactance (either inductive reactance, or capacitive reactance, or both).

Individual Branch Circuit   A household branch circuit supplying power to a single 120-volt appliance, such as a washing machine or dishwasher.

Inductance   The ability of a conductor to produce a voltage in another conductor when the current flow in the first conductor is varied.

Inductive Reactance   The opposition to the flow of alternating current due to the inductance of a circuit; it is expressed in ohms.

Insulator   A material of low conductivity, used to cover conductors.

Interlock   A switch that automatically shuts off power in a cabinet or other enclosure when the door is opened or panels are removed.

Jamb   A building term that describes the side post or lining of a door or window.

Joist   A building term meaning the timber that supports the floor or ceiling of a building.

Kilo-   A prefix meaning one thousand, as in kilovolt (1000 volts) or kilohm (1000 ohms).

Knockout   A section of metal in the wall of an electrical box that can be knocked out easily to form an opening for cable to enter.

Lamp Cord   A type of two-conductor wire commonly used for lamps and small appliances, but also suitable for low-voltage wiring.

Line Voltage Drop   This is the drop in voltage that occurs over long cable runs because of the resistance of the wires.

Lines of Force   Lines which represent the pattern made by iron filings when the filings are near a magnet. See Figure 3-3 in Chapter 3.

Load   A device connected to an electrical source to perform work; that is, to produce light, heat, or motion.

Low-Voltage Transformer   A transformer that steps down 120-volt power to 6 to 30 volts for low-voltage circuits.

Magnetism   A force that causes materials having magnetic qualities to be attracted or repelled in accordance with a definite set of rules.

Mapping A procedure (described in Chapter 15) for identifying all circuits in a building.

Matching Transformer A device used in TV antenna installations to match 300-ohm inputs to 75-ohm cable.

Meg- A prefix meaning one million, as in megohm (1,000,000 ohms).

Micro- A prefix meaning one-millionth, as in microfarad (one-millionth of a farad).

Middle-of-Run (MOR) This term describes an outlet on a branch circuit through which power must continue to supply one or more additional outlets.

Mil A unit of length which is one-thousandth of an inch.

Milli- A prefix meaning one-thousandth, as in milliampere (one-thousandth of an ampere).

Multiwire Branch Circuit This is a three-wire household circuit for units requiring 120/240-volt power.

National Electrical Code (NEC) A publication of the National Fire Protection Association. In combination with local building codes, this code governs virtually all electrical wiring in the United States.

New Work This term describes any electrical installation that is done before finished walls are in place.

Ohm The unit of electrical resistance.

Old Work This term describes any electrical installation that must be done in buildings having finished walls, ceilings, and floors.

On-Center A term that defines measurement points. It describes a measurement taken from the center of one item to the center of another.

Open Circuit A circuit which does not have a complete path for the flow of current.

Outlet A term meaning any point in a circuit at which an electrical box is installed.

Overcurrent A condition in which a circuit is carrying more than its rated current flow.

Overload A condition, applying principally to motors, of abnormally high flow of current in the motor windings.

Parallel Circuit A circuit in which all branches (loads) have the same applied voltage. Branch circuit currents can be the same or different, depending on the resistance in each branch.

Plot Plan A basic drawing used to start construction that shows how the house will be situated on lot.

Polarity Points in a material or a device at which opposite forces exist.

Polarized Receptacle Any receptacle having slots of different size or shape so that only one type of plug can be inserted.

Pole When applied to a switch, the term refers to the number of conductors that can be switched. When applied to a magnet, it refers to the points (north pole, south pole) around which magnetic lines of force are concentrated.

Pull Box A box installed in a conduit system solely for the purpose of inserting fish tape and pulling conductors back through the conduit.

Reactance The opposition offered to the flow of an alternating current by the inductance, capacitance, or both, in any circuit.

Receptacle A device mounted in an electrical box to provide a connection for cord and plug appliances and lights.

Relay An electrical device used primarily for remote switching.

Remote Control Circuit Any circuit in which low voltage is used to control full voltage at a location removed from the low-voltage control.

Resistance The opposition to the flow of current caused by the nature and physical dimensions of a conductor. Resistance is measured in ohms.

Riser A building term that means the vertical part of a step that supports the step tread.

Romex This is one manufacturer's trade name for non-metallic, sheathed cable. The term is widely used to mean any brand of nonmetallic cable.

Run Branch circuit wiring which begins at the overcurrent protection device in the service panel, and ends at the last outlet on the circuit.

Schematic Diagram This diagram employs symbols to show the electrical relationship of all devices on a circuit. It contains enough information to make circuit calculations.

Sensor A special type of switch used in alarm systems. The sensor makes or breaks a circuit when a door or window is opened, for example.

Series Circuit A circuit in which all loads are connected one to the other in a continuous loop, and the same current flows through all loads. The voltage across each load can be the same or different, depending on resistance.

Series-Parallel Circuit A circuit in which both series and parallel load arrangements are used.

Service Drop The overhead power lines that run from a utility pole to a building.

Service Entrance The place where utility company power lines are taped for connection to a building's electrical system. Altogether it consists of a service head, a meter, a main disconnect switch, and a service panel.

Service Head The point at which utility company power lines are attached to a building. The service head is sometimes called the entrance cap or weatherhead.

Service Lateral The underground power lines that run from a utility pole to a building.

Shock People experience shock when their bodies become a path for current flow. Another form of shock results from serious injury.

Short Circuit A low-impedance or zero-impedance path between two points, characterized by extremely high current flow.

Sine Wave A type of curve that represents a quantity constantly changing in magnitude and periodically changing in direction. Alternating voltage and current vary in accordance with this curve.

Slipring A device used to conduct energy in or out of rotating machinery.

Solenoid A multiturn coil of wire wound in a uniform layer or layers on a hollow cylindrical form. When energized, the coil creates a magnetic field that will attract an iron box and pull it into the coil.

Split Receptacle A double, or duplex, receptacle in which each half is wired on a different circuit.

Static Electricity The charge that results when electrons are transferred by heat or motion from one nonconductor (such as paper) to another nonconductor (such as rubber).

Three-Switch Circuit A circuit employing four-way switches to provide control of a fixture or receptacle from three (or more) switches at different locations.

Throw The number of positions to which a switch can be set.

Torque A force that produces rotation. The ability of a motor to overcome turning resistance.

Transformer An ac device composed of two or more coils, linked by magnetic lines of force. It is usually used to change the ratio of voltage to current between the coils.

Two-Switch Circuit A circuit employing three-way switches to provide control of a fixture or receptacle from two switches at different locations.

Underwriters' Laboratories (UL) A private testing organization that tests materials and devices having potential safety hazards. Satisfactory products are listed in a UL directory.

Vibrator Similar in principle to the solenoid, but designed to produce rapid, short-stroke movement.

Volt The unit of electrical potential; the amount of pressure required to cause 1 ampere of current to flow through a resistance of 1 ohm.

Voltage An electromotive force or potential difference, expressed in volts.

Voltampere The product of the applied voltage and the current as measured by a meter.

Watt The unit of electrical power; the actual work being done at any given moment.

Wire A single electrical conductor, covered by an insulator.

Wiring diagram This diagram shows how circuits are actually wired and at what points connections are made.

INDEX

SPECIAL OFFER FOR BOOK CLUB MEMBERS

Save $10 on these versatile Stellar 7 X 35 Binoculars

They're ideal all-purpose binoculars — good for a wide range of outdoor activities from football games to bird watching.

Look at these features:

Suggested Retail Price $49.95. Your Club Price only

$39.95

plus delivery and handling

- ☐ **Fully coated optics.** Both lenses and prisms are coated to give them maximum light-gathering power and to insure bright, clear, sharp images.
- ☐ **Quick, accurate focusing.** A right-eye adjustment compensates for differences in vision of the two eyes. A center focusing wheel permits fast adjustment.
- ☐ **Magnification.** "7 X" refers to the magnifying power of the binoculars. It means an object 700 feet away will appear to be only 100 feet away. "35" refers to the diameter in millimeters of the objective lenses, which determines the amount of light admitted. The larger the lenses, the greater the amount of light and the later in the evening you can use the binoculars.
- ☐ **Field of View.** The Stellar Binoculars provide a 393-foot field of view at 1000 yards.
- ☐ **Weight.** 21½ ounces.

The binoculars come in a soft vinyl case with carrying strap. You also get a shoulder strap and four lens covers.

Stellar 7 X 35 Binoculars are fully guaranteed against any defects in workmanship.

TO GET YOUR BINOCULARS, JUST SEND YOUR ORDER TO:
BOOK CLUB P.O. BOX 2044, LATHAM, N.Y. 12111

Ask for STELLAR BINOCULARS, NO. 7000, and enclose your check or money order for $39.95 plus $3.10 for delivery and handling and we'll send you your binoculars right away.